Signal Transduction during Biomembrane Fusion

CELL BIOLOGY: A Series of Monographs

EDITORS

D. E. BUETOW
*Department of Physiology
and Biophysics
University of Illinois
Urbana, Illinois*

I. L. CAMERON
*Department of Cellular and
Structural Biology
The University of Texas
Health Science Center at San Antonio
San Antonio, Texas*

G. M. PADILLA
*Department of Cell Biology
Duke University Medical Center
Durham, North Carolina*

A. M. ZIMMERMAN
*Department of Zoology
University of Toronto
Toronto, Ontario, Canada*

Signal Transduction during Biomembrane Fusion

Edited by

Danton H. O'Day

Department of Zoology
Erindale College
University of Toronto
Mississauga, Ontario, Canada

ACADEMIC PRESS, INC.
Harcourt Brace & Company
San Diego   New York   Boston   London   Sydney   Tokyo   Toronto

Academic Press, Inc.
1250 Sixth Avenue, San Diego, California 92101-4311

United Kingdom Edition published by
Academic Press Limited
24–28 Oval Road, London NW1 7DX

Library of Congress Cataloging-in-Publication Data

Signal transduction during biomembrane fusion. / edited by Danton H. O'Day.
        p.   cm. – (Cell biology)
    Includes index.
    ISBN 0-12-524155-0
    1. Cellular signal transduction. 2. Membrane fusion.
    I. O'Day, Danton H. II. Series.
    QP517.C45S54   1993
    574.87'.5–dc20
                                                                        92-42885
                                                                        CIP

PRINTED IN THE UNITED STATES OF AMERICA
93 94 95 96 97 98   B B   9 8 7 6 5 4 3 2 1

To David and Laurie-Ellen Gwynne,
two reasons why the academic life is worthwhile

Contents

I Introduction

1 Signal Transduction during Biomembrane Fusion: An Overview

Danton H. O'Day

2 The "Focal Membrane Fusion" Model Revisited: Toward a Unifying Structural Concept of Biological Membrane Fusion

Helmut Plattner and Gerd Knoll

3 Osmotic Phenomena in Membrane Fusion

Jack A. Lucy

II Endocytosis, Exocytosis, and Vesicle Formation

4 Cell Signaling and Regulation of Exocytosis at Fertilization of the Egg

Douglas Kline

5 Signal Transduction during Exocytosis in Mast Cells

Witte Koopmann

6 Protein Kinase C and Granule Membrane Fusion in Platelets

Jon M. Gerrard, Archibald McNicol, and Satya P. Saxena

7 GTP-Binding Proteins and Formation of Secretory Vesicles

Anja Leyte, Francis A. Barr, Sharon A. Tooze, and Wieland B. Huttner

8 Signal Transduction during Phagocytosis

Keith E. Lewis, Darren D. Browning, and Danton H. O'Day

III Cell Fusion

9 Calcium Signal Transduction Pathway and Myoblast Fusion

Joav Prives

10 Phospholipid Metabolism during Calcium-Regulated Myoblast Fusion

Victor S. Sauro and Kenneth P. Strickland

11 Protein Kinase C, Membrane Protein Phosphorylation, and Calcium Influx in Chick Embryo Skeletal Myoblast Fusion

John D. David and Ann Fitzpatrick

12   Signal Transduction and Cell Fusion in *Dictyostelium*: Calcium, Calmodulin, and an Endogenous Inhibitor

Michael A. Lydan and Danton H. O'Day

Contributors

Numbers in parentheses indicate the pages on which the authors' contributions begin.

Francis A. Barr (147), Institute for Neurobiology, University of Heidelberg, D-6900 Heidelberg, Germany

Darren D. Browning (163), Department of Zoology, Erindale College, University of Toronto, Mississauga, Ontario, Canada L5L 1C6

John D. David (223), Division of Biological Sciences, University of Missouri, Columbia, Missouri 65211

Ann Fitzpatrick (223), Division of Biological Sciences, University of Missouri, Columbia, Missouri 65211

Jon M. Gerrard (121), Manitoba Institute of Cell Biology, University of Manitoba, Winnipeg, Manitoba, Canada R3E 0V9

Wieland B. Huttner (147), Institute for Neurobiology, University of Heidelberg, D-6900 Heidelberg, Germany

Douglas Kline (75), Department of Biological Sciences, Kent State University, Kent, Ohio 44242

Gerd Knoll (19), Faculty of Biology, University of Konstanz, D-7750 Konstanz, Germany

Witte Koopmann[1] (103), Department of Biochemistry, Dartmouth Medical School, Hanover, New Hampshire 03755

Keith E. Lewis (163), Department of Zoology, Erindale College, University of Toronto, Mississauga, Ontario, Canada L5L 1C6

Anja Leyte (147), Institute for Neurobiology, University of Heidelberg, D-6900 Heidelberg, Germany

Jack A. Lucy (47), Department of Biochemistry and Chemistry, Royal Free Hospital School of Medicine, London NW3 2PF, England

[1]*Present address:* Department of Immunology, Division of Immunology, Box 3010, Duke University Medical School, Durham, North Carolina 27710.

Michael A. Lydan (245), Department of Zoology, Erindale College, University of Toronto, Mississauga, Ontario, Canada L5L 1C6

Archibald McNicol (121), Manitoba Institute of Cell Biology, University of Manitoba, Winnipeg, Manitoba, Canada R3E 0V9

Danton H. O'Day (3, 163, 245), Department of Zoology, Erindale College, University of Toronto, Mississauga, Ontario, Canada L5L 1C6

Helmut Plattner (19), Faculty of Biology, University of Konstanz, D-7750 Konstanz, Germany

Joav Prives (181), Department of Pharmacological Sciences, State University of New York at Stony Brook, Stony Brook, New York 11794

Victor S. Sauro (197), Department of Biochemistry, The University of Western Ontario, London, Ontario, Canada N6A 5C1

Satya P. Saxena (121), Manitoba Institute of Cell Biology, University of Manitoba, Winnipeg, Manitoba, Canada R3E 0V9

Kenneth P. Strickland (197), Department of Biochemistry, The University of Western Ontario, London, Ontario, Canada N6A 5C1

Sharon A. Tooze (147), Cell Biology Program, European Molecular Biology Laboratory, D-6900 Heidelberg, Germany

Preface

Although a number of books and review articles on signal transduction and on biomembrane fusion attest to the importance of these two subjects, to date the interrelationship between them has not been detailed beyond the primary literature. In this volume we attempt to consummate the relationship by focusing on a number of systems that have received attention recently. By documenting the current state of the art, I hope this book will set the stage for new research into the important and rapidly growing area of signal transduction during biomembrane fusion.

This volume is comprised of chapters contributed by an internationally recognized group of over twenty researchers. Essentially every aspect of biomembrane fusion is covered, including cell fusion, exocytosis, phagocytosis, secretion, intracellular membrane fusion, and vesiculation. The whole gamut of signal transduction events is detailed, from receptor stimulation through GTP-binding protein activation of effector enzymes. The generation of second messengers and subsequent cascade mechanisms involving protein kinases and intracellular modulators, such as calmodulin, are also emphasized.

The book begins with three review articles that put the problem of signal transduction and biomembrane fusion into a general perspective. Each subsequent chapter begins with an introduction which reviews past work on a specific biological system. The authors' current research is then detailed. The chapters conclude with final comments wherein the contributors express viewpoints about the general significance and progression of their work. In addition to emphasizing the biological and medical significance of their research, many of the authors provide detailed models of the cellular system being reviewed.

This book will meet the needs of workers in the areas of signal transduction and biomembrane fusion. It is also designed to appeal to everyone interested in the basic operation of cells, including biochemists, cell biologists, cell physiologists, developmental biologists, and pharmacologists. Because extensive background material is presented as a prelude to current research, the material will also be of interest to and could serve as a text for senior undergraduates and graduate students in cell biology and cell physiology.

These are exciting times. As scientists try to unravel the workings of the cell, I hope this book will meet the needs of some of them. Possibly, to some degree, it will "tie a tighter knot" between the two important areas of signal transduction and biomembrane fusion.

Danton H. O'Day

I

Introduction

1

Signal Transduction during Biomembrane Fusion: An Overview

DANTON H. O'DAY

Department of Zoology
Erindale College
University of Toronto
Mississauga, Ontario, Canada

I. INTRODUCTION

The fusion of biological membranes is a fundamental cellular process. During the origin of life, random fusions between primordial membranes surrounding mixtures of simple polynucleotides and polypeptides likely permitted the evolution of more complex "cellular" entities. Reproduction of these first "cells" undoubtedly occurred by random fission of their primitive surface membranes. As new species of cells evolved, unknown selection pressures must have operated to keep some of them apart. Thus with time, as

3

cells maintained their individuality, future diversity resulted from mutation and the more-or-less controlled intermixing of genetic material rather than by the continued, indiscriminate coalescence of masses of primeval genomes. Thus, one stage of cellular evolution must have involved the appearance of regulatory systems such as surface molecules that kept membranes from fusing randomly. As intracellular membrane systems formed, the subcompartmentalization of cells further required that intracellular membrane fusions be controlled rather than random.

Coupled with this need to maintain individuality or distinctiveness of cells and their compartments was the simultaneous need for membranes to be competent to fuse at certain times. From the beginning, asexual reproduction required some form of cytokinesis and, as sexuality evolved, specialized fusogenic gametes became essential for fertilization. As intracellular membrane systems became functionally more complex than basic barriers separating different cellular compartments, certain membrane fusions had to be permitted while others were prevented. From such humble beginnings arose the present regulatory systems that oversee the diverse examples of biomembrane fusion that occur within and between cells today.

Biomembrane fusion has been the subject of numerous books and reviews, and signal transduction has become one of the most dynamic areas of cell biology in the past decade. In spite of this and in spite of the critical interdependence of the two, it is only recently that enough material has been published for a diverse number of cell types to warrant a book interrelating the two topics. In this chapter, the contents of the volume are put in general perspective. After a short introduction to the topics of biomembrane fusion and signal transduction in eukaryotic cells, there is an overview of the entire book. The chapters that follow begin with a review of the system under analysis, followed by a summary, often provocative, of the current state of the authors' research. In the end, both the universal and the unique aspects of each cellular system should be evident.

II. BIOMEMBRANE FUSION

A. Some Examples

Biomembrane fusion lies at the heart of almost every biological process (e.g., Fig. 2, Chapter 2). In eukaryotic cells the final step in mitotic or meiotic cell division is the fission of the plasma membrane during cytokinesis. Prior to this, depending on the organism, fusion of intact or vesiculated nuclear envelope membranes is also essential. During cell division many cytoplasmic organelles undergo fragmentation and reformation to ensure their more-or-less equal distribution to the daughter cells.

The fusion of cell membranes mediates the extensively studied amalgamation of gametes during fertilization (Chapters 4 and 12) and the coalescence of myoblasts during muscle formation (Chapters 9, 10, and 11). The fusion of the plasma membrane is a fundamental step in the basic cellular events of exocytosis (Chapters 4, 5, and 6), endocytosis (Chapter 8), and viral penetration and release (e.g., Chapters 2 and 3).

Once inside the cell, the endosomal membranes are not exempt from biomembrane fusion. After their incorporation, endosomes follow specific intracellular routes based on their fates and functions (e.g., Brown and Greene, 1991). Like fusions of the plasma membrane, these events are precisely controlled. Phagosomes typically are targeted to and fuse with lysosomes to form digestive vacuoles. After cellular digestion, the residual vacuole is commonly directed to the cell surface where exocytosis occurs. Other endosomes (Chapter 7) may follow a similar route leading to the digestion of their contents or to alternative pathways where their fusion with intracellular vesicles leads to processing of the ingested molecules plus digestion or recycling of surface receptors.

Other events of intracellular membrane traffic involve the sequential, regulated budding and fusion of vesicles as molecules move from their sites of synthesis on the endoplasmic reticulum through regions of processing, packaging, and concentration within the Golgi. The fate of vesicles produced by this sequence depends on their role as constitutive or regulated secretory vesicles (Chapters 6, 7, and 9), as digestive vacuoles (e.g., lysosomes), as sources of membrane constituents or as intracellular sites of localized metabolic processes (e.g., glyoxysomes and peroxisomes). Most of these intracellular membrane pathways are still being characterized (e.g., Rothman and Orci, 1992).

In every cell, certain biomembrane fusion events must operate routinely. These include, but are not limited to, those involved in intracellular membrane traffic, in intracellular digestion, in the secretion of waste products, and in the uptake of essential nutrients from the extracellular milieu. In specialized secretory cells, this complexity is compounded by the need to be competent to simultaneously deliver on demand a specific hormone, factor, enzyme, or neurotransmitter.

If the cell, as an individual and as part of a multicellular organism, is to perform efficiently, each set of biomembrane fusion events must be kept functionally separate while often operating simultaneously. Spatial separation is possible, but more commonly they are kept apart by biochemical and molecular mechanisms. As a result, within each cell a diverse number of regulatory phenomena must operate to oversee the fundamental events of biomembrane fusion. A difficult problem is sifting out the constituents common to all examples of biomembrane fusion events from the equally critical regulatory elements.

B. The Sequence of Events

Whereas it is one thing to offer a short, speculative viewpoint on the role of biomembrane fusion in the evolution of the cell and to reflect on its importance in the cell's day-to-day survival, it is quite another to effectively summarize all of the information that exists on this subject. However, it is possible to set the stage for what follows in this volume by beginning with a short, generic summary of the events that occur during biomembrane fusion. More insight into the complexity of each stage is provided in Chapters 2 and 3.

The interaction between fusogenic membranes requires that they first come close enough to contact and then to recognize each other (e.g., via protein–protein interactions). The events leading to contact may be random but often they are directed (e.g., pheromones direct gamete cells toward each other; cytoskeleton directs movements of some intracellular vesicles). The importance of cytoskeletal constituents in the steps leading up to exocytosis of secretory vesicles is emphasized by several groups (Burgoyne, 1991; Monk *et al.*, 1990). In many instances, cytoskeletal elements (e.g., actin) in the cortex of the cell are believed to anchor secretory vesicles as a prelude to exocytosis; whereas the annexins, a group of calcium-dependent membrane-aggregating proteins, are prime candidates for a more dynamic mediatory role (Burgoyne, 1991; Zaks and Creutz, 1990). Annexins (e.g., calpactin and synexin) are not fusogenic, but instead they have been demonstrated to promote the close apposition or "docking" between fusogenic membranes leading to their fusion.

Recognition or adhesion typically involves proteins embedded in the membranes (e.g., Chapter 2). One of the best-characterized examples of an integral membrane protein that mediates membrane fusion is influenza virus hemagglutinin. The role of specific proteins in the endocytosis and intracellular release of enveloped viruses has recently been reviewed, but the signal transduction events involved have yet to be studied in enough detail to yield a comprehensive model (Hoekstra and Kok, 1989).

In Chapter 2, the role of proteins in biomembrane fusion is detailed by Plattner and Knoll. Recently, Blobel *et al.* (1992) presented data suggesting that fusion proteins that mediate both fertilization and viral incorporation are integral membrane glycoproteins that contain similar peptide domains involved in binding and membrane fusion. Such similarities may open the door for more unified molecular models of protein involvement in membrane fusion.

Once the biomembranes are juxtaposed, local alterations in the membrane occur. These local changes have been suggested to involve everything from calcium-mediated phase shifts in the lipid bilayer, to lipid digestion, and to the local formation of normal or inverted micelles. Sauro and Strickland (Chapter 10) discuss changes in the lipid bilayers associated with and possibly central to the fusion of myoblasts. Lucy (Chapter 3) details four successive stages of membrane coalescence in terms of the lipid bilayer, close apposition, hemifusion, fusion pore formation, and widening of fusion pore (Fig. 2, Chapter 3). In Lucy's view, the actual fusion process is driven by osmotic phenomena rather than by protein–lipid interactions.

The formation of a fusion pore by proteins occurs in several systems (Chapter 2). This local disruption in the bilayer or formation of a protein pore is followed by the amalgamation of the two biological membranes. Drawing on extensive data in a diverse number of systems, Plattner and Knoll (Chapter 2) review information about numerous systems to demonstrate that a previously proposed "focal membrane fusion" model can serve as a unifying concept for all forms of biomembrane fusion.

Membrane fusion mediates innumerable processes, all of which are essential to the survival and function of all eukaryotic cells. In Chapter 3, Lucy states, "there may well be a common molecular mechanism that underlies . . . membrane fusion events." The findings of Blobel *et al.* (1992), Plattner and Knoll (Chapter 2), and others also suggest this might be the case. If it is true, the current lack of concensus on a single, complete model for biomembrane fusion might be due in part to variations on a theme. Since several biomembrane fusion events can occur simultaneously in any single cell, it is unlikely that one type of regulatory system exists. To maintain cellular functions and avoid chaos, these processes must be regulated by different mechanisms. At the heart of many of these regulatory systems is signal transduction.

III. SIGNAL TRANSDUCTION: THE BASICS

Currently, the widely accepted view of signal transduction focuses on the binding of an extracellular ligand to a surface receptor that leads to the production of an intracellular signal that directs an immediate cellular response (Fig. 1). While hormonal interactions that alter transcriptional events are critical in long-term cellular processes, they do not appear to be involved

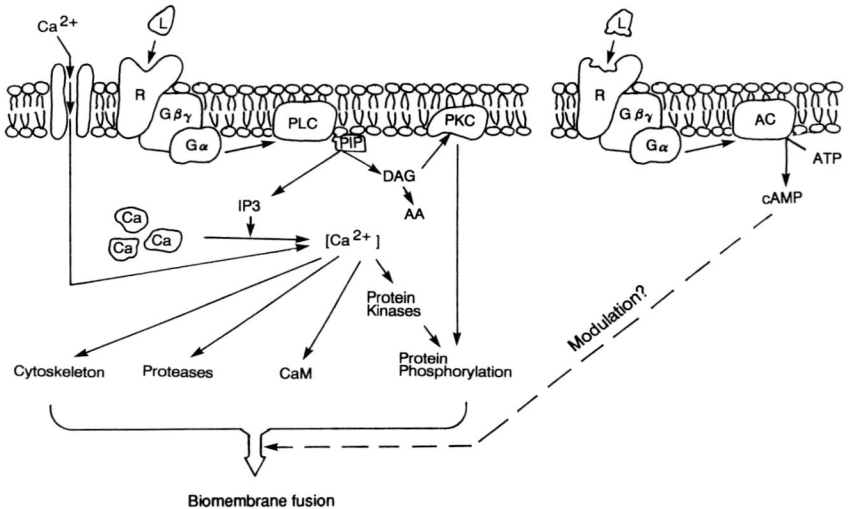

Fig. 1. A diagrammatic summary of some typical signal transduction events leading to biomembrane fusion. Calcium ions play a central role in most types of biomembrane fusion. Intracellular calcium levels can be raised by influx of calcium through channels or by release from intracellular stores. Inositol 1,4,5-trisphosphate (IP_3) can stimulate the release of calcium from such stores (e.g., in calciosomes). The intracellular IP_3 signal is typically generated by a ligand (L) binding to a membrane receptor (R). This binding activates GTPases, in this case heterotrimeric or G proteins. The α-subunit of the G protein then activates a membrane-bound phospholipase C (PLC), which hydrolyzes inositol 4,5-bisphosphate (PIP_2) to yield IP_3 and diacylglycerol (DAG). Calcium has numerous downstream effects, notably on the cytoskeleton, proteases, calmodulin (CaM), and protein kinases. Working in concert with DAG, calcium ions also activate protein kinase C (PKC). DAG may be further hydrolyzed to form arachidonic acid (AA). Various combinations of these downstream effects lead to biomembrane fusion in different systems. cAMP generated via signal transductive mechanisms has been implicated as an intracellular messenger in different aspects of biomembrane fusion, but usually functions to modulate the process.

in the immediate events of the regulation of biomembrane fusion. This book will mainly focus on membrane-receptor, G protein-mediated signal transduction mechanisms because they are central to the regulation of biomembrane fusion in all systems studied to date. However, as will become increasingly clear, our understanding of signal transduction mechanisms, especially as they relate to biomembrane fusion, is still in a state of flux.

In the traditional signal transduction pathway there is a well-known list of constituents (Fig. 1; see also Figs. 2 and 3, Chapter 4; Fig. 7, Chapter 11; Fig. 8, Chapter 12). A surface protein receptor exists that binds to specific ligands. The diverse nature and role of cell surface receptors and many of the component events involved in signal transduction in eukaryotes have recently been reviewed in detail (*Trends in Genetics,* Vol. 7, Issue 11/12, pp. 343–418, 1991). Once the ligand is bound to its receptor, the interaction activates intermediary guanosine triphosphate (GTP)-binding proteins, or G proteins, that are composed of three subunits (α and $\beta\gamma$). Leyte *et al.* (Fig. 1, Chapter 7) detail the events of the trimeric G protein cycle during signal transduction and the requisite role of GTP hydrolysis.

Typically, the α subunit of the heterotrimeric G proteins subsequently activate specific membrane-bound enzymes. A diverse number of different G proteins (e.g., G_o, G_i, G_s, G_t, and G_E) are known to exist that are usually defined on the basis of known functions or on specific inhibition/stimulation by certain agents (e.g., GTPγS, aluminum fluoride, and pertussis and cholera toxins; Freissmuth *et al.*, 1989). Some of these G proteins are detailed in the following chapters. Recently, other GTPases in addition to the heterotrimeric G proteins have been implicated in aspects of biomembrane fusion.

Small, ras-like GTPases show organelle-specific localization and are believed to function during vesicle formation and synaptic vesicle exocytosis, as well as during recognition and possible fusion between vesicles and their target compartment (Barr *et al.*, 1992; von Mollard, 1991). Future work may verify that the fusion of vesicles for nuclear envelope assembly also involves unique GTP-binding proteins (Boman *et al.*, 1991). To futher add to the complexity of the G protein story, Federman *et al.* (1992) have demonstrated that two mammalian adenylyl cyclases can be stimulated by hormone action via G_i protein $\beta\gamma$ subunits. Future work may further reveal that this alternative signaling mechanism, thought previously only to operate in yeast cells, may also function during some types of biomembrane fusion. Certain receptors have been shown to work without G protein mediation but, to date, biomembrane fusion events regulated directly by receptor tyrosine kinases (RTKs) without G protein mediation have not been reported.

Phospholipase C (PLC) and adenylate cyclase (AC) are two of the most widely studied enzymes to be regulated directly by G proteins (Fig. 1). As seen throughout this volume, numerous other enzymes and membrane proteins also have been shown to be regulated in response to hormone action, including phospholipase A_1, phospholipase D, cGMP phosphodiesterase, K^+ channels, and Ca^{2+} channels (Freissmuth *et al.*, 1989).

Once activated, phospholipase C hydrolyzes inositol 1,4-bisphosphate (PIP_2) to produce inositol 1,4,5-trisphosphate (IP_3) and diacylglycerol (DAG) (Fig. 1). Subsequently, the dual pathway effect of these products results in the release of calcium from intracellular vesicular stores (e.g., calciosomes) by IP_3 and the activation of protein kinase C (PKC) by DAG acting in concert with calcium (Berridge, 1987; O'Day, 1990; see also Chapters 4, 5, 6, 9, 10, and 11). The release of calcium and the activation of PKC result in an intracellular cascade of subsequent events, many of which involve protein phosphorylations (kinases) and dephosphorylations (phosphatases) (e.g., Chapter 2). As touched on in many of the chapters, the activation of numerous calcium-binding proteins (CaBPs), such as calmodulin (CaM), also occurs. Calmodulin, for example, then modulates the functions of calmodulin-binding proteins (CaMBPs, e.g., Chapter 12) such as myosin light chain kinase (MLCK; Chapter 10). These and other downstream regulatory elements are discussed to varying degrees in chapters throughout the book.

In recent years, other polyphosphoinositides in addition to IP_3 have been implicated in diverse aspects of intracellular signaling (e.g., Stephens *et al.*, 1991). The metabolism of DAG leads to the production of other lipids, some of which have been found to be fusogenic [e.g., arachidonic acid (AA)] in myoblasts and in macrophages (Chapter 8). This dual pathway is fundamental to a large number of regulated biomembrane fusion events.

When AC is activated, adenosine triphosphate (ATP) is hydrolyzed to yield cyclic AMP (cAMP), which subsequently activates cAMP-dependent protein kinase A (PKA), among other things, to set up an intracellular cascade mechanism leading to a specific cellular response. In most cases, the role of cAMP and other nucleotides in biomembrane fusion remains enigmatic. For example, cAMP inhibits cell fusion in *Dictyostelium* (Chapter 12), but later in the development of the same cells, it appears to enhance phagocytosis (Chapter 8). In muscle development, cAMP and PKA function during myogenesis, but it is unlikely they play a role in the fusion process itself (Chapter 10). The accumulated data suggest that, in many systems, cAMP plays a secondary or modulatory role in biomembrane fusion (e.g., during exocytosis; Chapter 5) but much remains to be learned about these functions. There are many other examples where the cAMP and inositol phosphate pathways intersect during various cellular processes, so it is likely that such cross-talk will also occur during many biomembrane fusion events (O'Day, 1990).

IV. SIGNAL TRANSDUCTION DURING BIOMEMBRANE FUSION

The means by which signal transduction is initiated as a prelude to bio-membrane fusion varies. Hormonal binding to a surface receptor clearly mediates most examples of regulated secretion (Chapters 5 and 6) and possibly chick myoblast fusion (Chapter 11). It is likely cell–cell contact sets in motion signal transductive events leading to phagocytosis (Chapter 8) and fertilization in diverse species (Chapters 4 and 12). In many of these initial cell–cell interactions, calcium ions play a role.

Calcium is a universal mediator of biomembrane fusion. During cell fusion, calcium ions function both intracellularly during the coalescence of the lipid bilayers and extracellularly during the initial cell adhesion/recognition step (Chapters 9–12). This first, often calcium-dependent, interaction between fusogenic membranes is clearly protein-mediated (Chapter 2). Once the proteins have done their work, the actual amalgamation of the bilayers remains to be completed. Lucy (Chapter 3) presents extensive data arguing that osmotic phenomena play an important role in this event.

Since calcium serves as the primary trigger in a large number of bio-membrane fusion events, it follows that signal transduction via the IP_3 pathway or via the opening of Ca^{2+} channels in the cell membrane will be critical. In both cases, G proteins are known to mediate enzyme activation or ion channel opening. The importance of calcium channels is discussed in several chapters. Specifically, GTP binding to the α subunit of heterotrimeric G proteins allows the α GTP complex to dissociate from the $\beta\gamma$ subunits and subsequently bind to its target protein. Hydrolysis of the bound GTP is essential for the cycle to function effectively in the signal transduction process.

In Chapter 7, Leyte *et al.* examine this cycle and the role of trimeric G proteins during secretory vesicle formation. Using a cell-free system, they reveal that secretory vesicle formation from Golgi membranes is under negative control mediated by an isomer of a pertussis-sensitive family of G proteins called G_{i3}. These data are discussed in terms of the current understanding of intracellular membrane traffic and of signal transduction in general. The role of G proteins is also discussed, to varying degrees, in many of the other chapters. For example, in Chapter 8, Lewis *et al.* reveal that the developmental loss of a 52-kDa Gα subunit correlated with the end of vegetative phagocytosis, followed subsequently by its cell-specific accumulation in association with the onset of cannibalistic phagocytosis. These authors have also shown a developmental increase in several low-molecular weight proteins that bind antibodies directed against Gα subunits, but the developmental role of the proteins remains speculative. Low-molecular weight or ras-like GTP-binding proteins have been previously implicated in the regulation of intracellular membrane traffic (Barr *et al.*, 1992; Pfeffer, 1992).

The identification of G protein-mediated fusion of intracellular vesicles raises the question of conservation of signal transduction components. Do intracellular fusion processes use similar transduction mechanisms as those that mediate cell surface membrane fusion events? If so, what is the arrangement of the signal transduction components within intracellular membranes? With the progress being made on intracellular membrane trafficking, this question may soon be answered (Chapter 7; see also the following discussion).

Several chapters cover the activation of PLC, which results in IP_3 and DAG production as preludes to intracellular calcium release and PKC activation during biomembrane fusion. After reviewing the events of myoblast fusion and the role of signal transduction in the component processes, Sauro and Strickland (Chapter 10) summarize their current research on the regulation of lipid metabolism during skeletal myogenesis. It is clear that PIP_2 breakdown functions to maintain membrane-associated PKC activity, which is essential in the steps leading to fusion. After fusion, PIP_2-specific PLC is down-regulated with a concomitant attenuation of PKC activity. The authors incorporate this data into a revised mechanism for myoblast fusion. PKC also comes under scrutiny in Chapter 11.

Extensive pharmacological studies on chick myoblast fusion reported by David and Fitzpatrick (Chapter 11) give rise to a working model for the signal transduction events leading to prostaglandin-mediated fusion of myoblasts. Central to this model is the binding of prostaglandin E_1 (PGE_1) to its receptor, leading to an increase in phosphoinositide metabolism and an associated signal cascade mechanism primarily involving Ca^{2+} and PKC.

The phosphoinositides and calcium have been shown to play several critical roles during fertilization in many animal species. During fertilization in animals, calcium fulfills its role as a central mediator of biomembrane fusion because it is involved in everything from gamete recognition to exocytosis of both sperm (acrosome) and egg (cortical granule) components. In Chapter 4, Kline provides an extensive review of the literature that culminates with two alternative signal transduction mechanisms that could lead to the intracellular calcium transient. This is coupled with two models that explain how the release of calcium ions could lead to cortical granule exocytosis. Several chapters focus on exocytosis, continuing to expand an understanding of the way the calcium signal is generated and how it serves as an intracellular messenger.

The pathway of exocytosis is one of the most well-defined biological processes (Butcher, 1978). In many examples of exocytosis, including the release of cortical granules from eggs, trichocysts from *Paramecium,* mast cell degranulation, and neurotransmitter release from nerve cell synaptosomes, the secretory organelles are docked at the cell surface. These exocytotic events are like a cocked gun, with calcium as the trigger.

Examining signal transduction during exocytosis in permeablized mast cells, Koopmann has detailed the interdependent roles of calcium, G proteins, PLC, and PKC (Chapter 5). After reviewing the critical role of calcium, attention turns to permeabilized cell systems and the development of an operational mast cell model. Comparing his results with others, Koopmann then analyzes the requirement for G proteins, concluding that their precise role during exocytosis remains enigmatic. Pharmacological perturbations strongly argue that PKC plays a modulatory rather than primary role in exocytosis from mast cells. Koopmann's work reveals the importance of comparative analyses between both nonpermeablized and permeablized cells in the search for the critical events of signal transduction leading to exocytosis in all systems.

To be effective, calcium needs intracellular targets in order to carry out specific cellular functions. Working in concert with DAG, calcium ions alter the activity of protein kinase C. Protein kinase C is focused on in detail for numerous types of biomembrane fusion. After reviewing the structure and function of platelets, Gerrard et al. (Chapter 6) address the importance of PKC in two principal processes: (1) platelet granule movement and (2) granule membrane fusion with the canalicular and plasma membranes. Calcium-mediated cytoskeletal activity results in granule redistribution within the platelet, whereas activation of PKC, probably via histamine and PLA_2, leads to membrane fusion. The recent identification of specific molecules (e.g., MARCKS) is helping to clarify the linkage of events between calcium function, protein kinase activation, and cytoskeletal regulation that are involved in biomembrane fusion and other cellular processes (e.g., Hartwig et al., 1992).

In Chapter 9, Prives also focuses on the calcium signal transduction pathway leading to protein phosphorylation and cytoskeletal operation. Prives begins his chapter with a review of the developmental events of primary cultures of chick skeletal muscle and the essential requirement for calcium ions. Operating at both extracellular and intracellular levels, this divalent cation functions in the initial events of cellular adhesion between myoblasts and in the subsequent steps resulting in cell fusion. At least some of calcium's intracellular functions are mediated by calmodulin and its binding proteins, especially myosin light chain kinase. Protein kinase C also phosphorylates myosin but at sites that differ from MLCK and with opposing effects. The role of these phosphorylations in the actin–myosin interactions involved in myoblast fusion is detailed in a way that must be done for other biomembrane fusion events to achieve meaningful and universal hypotheses.

Calcium also operates independently on a number of other intracellular CaBPs including Ca-dependent PK, cytoskeletal constituents, and calmodulin. Calmodulin is a universal mediator of calcium function in eukaryotic cells and is involved in regulating diverse types of biomembrane fusion from fertilization to exocytosis.

In Chapter 12, Lydan and O'Day review the importance of calcium and calmodulin during cell and pronuclear fusion in the model eukaryote *Dictyostelium*. Extensive work has demonstrated the importance of calcium and calmodulin during these events. Specific CamBPs have recently been identified that are associated with the events of gamete differentiation (CaMBP 91 kDa) and cell fusion (CaMPBs 48 and 38 kDa). In keeping with the critical role of calmodulin as a positive regulator of cell fusion, a powerful, low-molecular weight, hydrophobic, endogenous inhibitor of calmodulin activity is produced and secreted by zygotes to prevent other cells from fusing. Cyclic AMP, a chemoattractant secreted by the giant cells, also inhibits cell fusion. A dynamic, working model incorporating this and other known information is presented for cell fusion during sexual development of this model organism.

The importance of down-regulation in the control of biomembrane fusion is also discussed in other chapters. Whereas protein phosphorylation has clearly been shown to be important during biomembrane fusion, dephosphorylation of phosphoproteins is also implicated. For example, Plattner (Chapter 2) suggests that dephosphorylation of an intramembrane particle (IMP) protein might be critical to exocytosis. A 63-kDa phosphoprotein associated with the cortex of *Paramecium* is dephosphorylated in association with trichocyst discharge and thus is implicated in the process. Finally, the modulatory role of cAMP is mentioned in several chapters (e.g., Chapters 5, 10, and 12), but the accumulating evidence indicates this cyclic nucleotide may play a more central role in phagocytosis.

Although phagocytosis is a fundamental process that is essential for everything from food uptake in protozoans to the removal of infectious particles by macrophages, the accumulated information on signal transduction during phagocytosis in any one system is sketchy at best. Lewis *et al.* (Chapter 8) have begun investigating the regulation of amebal uptake by zygote giant cells in *Dictyostelium*. Having previously characterized the fundamental events of the process and the importance of specific cell surface glycans as mediators of cannibalistic phagocytosis by giant cells, they have recently turned their attention to signal transductive events. Although the importance of signal transduction involving calcium remains enigmatic, their data suggest that a G protein-mediated signaling sequence involving adenylyl cyclase is importance for the species-specific ingestion of amebas by zygote giant cells. The simplicity of the system and the ability to purify zygote giant cells and amebas facilitates continued studies in this model system.

V. FINAL COMMENTS

In spite of its central role in every aspect of cell life, the precise events of biomembrane fusion remain a mystery. Although still in its comparative infancy, the study of signal transduction during biomembrane fusion has already yielded some interesting results. Once more is known about this process, it may be feasible to use the information in diverse ways.

The ability to enhance biomembrane fusion events involved in secretion would be useful for increasing the biotechnological production of specific secretory molecules. The ability to selectively inhibit biomembrane fusion could provide routes to developing new contraceptives (i.e., inhibiting fertilization), to producing antiviral products (e.g., inhibit uptake, spreading, and release of enveloped viruses), or to reducing the debilitating effects of high-levels of secreted molecules (e.g., preventing the release of enzymes associated with arthritis and metastasis). The specific targeting of cells for the incorporation of drugs or genes might be enhanced with a better knowledge of signal tranduction events involved in biomembrane fusion in specific cell types. Thus, a targeted liposome's efficacy might be enhanced if it was presented at the same time as a specific pharmacolgical agent against nontarget cells or a signal transduction pathway component of target cells. Similarly, it might be possible to enhance phagocytic removal of tumor cells and infectious agents by selective induction of phagocytosis in macrophages.

Whereas it may be difficult in the short-term to identify the precise sequence of events that occur when the bilayers coalesce, it should be relatively easy to characterize the events of signal transduction that lead up to these events. From this will come insight into the receptors, G proteins, and effector molecules that drive this fundamental but universally important process. In this regard, the accumulating evidence suggests some biochemical and molecular similarities between intracellular membrane fusions and those occurring at the cell surface. The involvement of diverse types of GTPases, especially trimeric G proteins, in membrane traffic might be reflective of an evolutionary conservation of fundamental transduction processes with modifications that enable them to function in diverse cellular environments.

ACKNOWLEDGMENTS

I especially thank Darren Browning for his critical and helpful suggestions for improving this work. Kim Gunther, Keith Lewis, Mike Lydan, and Raymund Rama are thanked for their comments on the original draft of this chapter. This work was supported by a grant from the Natural Sciences and Engineering Research Council of Canada.

REFERENCES

Barr, F. A., Leyte, A., and Huttner, W. B. (1992). Trimeric G proteins and vesicle formation. *Trends Cell Biol.* **2,** 91–94.

Berridge, M. J. (1987). Inositol trisphosphate and diacylglycerol: Two interacting second messengers. *Annu. Rev. Biochem.* **56,** 159–193.

Blobel, C. P., Wolfsberg, T. G., Turck, C. W., Myles, D. G., Primakoff, P., and White, J. M. (1992). A potential fusion peptide and an integrin ligand domain in a protein active in sperm–egg fusion. *Nature* **356,** 248–252.

Boman, A. L., Delannoy, M. R., and Wilson, K. L. (1991). GTP hydrolysis is required for vesicle fusion during nuclear envelope assembly *in vitro. J. Cell Biol.* **116,** 281–294.

Brown, V. I., and Greene, M. I. (1991). Molecular and cellular mechanisms of receptor-mediated endocytosis. *DNA Cell Biol.* **10,** 399–409.

Burgoyne, R. D. (1991). Control of exocytosis in adrenal chromaffin cells. *Biochim. Biophys. Acta* **1071,** 174–202.

Butcher, F. R. (1978). Regulation of exocytosis. *In* "Biochemical Actions of Hormones" (G. Litwack, ed.), Vol. 5, Ch. 2, pp. 53–99. Academic Press, New York.

Federman, A. D., Conklin, B. R., Schrader, K. A., Reed, R. R., and Bourne, H. R. (1992). Hormonal stimulation of adenylyl cyclase through G_i protein beta subunits. *Nature* **356,** 159–161.

Freissmuth, M., Casey, P. J., and Gilman, A. G. (1989). G proteins control diverse pathways of transmembrane signaling. *FASEB J.* **3,** 2125–2131.

Hartwig, J. H., Thelen, M., Rosen, A., Janmey, P. A., Nairn, A. C., and Aderem, A. (1992). MARCKS is an actin filament, cross-linking protein required by protein kinase C and calcium–calmodulin. *Nature* **356,** 618–622.

Hoekstra, D., and Kok, J. W. (1989). Entry mechanisms of enveloped viruses. Implications for fusion of intracellular membranes. *Biosci. Rep.* **9,** 273–305.

von Mollard, G. F., Südhof, T. C., and Jahn, R. (1991). A small GTP-binding protein dissociates from synaptic vesicles during exocytosis. *Nature* **349,** 79–81.

Monk, J. R., De Toledo, G. A., and Fernandez, J. M. (1990). Tension in secretory granule membranes causes extensive membrane transfer through the exocytotic fusion pore. *Proc. Natl. Acad. Sci. USA* **87,** 7804–7808.

O'Day, D. H., ed. (1990). "Calmodulin as an Intracellular Messenger in Eucaryotic Microbes," 22 chapters, 418 pages. American Society for Microbiology, Washington, D.C.

Pfeffer, S. R. (1992). GTP-binding proteins in intracellular transport. *Trends Cell Biol.* **2,** 41–45.

Rothman, J. E., and Orci, L. (1992). Molecular dissection of the secretory pathway. *Nature* **335,** 409–415.

Stephens, L. R., Hughes, K. T., and Irvine, R. F. (1991). Pathway of phosphatidyl-inositol(3,4,5)trisphosphate synthesis in activated neutrophils. *Nature* **351,** 33–39.

Zaks, W. J., and Creutz, C. E. (1990). Evaluation of the annexins as potential mediators of membrane fusion in exocytosis. *J. Bioenerg. Biomembr.* **22,** 97–120.

2

The "Focal Membrane Fusion" Model Revisited: Toward a Unifying Structural Concept of Biological Membrane Fusion

HELMUT PLATTNER AND GERD KNOLL

Faculty of Biology
University of Konstanz
Konstanz, Germany

I. INTRODUCTION

The "focal membrane fusion" concept (Fig. 1) was originally formulated to characterize structural events occurring during membrane fusion in the course of exocytosis (Plattner, 1981). Its essentials are that biological membrane fusion, in most cases, is rapid and occurs on a very small focus in the presence

19

SIGNAL TRANSDUCTION DURING
BIOMEMBRANE FUSION

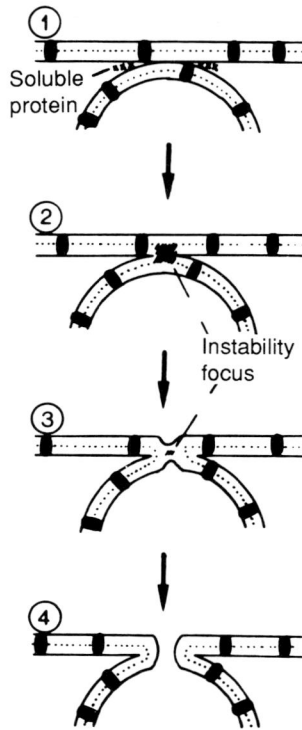

Fig. 1. The "focal membrane fusion" concept as presented for exocytosis by Plattner (1981). Note the occurrence of integral and membrane-associated proteins at the contact site, as well as the occurrence of a small instability focus engaged in lipid rearrangement. Dotted lines indicate lipid bilayer halves, which are separated by freeze-fracturing, thus exposing IMPs. Reproduced with permission.

of membrane-intrinsic and membrane-associated proteins. This chapter documents how this scheme might be applicable to all different kinds of biological membrane fusion.

A. Occurrence and Type of Biological Membrane Fusion

Figure 2 summarizes the numerous types of membrane fusion occurring in eukaryotic cells. This mainly includes cell–cell fusion, cell division, endo-

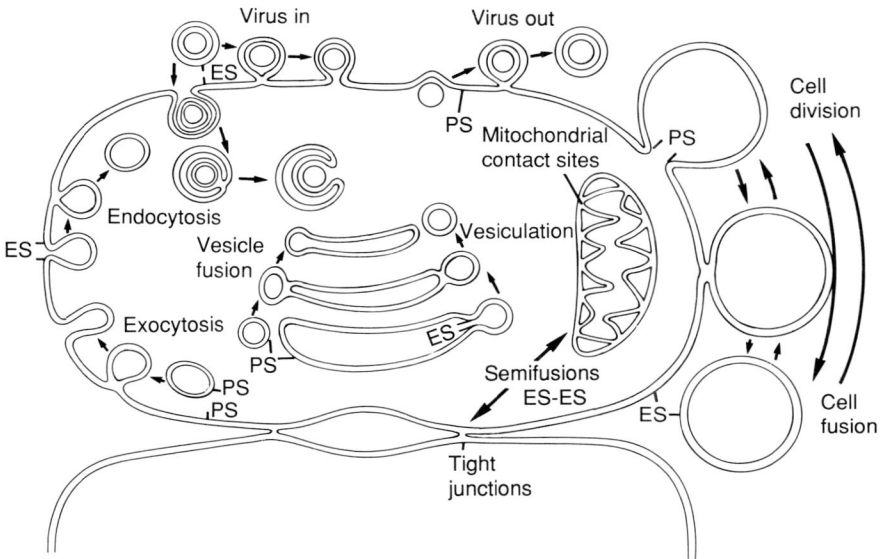

Fig. 2. Scheme of membrane fusion processes. ES, exoplasmic surface; PS, plasmatic surface involved in E- or P-type membrane fusions. For detail, see text. From Knoll and Plattner (1989), reproduced with permission.

cytosis, exocytosis, cytoplasmic vesicle fusion, viral membrane–cell membrane fusion and virus–lysosomal membrane fusion, as well as virus budding (Knoll and Plattner, 1989). Other types may be more or less persistent "hemifusion" stages, such as the "contact sites" between the inner and outer membranes of mitochondria or chloroplasts, or possibly also the cell membrane contacts of "tight junctions," that is, between adjacent epithelial cells. Hemifusions will not be considered further (for detail, see Knoll and Plattner, 1989).

These different types of membrane fusions can be divided into two categories (Stossel *et al.,* 1978; Knoll and Plattner, 1989): (1) Plasmatic faces fuse (PF-type of membrane fusion) as in cell division, exocytosis, cytoplasmic vesicle fusion, and virus budding; and (2) exoplasmic faces fuse (EF-type) in cell–cell fusion, endocytosis, and virus–lysosome fusion. This terminology is based on the following. Some subcellular components are—via "membrane flow"—potentially connected to the extracellular space, and thus are "exoplasmic." This holds true for intracellular vesicle transport (and, according to the endosymbiont hypothesis, also for the organelles with a double membrane, i.e., mitochondria and chloroplasts).

PF- and EF-type fusion processes might have quite different requirements, considering, for example, the difference in free calcium concentrations (pCa) available, asymetry of lipids, and the presence or absence of a glycokalix. Intracellular PF-type fusions entail the coalescence of two membranes to a continuum, while the opposite holds true for virus budding and fusion during cell division. EF fusions again result in a membrane continuum. This implies that membrane fusion can mean the merger of two membranes to form one continuum ("fusion" proper) or a separation process, for example, cell division or the pinching off of a membrane enveloped vesicle ("fission"). For detail, see Fig. 2.

B. Problems and Methods in Membrane Fusion Analysis

Most membrane fusions are now generally considered extremely fast and restricted in space. Assuming a duration of approximately 1 msec and a spatial extension of about 10 nm or even less as typical values, for example, for membrane fusion during exocytosis (Almers, 1990; Almers, *et al.,* 1989; Almers and Tse, 1990; Knoll and Plattner, 1989; Plattner, 1989a,b, 1991), this would imply the following. (1) Only a very small fraction of membrane lipids would be involved. (2) It would be more difficult to obtain relevant biochemical data, for example, on lipid modifications, than to obtain insight into structural rearrangements by methods with high-temporal and spatial resolution, that is, by electron microscopy after fast-freezing or, for the precise time sequence, by electrophysiology (in the absence of any precise topological correlation).

Electron microscope analysis of biomembrane fusion has been hampered for a long time by artifacts induced by chemical fixatives and antifreeze impregnation (Plattner, 1981; Plattner and Bachmann, 1982; Chandler, 1988; Knoll and Plattner, 1989). Physical fixation (cryofixation) is enormously faster and operates with a time resolution in the submillisecond range (Plattner and Bachmann, 1982; Chandler, 1988). Subsequent freeze-fracturing allows the visualization of intramembranous differentiations, such as IMPs (intramembrane particles, in most cases equivalent to intrinsic membrane proteins; Zingsheim and Plattner, 1976; Plattner and Zingsheim, 1983). With freeze-fracture replica techniques, spatial resolution is restricted to the nm range. Restrictions in resolution are imposed by more factors than generally assumed, in particular by material dislocation during and after freezing, by grain size of the replication material, and by capillary condensation (Zingsheim and

Plattner, 1976). This is to be considered particularly for visualizing small fusion pores (see Section II,A,1).

Electrophysiological measurements now allow the kinetics of single membrane fusion events to be registered with even higher precision by following membrane capacitance changes (Neher and Marty, 1982; Almers and Neher, 1987; Zimmerberg, 1987; Almers, 1990; Spruce *et al.,* 1990). For instance, with mutant beige mouse mast cells, Breckenridge and Almers (1987), as well as Zimmerberg *et al.* (1987), could thus record exocytotic membrane fusion within a range of 1 msec and 2 nm. Also, it is easier to analyze non-synchronous events in this way than with electron microscopic methods. It must be said, however, that the "focal membrane fusion" concept was elaborated on much earlier, with the use of advanced electron microscopic techniques, primarily with correct figures on time course and spatial extension (see Plattner, 1981, 1989a,b, 1991). The essentially new contribution was the hypothetic assumption of fusion pore forming proteins (Almers *et al.,* 1989; Almers, 1990; Almers and Tse, 1990).

This chapter shows how data obtained by appropriate electron microscopic analysis and by recent electrophysiological recordings all converge for widely different types of biological membrane fusion and that they all support the "focal fusion concept" formulated a decade ago (Plattner, 1981). Interestingly, this concept may hold true for both PF- and EF-type fusions. Only in some cases may fusion intermediates be long-lived. Combining the widely different methods now available will also allow the specification of the original concept of the involvement of membrane-associated and membrane-integral proteins in biomembrane fusion (Plattner, 1981), which at the time it was proposed, was also quite heretical.

II. CURRENT RESEARCH

Considerable efforts are currently being undertaken toward an understanding of the mechanisms of the kinds of membrane fusion listed in Section I (Fig. 2). It is also desirable to analyze the actual kinetics and spatial extension for all types of biomembrane fusions. However, both ultrastructural and electrophysiological analysis still frequently encounter insurmountable difficulties, such as low frequency of fusions (Knoll *et al.,* 1987) and the small size of the organelles involved, for example, in intracellular vesicle fusion.

The systems analyzed in most detail up to now are viral fusion processes and exocytosis.

A. General Aspects of Exocytosis and Endocytosis

On the ultrastructural level it is difficult in most systems to differentiate unequivocally between exocytotic and endocytotic processes, because in most cases triggered (regulated) exocytosis is rapidly coupled to endocytosis, whereas constitutive (nonregulated) exocytosis is difficult to pinpoint. Therefore, examples of a clear-cut distinction between exocytosis and endocytosis are discussed separately in Sections II,B and II,C.

Different strategies have been employed for a fast-freezing, freeze-fracture analysis. Massive stimulation, for example, within ≤1 min, of mast cells (Chandler and Heuser, 1980), sea urchin egg cells (Chandler and Heuser, 1979; see also Chandler, 1988), or *Limulus* amebocytes (Ornberg and Reese, 1981) allows the analysis of preferably or exclusively exocytotic membrane fusion. This also works for gland cells such as the chromaffin cells from the adrenal medulla (Fig. 3). They also have been stimulated for short periods and the preponderance of exocytosis has been ascertained by registering secretory product release (Schmidt *et al.*, 1983), although some endocytosis certainly goes on at the same time. Highly synchronous exocytosis has been achieved in

Fig. 3.   Focal membrane fusion occurring during exocytotic catecholamine release from a bovine chromaffin cell (freeze-fracture). Note the volcanolike deformation of the plasma membrane, the craterlike pit is the fusion spot. No IMP rearrangement or diaphragm formation occurs. Shadowing from bottom to top. Bar = 0.1 μm. Reproduced from Schmidt *et al.* (1983) with permission.

nerve terminals by electrical stimulation [with chemically (4-aminopyridine) induced pileup of opening stages] when triggering exocytosis was combined with the cryofixation process (Heuser et al., 1979; Torri-Tarelli et al., 1985). A considerable degree of synchrony (<1 sec), combined with high-efficiency (up to 95% of all secretory organelles), was achieved with *Paramecium* cells, a ciliated protozoan (Plattner et al., 1985; Plattner, 1987; Knoll et al., 1991a). With nerve terminals and paramecia, the occurrence of predictable, morphologically defined fusogenic sites further facilitates the analysis of fusion processes. However, only with *Paramecium* cells can the potential fusion zone be pinpointed within an area of several times 10 nm, where it contains an IMP aggregate ("fusion rosette") that is surrounded by concentric IMP arrangements ("rings") (Plattner et al., 1973; Beisson et al., 1976; see Plattner, 1987 for a review).

Another important contribution comes from the development of a new quenched-flow procedure (rapid mixing of cells with trigger agent and freezing at defined time points), applicable even to sensitive cells (Knoll et al., 1991a). This method has been successfully applied to analyze the secretory process in paramecia with an even higher degree of synchrony. It can be shown that synchronous exocytotic membrane fusion is accomplished within 80 msec, and that contents extrusion is rapidly followed by synchronous (80–350 msec) endocytotic membrane resealing and subsequent detachment of empty ghost membranes. Furthermore, stepwise multiple mixing of cells with a stimulant or with inhibitors such as Ca^{2+} chelators, are possible (Knoll et al., 1991a). This now allows dissociation of endo- from exocytotic membrane fusion, which otherwise would not easily be discernible as discrete steps.

1. Exocytosis

Ultrastructural analysis of a variety of regulated exocytotic systems has shown that early fusion stages are represented by an especially small focus (approximately 10 nm or sometimes slightly larger, depending on the precision of the method; Fig. 3), sometimes seen as an IMP or as a corresponding pit (egg cells: Chandler and Heuser, 1979; mast cells: Chandler and Heuser, 1980; amebocytes: Ornberg and Reese, 1981; chromaffin cells: Schmidt et al., 1983; nerve terminals: Torri-Tarelli et al., 1985; *Paramecium:* Momayezi et al., 1987). This fusion intermediate must contain membrane lipids because it gradually expands later to a membrane continuum with an exocytotic opening the size of the vesicle diameter. (The likely involvement of proteins in fusion pore formation will be discussed later.) As first shown by Heuser et al. (1979), IMPs present in membranes before fusion are not rearranged before fusion and no lipidic diaphragm is formed during fusion (see Plattner, 1981, in contrast to sporadic, even quite recent claims for some systems, e.g., by Creutz et al., 1987; Pinto Da Silva, 1988; and in some recent cell biology textbooks).

Another general aspect of fusion (except for nerve terminals) is the formation of a funnel-like protrusion, at least on one of the membranes (Fig. 3). This was found with mast and egg cells (Chandler, 1988), amebocytes (Ornberg and Reese, 1981), chromaffin cells (Schmidt et al., 1983), and paramecia (Knoll et al., 1991a; Fig. 4). It might indicate the relevance of a small curvature for fusion capacity (see Plattner, 1989a,b). Its inconspicuous

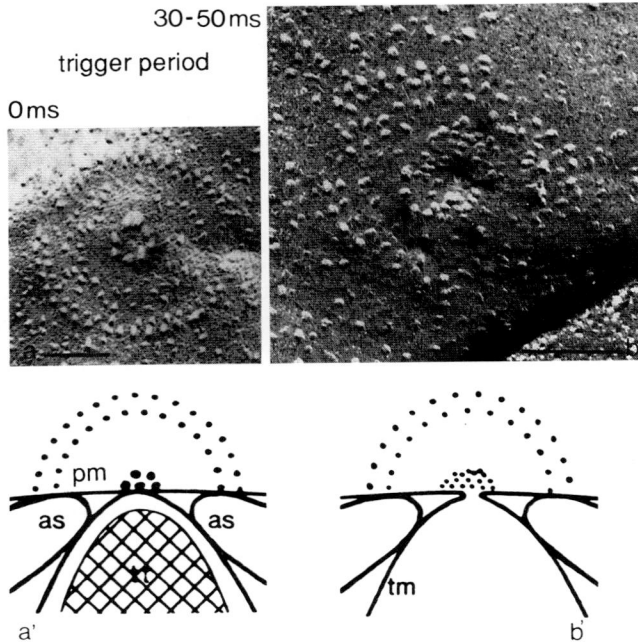

Fig. 4. In freeze-fracture replicas of *Paramecium* cells, the fusogenic zone of exocytotic trichocyst release is surrounded by a double ring of IMPs (its diameter corresponding to the final size of an exocytotic opening). Large rosette IMPs appear to disintegrate into subunits during focal membrane fusion. Whereas only one fusion is formed (arrowhead) in the protruding plasma membrane, the number of rosette IMPs might represent multiple potential "ignition" sites to ensure a quick trichocyst release response. Shadowing from bottom to top. as, alveolar sacs (subplasmalemmal Ca^{2+} stores); tm, trichocyst membrane; tt, trichocyst tip; pm, plasma membrane. Bar = 0.1 μm. Reproduced from Knoll et al. (1991a) with permission.

appearance in nerve terminals (Heuser *et al.*, 1979; Torri-Tarelli *et al.*, 1985) might be compensated by the small size of neurotransmitter vesicles.

Another feature that different kinds of exocytotic membrane fusion have in common is the presence of electron dense, amorphous to fibrillar materials ("membrane connecting material") at the fusogenic zone, for example, in protozoa, egg cells, some gland cells, and along "active zones" of nerve terminals (see Plattner, 1981, 1987, 1989a,b, 1991). In *Paramecium* cells the occurrence of calmodulin (CaM) and of CaM binding at preformed sites of trichocyst exocytosis has been established (Plattner, 1987). The *Paramecium* cell cortex also contains a phosphoprotein of 63 kDa, "PP63" (Stecher *et al.*, 1987), which becomes selectively and rapidly (<1 sec) dephosphorylated in the course of synchronous (1 sec) exocytosis and then rephosphorylated (5–40 sec) (Zieseniss and Plattner, 1985). Quenched-flow analysis shows convincingly the coincidence of the dephosporylation step with membrane fusion, both within 80 msec (Höhne-Zell *et al.*, 1992). It has to be assumed that the corresponding protein phosphatase also occurs in the *Paramecium* cortex to regulate PP63 dephosphorylation. Because antibodies against any of these components inhibit exocytosis, it can be concluded that these soluble proteins form part of the "connecting material" at fusogenic sites in paramecia and also form part of the stimulus–secretion coupling machinery (Plattner *et al.*, 1991).

The absence of any ultrastructurally recognizable "connecting material," as well as of fusion rosette IMPs from secretory organelle docking sites in "nondischarge" mutants (Beisson *et al.*, 1976; Pouphile *et al.*, 1986), underscores their relevance for membrane fusion. Some fusion-relevant components can be transferred from normal donor cells to nondischarge mutants, which then can be cured (Beisson *et al.*, 1980, Lefort-Tran *et al.*, 1981). This involves assembly of fusion rosettes and of connecting materials (Pape *et al.*, 1988). Double-trigger experiments with wild-type cells support this conclusion, since the capacity to dephosphorylate PP63 is paralleled by structural assembly of the fusogenic site (Zieseniss and Plattner, 1985). CaM is also quite distinctly localized at exocytotic sites in other systems (for review, see Plattner, 1989a). All this points to the importance of local assembly of these protein components (and probably of some additional binding proteins), thus, providing the capacity for focal membrane fusion.

Dephosphorylation, rather than phosphorylation, of some proteins might be a crucial step for actual membrane fusion (Plattner, 1989a) (if preparatory steps such as phosphorylation of ion channel and cytoskeletal elements are disregarded; see Plattner, 1989a). This is now becoming increasingly accepted as a general principle (Gomperts *et al.*, 1990; Whalley *et al.*, 1991). The role of PP63 is not known precisely in any system; but in *Paramecium*, PP63 is not an intrinsic, but rather a surface structure-bound protein (Höhne *et al.*, 1992). One might now speculate (and experiment further) along the following lines (Fig. 5). (1) Phosphorylation might be required for the assembly of rosette IMPs, for example, to regulate *Paramecium* surface pattern formation (Keryer *et al.*, 1987). This is not sufficient for the assembly of trichocyst docking sites because rephosphorylation occurs significantly faster than rosette reassembly (Zieseniss and Plattner, 1985). It might be, however, a prerequisite for the docking of newly formed secretory organelles. (2) PP63 dephosphorylation might be necessary for membrane fusion. It could transfer a membrane-associated phosphoprotein, like PP63, from a hydrophilic to a hydrophobic state and thus mediate a hydrophobic bridge between membranes achieving close contact and ultimately fusion (Plattner, 1989a). This mechanism could operate mainly on the organelle membrane because the cell membrane contains intrinsic (rosette) proteins at this site. (3) Alternatively, PP63 dephosphorylation might allow attached rosette IMPs to induce membrane fusion, if their cytoplasmic regions come in direct contact with the secretory organelle membrane, before they disperse as monomers. The precise mechanism is not yet known.

Along these lines, the following is an important recent finding. One rosette IMP is possibly made up of several subunits because during synchronous exocytosis (as analyzed by quenched-flow) rosette IMPs disappear more rapidly than can be accounted for by lateral diffusion in the cell membrane, whereas the number of much smaller IMPs in the fusogenic zone of the cell membrane increases with about sixfold stoichiometry (Knoll *et al.*, 1991a;

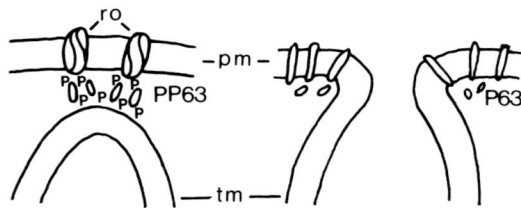

Fig. 5. Possible sequence of events during exocytotic membrane fusion in *Paramecium*. ro, rosette IMPs; pm, plasma membrane, tm, trichocyst membrane. PP63, P63, 63-kDa protein in its phosphorylated and dephosphorylated forms, respectively. P, phosphorylation sites. The scheme assumes occurrence of PP63 close to the membrane interaction zone, where it may serve for rosette assembly and/or reversible interaction with the trichocyst membrane and subsequent disintegration of rosette IMPs into subunits (compare with Fig. 4), thus allowing for lipid rearrangement in the course of membrane fusion.

Fig. 4). If a rosette IMP were an oligomeric protein, it could mediate, before total dissociation, the formation of a hydrophilic fusion channel (as postulated by electrophysiologists for other secretory systems; see Fig. 6). However, any attempt to directly visualize such a channel by freeze-etching would be difficult because of capillary condensation (Zingsheim and Plattner, 1976). Theoretically, one oligomeric rosette IMP would then suffice for membrane fusion initiation, but the availability of several "ignition sites" would make the exocytotic response safer (as proposed previously; Plattner, 1981), particularly in a system engaged in an instantaneous defense mechanism (Knoll *et al.*, 1991b).

Rosette IMPs in paramecia and in other protozoa recall IMP rows along "active zones" (also with attached electron-dense materials) in different types of nerve terminals (as reviewed by Harris and Landis, 1986; Rash *et al.*, 1988). Speculations on dense materials in neurons similar to ours can be found in Kelly (1988). For a more recent survey of components of presynaptic "dense materials" and their possible implications in vesicle docking and fusion, see Trimble *et al.* (1991).

Although no such conspicuous IMP aggregates are recognized with other exocytotic systems, the presence of single IMPs close to fusion sites (together with membrane-attached materials) might have the same bearing on fusion regulation as previously outlined.

Patch-clamp analysis of the early fusion events during neurotransmitter release has shown reversible conductance "flickering" (Almers *et al.*, 1989; Almers, 1990; Almers and Tse, 1990). This has been interpreted as a reversible formation of a small aqueous pore, possibly caused by a conformational change of an oligomeric fusogenic protein (Fig. 6). The recent discovery that synaptophysin (p38), an intrinsic protein of neurotransmitter vesicle membranes, forms hexamers (Thomas *et al.*, 1988) has led to widespread support of the implication that oligomeric protein aggregates may form pores in different kinds of membrane fusion events.

It then becomes necessary to look for matching oligomeric proteins in the cell membrane. Using aldehyde fixation and glycerol impregnation, Pfenninger and colleagues (1972) observed the occurrence of fusion "craters" and pores, studded with hexagonally arranged IMPs. Although this is now interpreted as an equivalent of pore-forming proteins (Südhof and Jahn, 1991), it remains problematic considering the preparation schedule used (see Section I,B). The "mediatophore" protein (Morel *et al.*, 1987) was considered a candidate (Thomas *et al.*, 1988; Morel *et al.*, 1991), although the function assigned by these authors was thought to be different, namely the release of acetylcholine from the cytosolic compartment rather than from vesicles (Israel

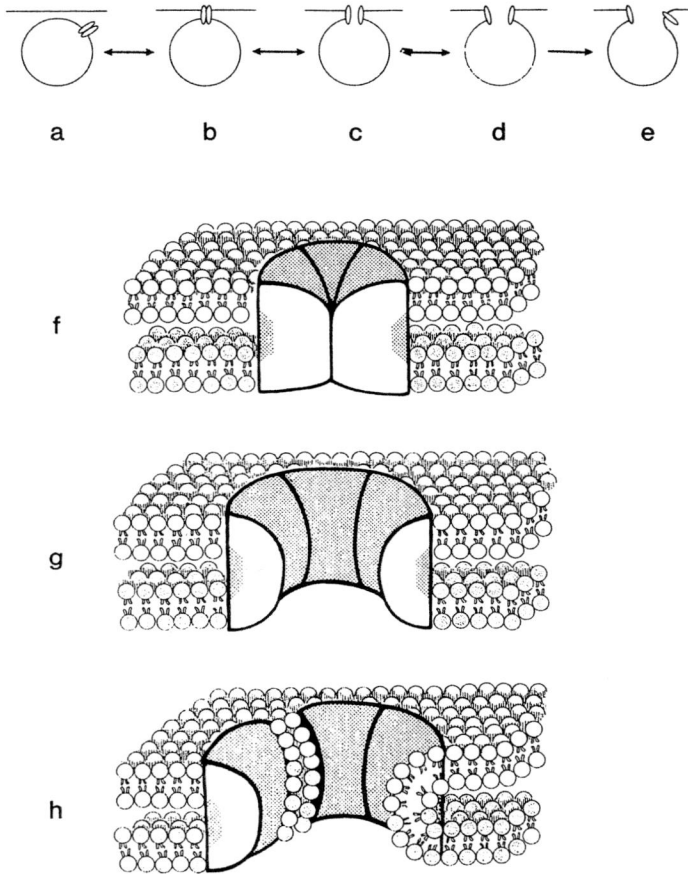

Fig. 6. This scheme by Almers (1990) characterizes current discussions on the formation of a fusion focus during exocytosis by neurotransmitter vesicles. This hypothesis assumes subunits of fusogenic proteins to undergo a conformation change (or other rearrangement) during fusion induction, thus allowing for formation of a hydrophilic pore and the rearrangement of lipids to form a membrane continuum. The scheme presented assumes one type of fusogenic protein penetrating both the membrane of a neurotransmitter-storing organelles and the cell membrane (see text, however). Reproduced with permission.

et al., 1989). Later on physophilin (36 kDa) was assumed to be a probably oligomeric equivalent (Thomas and Betz, 1990). The occurrence of matching oligomers in both membranes to be fused is reminiscent of gap junctions, although synaptophysin is genetically different from connexins (Bennett *et al.,* 1991). Synaptoporin, largely homologous to synaptophysin, is now considered by some as another candidate for a fusion-mediating protein in synaptic vesicles (Knaus *et al.,* 1990), and oligomeric SVp25 is nominated by others (Volknandt *et al.,* 1990). In a recent review, Südhof and Jahn (1991) advocate synaptophysin, whereas synaptotagmin might operate in vesicle docking and/or fusion.

The actual engagement of any of these oligomeric proteins in exocytotic membrane fusion still needs to be shown, although the concept is supported by the electrophysiological data and by the decay of rosette IMPs observed by the authors during trichocyst exocytosis. For reasons discussed in Section II,C, it is not imperative, however, that matching oligomeric IMP sets on both membranes be mandatory, and no IMPs are seen on the tips of trichocysts (Plattner, 1987). Furthermore, from the absence of synaptophysin from cell types other than neurons, one would have to postulate the occurrence of quite different fusogenic proteins in other secretory systems. For further critical discussion, see Section III.

2. Endocytosis

For endocytosis, few fast-freezing analyses are available because timing and identification of the process in freeze-fracturing are even more difficult than with exocytosis.

Only in paramecia, did clear dissociation of endocytosis from trichocyst exocytosis (see discussion earlier) allow selective visualization of the ultrastructural fusion intermediate of endocytotic membrane fusion. Again, it is focal, represented by an IMP or a pit, respectively, located on a protrusion of the cell membrane (unpublished observations). Thus, this process has a morphology quite similar to fusion during exocytosis. This is remarkable, because we are dealing alternatively with PF- or with EF-type fusion.

B. Endocellular Membrane Fusions

Very little is known on an ultrastructural level—and nothing at all on an electrophysiological level—about endocellular membrane fusion, although the biochemical background on these fusion processes has increased enormously. Everything points toward the importance of a variety of membrane proteins for docking and fusion.

Some Golgi-associated vesicles are surrounded by a nonclathrin coat (Orci *et al.*, 1986). Their fusion capacity depends on adenosine triphosphate (ATP) and soluble proteins and is sensitive to *N*-ethylmaleimide (NEM), (Wilson *et al.*, 1989). A NEM-sensitive fusion factor (NSF) is required for fusion processes not only in the Golgi region, but also for endosome fusion (Diaz *et al.*, 1989). Other possible relevant factors are small GTP-binding proteins (G proteins), including different Sec gene products in yeast (Bowser and Novick, 1991). The mutated Sec17 gene in yeast allowed assessment of its identity with an NSF attachment protein (SNAP) operating in the endoplasmic reticulum–Golgi apparatus shuttle system (Clary *et al.*, 1990). Certain aspects of vesicle targeting and docking are difficult to discriminate from actual fusion; to this end, membrane fusion would probably have to operate at resting pCa levels. This, as well as stimulation by GTP, have been shown for the fusion of elements of the endoplasmic reticulum and of the Golgi apparatus (Paiement and Bergeron, 1985). With neutrophilic granulocytes, only phagocytosis, but not phagosome–lysosome fusion, can operate at resting free Ca^{2+} levels (Jaconi *et al.*, 1990).

A tetrameric structure, 6.3 nm in diameter, has been reported for an intracellular NEM-sensitive protein with alledged fusogenic properties (Block *et al.*, 1988). The NEM-sensitive fusion factor and Sec18p gene product, small G proteins in yeast, have similar structure and function as they form 12.5-nm large homo-hexamers, and may have NEM-sensitive ATPase activity and function in vesicle interaction (Peters *et al.*, 1990). Although these are oligomeric proteins, they are not membrane integrated. It remains open as to whether they are engaged in fusion per se.

Unfortunately, ultrastructural data on endomemembrane fusion based on fast-freezing are scarce. However, endocytic vesicles in renal tubule epithelial cells (Fig. 7) display focal fusion profiles that are thought to expand (Fujioka *et al.*, 1990).

C. Virus Fusion

Virus fusion involves (1) the budding of membrane-enveloped virus particles as well as (2) their fusion with the cell membrane and, once incorporated via endocytic vesicles, (3) fusion with the lysosomal membrane. Viral fusion

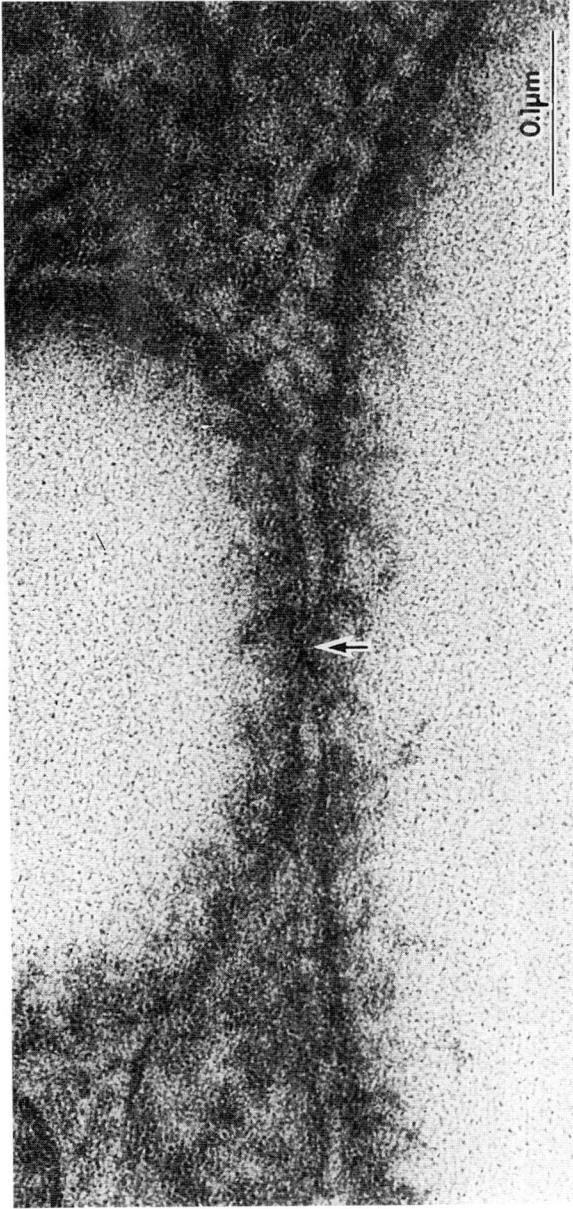

Fig. 7. Endosome fusion (or fission) in a mouse kidney tubule cell showing focal fusion (arrow) of membranes with attached electron-dense materials. Micrograph obtained by fast-freezing and freeze-substitution. Bar = 0.1 μm. Reproduced from Fujioka *et al.* (1990), with permission.

proteins, such as hemagglutinin (HA) and "G protein" (different from GTP-binding G proteins; compare White *et al.*, 1981; Schlegel, 1987) are the only fusion proteins (i.e., fusogenic proteins, different from genetic terminology) identified unequivocally so far, their oligomeric structure and conformational change during fusion being undebated (White *et al.*, 1981, 1983; Blumenthal, 1987; Schlegel, 1987; Stegmann *et al.*, 1989, 1990; Bentz *et al.*, 1990; White, 1990), except for some minor details. Fast-freezing and freeze-fracture analysis during budding (Fig. 8) reveals point fusion, as does the fusion of liposomes with the membrane of virus-infected cells (Knoll *et al.*, 1988). The 10-nm large IMP or pit at the fusion site probably represents lipids, possibly in an inverted micelle configuration. The fusion protein itself probably remains undetected for reasons indicated by Knoll *et al.* (1988). If so, one must not expect matching IMPs at exocytosis sites in both membranes in some other

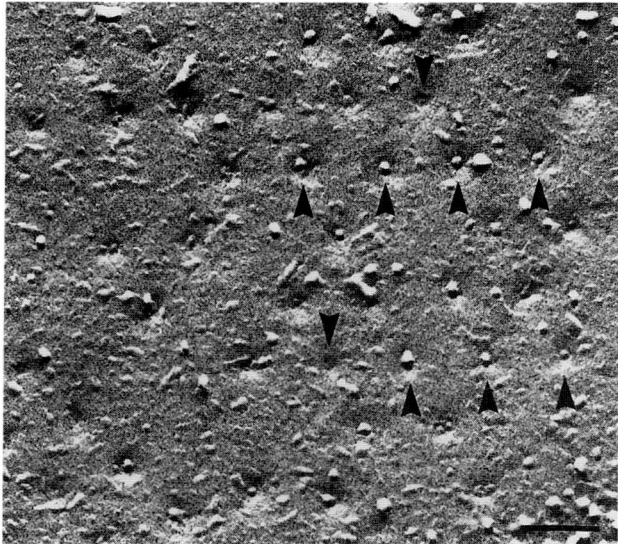

Fig. 8.   MDCK epithelial kidney cell line infected with influenza virus. This freeze-fracture replica of the EF-face of the apical cell membrane reveals multiple fusion spots represented by IMPs (upward arrowheads) or occasionally by pits (downward arrowheads), during synchronous virus budding. Shadowing from bottom to top. Bar = 0.1 μm. From Knoll *et al.* (1988). Reproduced with permission.

cases of biological membrane fusion, for example, during exocytosis. The absence of counterparts of rosette IMPs in trichocyst membranes of paramecia (Section II,A,1) could thus be explained, even if the fusogenic mechanism were quite similar.

For a long time, viral systems have been considered a paradigm of other biological membrane fusions (White *et al.*, 1983), which were therefore also expected to involve homo-oligomeric fusion proteins with the capability of undergoing conformational change to induce fusion. Partial penetration or very close apposition of the partner membrane, for example by HA, appears sufficient to induce fusion (Stegmann *et al.*, 1990; Bentz *et al.*, 1990) in accordance with the freeze-fracture morphology just described. Fusion can be achieved even with a hydrophobic 20-amino-acid long NH_2-terminal sequence (Murata *et al.*, 1987).

In contrast to other biomembrane fusions, viral fusion processes may last for a relatively long time because they are easier to capture by fluorescence assays (Sarkar *et al.*, 1989) or by freeze-fracturing (Burger *et al.*, 1988; Knoll *et al.*, 1988) than most other cellular fusion process, such as most cases of exocytosis (see Zimmerberg, 1987; Almers and Tse, 1990). Electrophysiological recordings of virus-mediated single cell fusion has confirmed initial point fusion with low-expansion velocity, that is, approximately 10^3-times slower than during exocytosis (see Section II,A,1; Spruce *et al.*, 1989). This would be compatible with inverted lipid micelle formation (Burger *et al.*, 1988; Knoll *et al.*, 1988), which can also be rather long-lived (Siegel *et al.*, 1989; Verkleij, 1989), similarly to that proposed recently by Bentz *et al.* (1990) and by White (1990; Fig. 9).

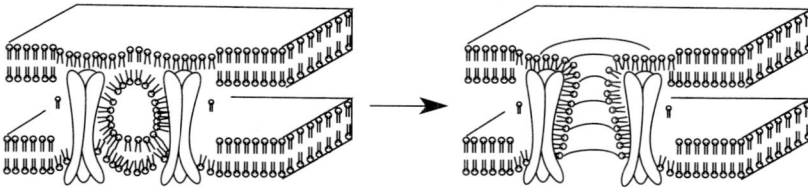

Fig. 9. One of the models currently under discussion for virus-membrane fusion. Oligomeric fusogenic proteins aggregate and undergo a small conformational change (not visible). This allows for partial penetration of the other membrane. This in turn entails the rearrangement of membrane lipids, possibly in an inverted micelle configuration, and finally membrane fusion. Reproduced from White *et al.* (1990) with permission.

D. Cell–Cell Fusion

1. Gamete Fusion

In *Chlamydomonas* (flagellate algae), isospores fuse at a preformed site studded with IMPs and with some subjacent electron-dense materials (Weiss *et al.*, 1977). IMP aggregates are absent from fusion defective mutants (Forest and Ojakian, 1989). This again points toward the importance of membrane protein assembly for membrane interaction and fusion (for exocytosis, see Section II,A,1). Such proteins have been analyzed in more detail with the slime mold, *Dictyostelium* (O'Day and Lydan, 1989; Ishikawa *et al.*, 1990), and in yeast (Trueheart *et al.*, 1987), although neither their role in actual fusion nor the mechanics of fusion itself have yet been analyzed.

For egg–sperm interaction, two fusogenic proteins are now known to occur in the sperm cell membrane: bindin in the sea urchin (Glabe, 1985) and, in mammals, a heteromeric antigen (Blobel *et al.*, 1990) that binds monoclonal antibodies (MAbs) of the type PH-30. Anti-PH-30 MAbs inhibit fusion and, by immunofluorescence, they particularly stain the postnuclear region of the sperm head (Primakoff *et al.*, 1987) where it fuses with the egg cell (Yanagimachi, 1978) and possibly where the sperm plasmalemma is studded with linear IMP aggregates (Koehler, 1966). Bindin was shown to cause fusion of different biomembranes and of liposomes *in vitro* (Glabe, 1985). No data are available, however, for any of these systems to show whether fusion would be the focal type.

2. Somatic Cell Fusion

Myoblast fusion, according to results obtained by fast-freezing and freeze-fracturing or freeze-substitution under considerably synchronized fusion conditions (Fig. 10), involves contacts by small protrusions on only small areas of cell membrane (Knoll and Plattner, 1989). The absence of IMP dispersal and of a diaphragm also recalls the process described in Sections II,A–C.

Electroporation induces small latent defects (Bates *et al.*, 1987) mediating EF-type fusion capacity to very different cells (Zimmermann, 1991). When subjected to fast-freezing within a msec time scale after poration and then freeze-fractured, erythrocytes immediately show volcanolike protrusions with the formation of a fusion focus in the absence of any IMP clearing, before the fusion pore rapidly expands to 20–120 nm (Chang and Reese, 1990). In a less synchronized assay, IMPs were seen to be cleared from protruding fusion

Fig. 10. Synchronized myoblast fusion *in vitro*. The freeze-fractured cell membranes of two adjacent cells (C1, C2) show several depressions, probably indicating small fusion areas (arrowheads). Shadowing from bottom right to top left. Bar = 0.1 μm. Reproduced from Knoll and Plattner (1989) with permission.

zones (Hui *et al.*, 1985). With polyethylene glycol, chemically induced erythrocyte fusion is accompanied by IMP clearing, while extended fusion areas are formed, even when analyzed by fast-freezing (Stenger and Hui, 1986). This clearly shows the importance of strict synchronization and, moreover, the fact that nonphysiological triggers (as applied to produce hybrid cells) may not always faithfully reflect natural biomembrane fusion.

E. Cell Surface "Blebbing"

Interestingly, vesicle "blebbing" from erythrocytes (Fig. 11) exposed to hypertonic media (PF-fusion) probably involves the clustering of intrinsic proteins, which may form a fusion pore (Lew *et al.*, 1988). This is reminiscent of the hypotheses currently under discussion for virus–membrane fusion (Section II,C) and for exocytotic membrane fusion (Section II,A,1).

III. FINAL COMMENTS

Electron microscopic analysis (based on rapid-freezing, combined with freeze-fracturing or freeze-substitution) has reavealed the following features in common to most, if not all, biological membrane fusion processes. (1) The earliest stages recognizable are approximately 10 nm in size. (2) IMPs, representing integral proteins in most cases, remain closely associated with the

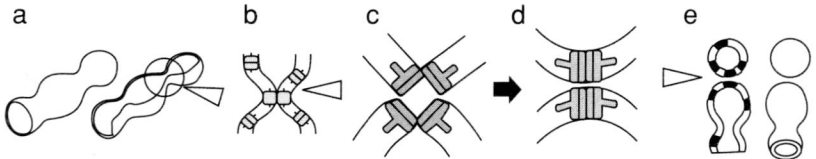

Fig. 11. Membrane "blebbing" (fission) in erythrocytes. Note that (a), (b), (c, d) and (e) are drawn on a different scale. In (b, c, d) membrane interaction is indicated to occur via aggregated intrinsic membrane proteins (dark). The clue of this scheme is that the reversible rearrangement of such proteins should allow lipid rearrangement and, thus, induce membrane fusion. Reproduced from Lew *et al.* (1988) with permission.

fusion site. (3) No lipidic diaphragm is formed and no IMP dispersal occurs before fusion. (4) Even chemical fixation, if appropriate, can demonstrate the occurrence of membrane-associated materials (proteins) on fusogenic zones. (5) Initial fusion can be fast, that is, in the msec range. (6) The fusion pore might expand at quite a different speed, depending on the system analyzed. (7) Some of these aspects might be different with viral fusion (e.g., the fusion intermediate and their lifetime). (8) The aspect recognized most recently is the possible decay of IMPs into subunits during membrane fusion. This has been postulated mainly for exocytosis, but implicated also for endocellular fusion processes. For exocytosis, some support for this hypothesis has been obtained, yet the occurrence of genetically unrelated oligomeric intrinsic-membrane proteins in the plasmalemma of many cells (Baker and Saier, 1990; Bennett et al., 1991), including endomembrane-free erythrocytes (Smith and Agre, 1991), casts some doubt on a straightforward interpretation by assuming that the mere occurrence of oligomeric protein assemblies would suffice to derive their potential engagement in biological fusion processes. Finally, beyond proteins, one should also keep in mind the local formation of fusogenic lipids by endogenous enzymes associated with fusable membranes (Karli et al., 1990; Hildebrandt and Albanesi, 1991).

There is now an impressive convergence of previous ultrastructural findings with more recent data obtained by patch-clamp analysis (however, this literature sometimes neglects reliable and concise ultrastructural data). Because this methodology provides improved time resolution (although in the absence of any spatial resolution), it was able to show in more detail the following aspects. (1) The initial conductance observed corresponds to a 2-nm large fusion pore. (2) Electrical conductance "flickering" phenomena indicate reversibility of the earliest fusion pore. This has been interpreted to mean that oligomeric transmembrane proteins would form a hydrophilic, expandable pore, and thus be engaged in membrane fusion. As already mentioned, we have recently found, by fast-freezing analysis, some ultrastructural data that could be compatible with this hypothesis for trichocyst exocytosis (Knoll et al., 1991a).

The dependency of different types of biological membrane fusions on a variety of parameters, such as pCa, protein (de-)phosphorylation, and phospholipid turnover, is well established (see Sections I and II). However, their actual contribution to membrane fusion, as well as the molecular identity of fusion-mediating proteins and the precise kind of lipid rearrangement, all remain to be elucidated. Therefore, one of the main goals in the future will be the identification of fusogenic proteins (which will probably differ from one system to another) and to analyze the modifications of such proteins or the proteins associated with them to understand their rearrangement in the course of membrane fusion.

For fusion to occur, the membrane lipids have to leave, within a small area and for a restricted time, their bilayer arrangement. Evidence available for virus-mediated fusion favors the formation of "inverted lipid micelles." Only the identification of fusogenic proteins (and possibly their reconstitution into artificial membranes) will also allow this hypothesis to be tested for other cellular fusion processes.

Altogether, fusion of biological membranes appears to be subject to finely tuned regulation processes and, hence, to differ from what occurs with artificial systems. The "focal fusion concept" proposed on the basis of the earliest reliable ultrastructural data (Plattner, 1981) can now be considered valid for practically all types of biomembrane fusion processes.

ACKNOWLEDGMENTS

We thank Mrs M. Harral-Schulz for secretarial help. Some of the authors' work cited was supported by SFB156 and DFG.

REFERENCES

Almers, W. (1990). Exocytosis. *Annu. Rev. Physiol.* **52,** 607–624.

Almers, W., and Neher, E. (1987). Gradual and stepwise changes in the membrane capacitance of rat peritoneal mast cells. *J. Physiol.* **386,** 205–217.

Almers, W., and Tse, F. W. (1990). Transmitter release from synapses: Does a preassembled fusion pore initiate exocytosis? *Neuron* **4,** 813–818.

Almers, W., Breckenridge, L. J., and Spruce, A. E. (1989). The mechanism of exocytosis during secretion in mast cells. *In* "Secretion and Its Control" (G. S. Oxford and C. M. Armstrong, eds.), pp. 269–282. Rockefeller University Press, New York.

Baker, M. E., and Saier, M. H. (1990). A common ancestor for bovine lens fiber major intrinsic protein, soybean modulin-26, and *E. coli* glycerol facilitator. *Cell* **60,** 185–186.

Bates, G. W., Saunders, J. A., and Sowers, A. E. (1987). Electrofusion. Principles and applications. *In* "Cell Fusion" (A. E. Sowers, ed.), pp. 367–395. Plenum Press, New York.

Beisson, J., Lefort-Tran, M., Pouphile, M., Rossignol, M., and Satir, B. (1976). Genetic analysis of membrane differentiation in *Paramecium*. Freeze-fracture study of the trichocyst cycle in wild-type and mutant strains. *J. Cell Biol.* **69,** 126–143.

Beisson, J., Cohen, J., Lefort-Tran, M., Pouphile, M., and Rossignol, M. (1980). Control of membrane fusion in exocytosis. Physiological studies on a *Paramecium* mutant blocked in the final step of the trichocyst extrusion process. *J. Cell Biol.* **85**, 213–227.

Bennett, M. V. L., Barrio, L. C., Bargiello, T. A., Spray, D. C., Hertzberg, E., and Saez, J. C. (1991). Gap junctions: New tools, new answers, new questions. *Neuron* **6**, 305–320.

Bentz, J., Ellens, H., and Alford, D. (1990). An architecture for the fusion site of influenza hemagglutinin. *FEBS Lett.* **276**, 1–5.

Blobel, C. P., Myles, D. G., Primakoff, P., and White, J. M. (1990). Proteolytic processing of a protein involved in sperm–egg fusion correlates with acquisition of fertilization competence. *J. Cell Biol.* **111**, 69–78.

Block, M. R., Glick, B. S., Wilcox, C. A., Wieland, F. T., and Rothman, J. E. (1988). Purification of an *N*-ethylmaleimide-sensitive protein catalyzing vesicular transport. *Proc. Natl. Acad. Sci. USA* **85**, 7852–7856.

Blumenthal, R. (1987). Membrane fusion. *Curr. Top. Membr. Transp.* **29**, 203–254.

Bowser, R., and Novick, P. (1991). Sec15 protein, an essential component of the exocytotic apparatus, is associated with the plasma membrane and with a soluble 19.5*S* particle. *J. Cell Biol.* **112**, 1117–1131.

Breckenridge, L. J., and Almers, W. (1987). Currents through the fusion pore that forms during exocytosis of a secretory vesicle. *Nature* **328**, 814–817.

Burger, K. N. J., Knoll, G., and Verkleij, A. J. (1988). Influenza virus-model membrane interaction. A morphological approach using modern cryotechniques. *Biochim. Biophys. Acta* **939**, 89–101.

Chandler, D. E. (1988). Exocytosis and endocytosis: Membrane fusion events captured in rapidly frozen cells. *Curr. Top. Membr. Transp.* **32**, 169–202.

Chandler, D. E., and Heuser, J. (1979). Membrane fusion during secretion. Cortical granule exocytosis in sea urchin eggs as studied by quick-freezing and freeze-fracture. *J. Cell Biol.* **83**, 91–108.

Chandler, D. E., and Heuser, J. E. (1980). Arrest of membrane fusion events in mast cells by quick-freezing. *J. Cell Biol.* **86**, 666–674.

Chang, D. C., and Reese, T. S. (1990). Changes in membrane structure induced by electroporation as revealed by rapid-freezing electron microscopy. *Biophys. J.* **58**, 1–12.

Clary, D. O., Griff, I. C., and Rothman, J. E. (1990). SNAPs, a family of NSF attachment proteins involved in intracellular membrane fusion in animals and yeast. *Cell* **61**, 709–721.

Creutz, C. E., Zaks, W. J., Hamman, H. C., and Martin, W. H. (1987). The roles of Ca^{2+}-dependent membrane-binding proteins in the regulation and mechanism of exocytosis. *In* "Cell Fusion" (A. E. Sowers, ed.), pp. 45–68. Plenum Press, New York.

Diaz, R., Mayorga, L. S., Weidman, P. J., Rothman, J. E., and Stahl, P. D. (1989). Vesicle fusion following receptor-mediated endocytosis requires a protein active in Golgi transport. *Nature* **339**, 398–400.

Forest, C. L., and Ojakian, G. K. (1989). Mating structure differences demonstrated by freeze-fracture analysis of fusion-defective *Chlamydomonas* gametes. *J. Protozool.* **36**, 548–556.

Fujioka, A., Ohtsuki, M., Nagano, M., and Mori, S. (1990). Membrane fusion events during endocytosis in mouse kidney tubule cells detected by rapid-freezing followed by freeze-substitution. *J. Electron Microsc.* **39**, 356–362.

Glabe, C. G. (1985). Interaction of the sperm adhesive protein, bindin, with phospholipid vesicles. II. Bindin induces the fusion of mixed-phase vesicles that contain phosphatidylcholine and phosphatidylserine *in vitro*. *J. Cell Biol.* **100**, 800–806.

Gomperts, B. D., Churcher, A., Koffer, A., Kramer, I. M., Lillie, T., and Tatham, P. E. R.

(1990). The role and mechanism of the GTP-binding protein G_E in the control of regulated exocytosis. *Biochem. Soc. Symp.* **56**, 85–101.

Harris, K. M., and Landis, D. M. D. (1986). Membrane structure at synaptic junctions in area CA1 of the rat hippocampus. *Neuroscience* **19**, 857–872.

Heuser, J. E., Reese, T. S., Dennis, M. J., Jan, Y., Jan, L., and Evans, L. (1979). Synaptic vesicle exocytosis captured by quick-freezing and correlated with quantal transmitter release. *J. Cell Biol.* **81**, 275–300.

Hildebrandt, E., and Albanesi, J. P. (1991). Identification of a membrane-bound glycol-stimulated phospholipase A2' located in the secretory granules of the adrenal medulla. *Biochemistry* **30**, 464–472.

Höhne-Zell, B., Knoll, G., Riedel-Gras, U., Hofer, W., and Plattner, H. (1992). A cortical phosphoprotein ('PP63') sensitive to exocytosis triggering in *Paramecium* cells. Immunolocalization and quenched-flow correlation of time course of dephosphorylation with membrane fusion. *Biochem. J.* **286**, 843–849.

Hui, S. W., Boni, I. L. T., and Sen, A. (1985). Action of polyethylene glycol on the fusion of human erythrocyte membranes. *J. Membr. Biol.* **84**, 137–146.

Ishikawa, T. O., Urushihara, H., Habata, Y., and Yanagisawa, K. (1990). Affinity purification of a 70K protein, a membrane protein relevant to sexual cell fusion in *Dictyostelium discoideum*. *Cell Differ. Dev.* **31**, 177–184.

Israel, M., Lesbats, B., and Suzuki, A. (1989). Characterization of a polyclonal antiserum raised against mediatophore, a protein that translocates acetylcholine. *Cell Biol. Int. Rep.* **13**, 1097–1107.

Jaconi, M. E. E., Lew, D. P., Carpentier, J. L., Magnusson, K. E., Sjögren, M., and Stendahl, O. (1990). Cytosolic free calcium elevation mediates the phagosome–lysosome fusion during phagocytosis in human neutrophils. *J. Cell Biol.* **110**, 1555–1564.

Karli, U. O., Schäfer, T., and Burger, M. M. (1990). Fusion of neurotransmitter vesicles with target membrane is calcium independent in a cell-free system. *Proc. Natl. Acad. Sci. USA* **87**, 5912–5915.

Kelly, R. B. (1988). The cell biology of the nerve terminal. *Neuron* **1**, 431–438.

Keryer, G., Davis, F. M., Rao, P. N., and Beisson, J. (1987). Protein phosphorylation and dynamics of cytoskeletal structures associated with basal bodies in *Paramecium*. *Cell Motil. Cytoskel.* **8**, 44–54.

Knaus, P., Marquèze-Pouey, B., Scherer, H., and Betz, H. (1990). Synaptoporin, a novel putative channel protein of synaptic vesicles. *Neuron* **5**, 453–462.

Knoll, G., and Plattner, H. (1989). Ultrastructural analysis of biological membrane fusion and a tentative correlation with biochemical and biophysical aspects. *In* "Electron Microscopy of Subcellular Dynamics" (H. Plattner, ed.), pp. 96–117. CRC Press, Boca Raton, Florida.

Knoll, G., Verkleij, A. J., and Plattner, H. (1987). Cryofixation of dynamic processes in cells and organelles. *In* "Cryotechniques in Biological Electron Microscopy" (R. A. Steinbrecht and K. Zierold, eds.), pp. 258–271. Springer-Verlag, Berlin/Heidelberg/New York.

Knoll, G., Burger, K. N. J., Bron, R., Van Meer, G., and Verkleij, A. J. (1988). Fusion of liposomes with the plasma membrane of epithelial cells: Fate of incorporated lipids as followed by freeze fracture and autoradiography of plastic sections. *J. Cell Biol.* **107**, 2511–2521.

Knoll, G., Braun, C., and Plattner, H. (1991a). Quenched-flow analysis of exocytosis in *Paramecium* cells: Time course, changes in membrane structure and calcium requirements revealed after rapid mixing and rapid freezing of intact cells. *J. Cell Biol.* **113**, 1295–1304.

Knoll, G., Haacke-Bell, B., and Plattner, H. (1991b). Local trichocyst exocytosis provides an efficient escape mechanism for *Paramecium* cells. *Eur. J. Protistol.* **27**, 381–385.

Koehler, J. K. (1966). Fine structure observations in frozen-etched bovine spermatazoa. *J. Ultrastruct. Res.* **16**, 359–375.

Lefort-Tran, M., Aufderheide, K., Pouphile, M., Rossignol, M., and Beisson, J. (1981). Control of exocytotic processes: Cytological and physiological studies of trichocyst mutants in *Paramecium tetraurelia. J. Cell Biol.* **88**, 301–311.

Lelkes, P. I., and Pollard, H. B. (1991). Cytoplasmic determinants of exocytotic membrane fusion. *In* "Membrane Fusion" (J. Wilschut and D. Hoekstra, eds.), pp. 511–551. Marcel Dekker, New York.

Lew, V. L., Hockaday, A., Freeman, C. J., and Bookchin, R. M. (1988). Mechanism of spontaneous inside-out vesiculation of red cell membranes. *J. Cell Biol.* **106**, 1893–1901.

Momayezi, M., Girwert, A., Wolf, C., and Plattner, H. (1987). Inhibition of exocytosis in *Paramecium* cells by antibody-mediated cross-linking of cell membrane components. *Eur. J. Cell Biol.* **44**, 247–257.

Morel, N., Israel, M., Lesbats, B., Birman, S., and Manaranche, R. (1987). Characterization of a presynaptic membrane protein ensuring a calcium-dependent acetylcholine release. *Ann. N. Y. Acad. Sci.* **493**, 151–154.

Morel, N., Synguelakis, M., and Le Gal La Salle, G. (1991). Detection with monoclonal antibodies of a 15-kDa proteolipid in both presynaptic plasma membranes and synaptic vesicles in *Torpedo* electric organ. *J. Neurochem.* **56**, 1401–1408.

Murata, M., Sugahara, Y., Takahashi, S., and Ohnishi, S. I. (1987). pH-dependent membrane fusion activity of a synthetic twenty amino acid peptide with the same sequence as that of the hydrophobic segment of influenza virus hemagglutinin. *J. Biochem.* **102**, 957–962.

Neher, E., and Marty, A. (1982). Discrete changes of cell membrane capacitance observed under conditions of enhanced secretion in bovine adrenal chromaffin cells. *Proc. Natl. Acad. Sci. USA* **79**, 6712–6716.

O'Day, D. H., and Lydan, M. A. (1989). The regulation of membrane fusion during sexual development in *Dictyostelium discoideum. Biochem. Cell Biol.* **67**, 321–326.

Orci, L., Glick, B. S., and Rothman, J. E. (1986). A new type of coated vesicular carrier that appears not to contain clathrin: Its possible role in protein transport within the Golgi stack. *Cell* **46**, 171–184.

Ornberg, R. L., and Reese, T. S. (1981). Beginning of exocytosis captured by rapid-freezing of *Limulus* amebocytes. *J. Cell Biol.* **90**, 40–54.

Paiement, J., and Bergeron, J. J. M. (1985). The specificity of fusion among endoplasmic reticulum and Golgi membranes. *J. Cell Biol.* **101**, 60a.

Pape, R., Haacke-Bell, B., Lüthe, N., and Plattner, H. (1988). Conjugation of *Paramecium tetraurelia* cells: Selective wheat germ agglutinin binding, reversible local trichocyst detachment, and secretory function repair. *J. Cell Sci.* **90**, 37–49.

Peters, J. M., Walsh, M. J., and Franke, W. W. (1990). An abundant and ubiquitous homo-oligomeric ring-shaped ATPase particle related to the putative vesicle fusion proteins Sec18p and NSF. *EMBO J.* **9**, 1757–1767.

Pfenninger, K., Akert, K., Moor, H., and Sandri, C. (1972). The fine structure of freeze-fractured presynaptic membranes. *J. Neurocytol.* **1**, 129–149.

Pinto Da Silva, P. (1988). Geometric topology of membrane fusion: From secretion to intracellular junctions. *In* "Molecular Mechanisms of Membrane Fusion" (S. Ohki, D. Doyle, T. D. Flanagan, S. W. Hui, and E. Mayhew, eds.), pp. 521–530. Plenum Press, New York.

Plattner, H. (1981). Membrane behaviour during exocytosis. *Cell Biol. Int. Rep.* **5**, 435–459.

Plattner, H. (1987). Synchronous exocytosis in *Paramecium* cells. *In* "Cell Fusion" (A. E. Sowers, ed.), pp. 69–98. Plenum Press, New York.

Plattner, H. (1989a). Regulation of membrane fusion during exocytosis. *Int. Rev. Cytol.* **119**, 197–286.

Plattner, H. (1989b). Freeze-fracture (-etching) analysis of exo-endocytosis. *In* "Freeze Fracture Studies of Membranes" (S. W. Hui, ed.), pp. 147–173. CRC Press, Boca Raton, Florida.

Plattner, H. (1991). Ultrastructural aspects of exocytosis. *In* "Membrane Fusion" (J. Wilschut and D. Hoekstra, eds.), pp. 571–598. Marcel Dekker, New York.

Plattner, H., and Bachmann, L. (1982). Cryofixation: A tool in biological ultrastructural research. *Int. Rev. Cytol.* **79**, 237–304.

Plattner, H., and Zingsheim, H. P. (1983). Electron microscopic methods in cellular and molecular biology. *In* "Subcellular Biochemistry" (D. B. Roodyn, ed.), Vol. 9, pp. 1–236. Plenum Press, New York.

Plattner, H., Miller, F., and Bachmann, L. (1973). Membrane specializations in the form of regular membrane-to-membrane attachment sites in *Paramecium*. A correlated freeze-etching and ultrathin-sectioning analysis. *J. Cell Sci.* **13**, 687–719.

Plattner, H., Stürzl, R., and Matt, H. (1985). Synchronous exocytosis in *Paramecium* cells. IV. Polyamino-compounds as potent trigger agents for repeatable trigger-redocking cycles. *Eur. J. Cell Biol.* **36**, 32–37.

Plattner, H., Lumpert, C. J., Knoll, G., Kissmehl, R., Höhne, B., Momayezi, M., and Glas-Albrecht, R. (1991). Stimulus-secretion coupling in *Paramecium* cells. *Eur. J. Cell Biol.* **55**, 3–16.

Pouphile, M., Lefort-Tran, M., Plattner, H., Rossignol, M., and Beisson, J. (1986). Genetic dissection of the morphogenesis of exocytosis sites in *Paramecium*. *Biol. Cell* **56**, 151–162.

Primakoff, P., Hyatt, H., and Tredick-Kline, J. (1987). Identification and purification of a sperm surface protein with a potential role in sperm–egg membrane fusion. *J. Cell Biol.* **104**, 141–149.

Rash, J. E., Walrond, J. P., and Morita, M. (1988). Structural and functional correlates of synaptic transmission in the vertebrate neuromuscular junction. *J. Electron Microsc. Tech.* **10**, 153–185.

Sarkar, D. P., Morris, S. J., Eidelman, O., Zimmerberg, J., and Blumenthal, R. (1989). Initial stages of influenza hemagglutinin-induced cell fusion monitored simultaneously by two fluorescent events: Cytoplasmic continuity and lipid mixing. *J. Cell Biol.* **109**, 113–122.

Schlegel, R. (1987). Probing the function of viral fusion proteins with synthetic peptides. *In* "Cell Fusion" (A. E. Sowers, ed.), pp. 33–43. Plenum Press, New York.

Schmidt, W., Patzak, A., Lingg, G., Winkler, H., and Plattner, H. (1983). Membrane events in adrenal chromaffin cells during exocytosis: A freeze-etching analysis after rapid cryofixation. *Eur. J. Cell Biol.* **32**, 31–37.

Siegel, D. P., Burns, J. L., Chestnut, M. H., and Talmon, Y. (1989). Intermediates in membrane fusion and bilayer/nonbilayer phase transitions imaged by time-resolved cryotransmission electron microscopy. *Biophys. J.* **56**, 161–169.

Smith, B. L., and Agre, P. (1991). Erythrocyte M_r 28,000 transmembrane protein exists as a multisubunit oligomer similar to channel proteins. *J. Biol. Chem.* **266**, 6407–6415.

Spruce, A. E., Iwata, A., White, J. M., and Almers, W. (1989). Patch clamp studies of single cell-fusion events mediated by a viral fusion protein. *Nature* 342, 555–558.

Spruce, A. E., Breckenridge, L. J., Lee, A. K., and Almers, W. (1990). Properties of the fusion pore that forms during exocytosis of a mast cell secretory vesicle. *Neuron* **4**, 643–654.

Stecher, B., Höhne, B., Gras, U., Momayezi, M., Glas-Albrecht, R., and Plattner, H. (1987). Involvement of a 65 kDa phosphoprotein in the regulation of membrane fusion during exocytosis in *Paramecium* cells. *FEBS Lett.* **223**, 25–32.

Stegmann, T., Doms, R. W., and Helenius, A. (1989). Protein-mediated membrane fusion. *Annu. Rev. Biophys. Biophys. Chem.* **18**, 187–211.

Stegmann, T., White, J. M., and Helenius, A. (1990). Intermediates in influenza induced membrane fusion. *EMBO J.* **9**, 4231–4241.

Stenger, D. A., and Hui, S. W. (1986). Kinetics of ultrastructural changes during electrically induced fusion of human erythrocytes. *J. Membr. Biol.* **93**, 43–53.

Stossel, T. P., Bretscher, M. S., Ceccarelli, B., Dales, S., Helenius, A., Heuser, J. E., Hubbard, A. L., Kartenbeck, J., Kinne, R., Papahadjopoulos, D., Pearse, B., Plattner, H., Pollard, T. D., Reuter, W., Satir, B. H., Schliwa, M., Schneider, Y. J., Silverstein, S. C., and Weber, K. (1978). Membrane dynamics. *In* "Transport of Macromolecules in Cellular Systems" (S. C. Silverstein, ed.), *Life Sci. Res. Rep.* **11**, pp. 503–516. Dahlem Konferenzen, Berlin.

Südhof, T. C., and Jahn, R., (1991). Proteins of synaptic vesicles involved in exocytosis and membrane recycling. *Neuron* **6**, 665–677.

Thomas, L., and Betz, H. (1990). Synaptophysin binds to physophilin, a putative synaptic plasma membrane protein. *J. Cell Biol.* **111**, 2041–2052.

Thomas, L., Hartung, K., Langosch, D., Rehm, H., Bamberg, E., Franke, W. W., and Betz, H. (1988). Identification of synaptophysin as a hexameric channel protein of the synaptic vesicle membrane. *Science* **242**, 1050–1053.

Torri-Tarelli, F., Grohovaz, F., Fesce, R., and Ceccarelli, B. (1985). Temporal coincidence between synaptic vesicle fusion and quantal secretion of acetylcholine. *J. Cell Biol.* **101**, 1386–1399.

Trimble, W. S., Linial, M., and Scheller, H. (1991). Cellular and molecular biology of the presynaptic nerve terminal. *Annu. Rev. Neurosci.* **14**, 93–122.

Trueheart, J., Boeke, J. D., and Fink, G. R. (1987). Two genes required for cell fusion during yeast conjugation: Evidence for a pheromone-induced surface protein. *Mol. Cell. Biol.* **7**, 2316–2328.

Verkleij, A. J. (1989). Lipid phase transitions and micro-domains in biomembranes and their possible relevance for biological processes. *In* "Electron Microscopy of Subcellular Dynamics" (H. Plattner, ed.), pp. 75–93. CRC Press, Boca Raton, Florida.

Volknandt, W., Schläfer, M., Bonzelius, F., and Zimmermann, H. (1990). Svp25, a synaptic vesicle membrane glycoprotein from *Torpedo* electric organ that binds calcium and forms a homo-oligomeric complex. *EMBO J.* **9**, 2465–2470.

Weiss, R. L., Goudenough, D. A., and Goodenough, U. W. (1977). Membrane differentiations at sites specialized for cell fusion. *J. Cell Biol.* **72**, 144–160.

Whalley, T., Crossley, I., and Whitaker, M. (1991). Phosphoprotein inhibition of calcium-stimulated exocytosis in sea urchin eggs. *J. Cell Biol.* **113**, 769–778.

White, J. W. (1990). Viral and cellular membrane fusion proteins. *Annu. Rev. Physiol.* **52**, 675–697.

White, J., Matlin, K., and Helenius, A. (1981). Cell fusion by Semliki Forest, influenza, and vesicular stomatitis viruses. *J. Cell Biol.* **89**, 674–679.

White, J., Kielian, M., and Helenius, A. (1983). Membrane fusion proteins of enveloped animal viruses. *Q. Rev. Biophys.* **16**, 151–195.

Wilson, D. W., Wilcox, C. A., Flynn, G. C., Chen, E., Kuang, W. J., Henzel, W. J., Block, M. R., Ullrich, A., and Rothman, J. E. (1989). A fusion protein required for vesicle-mediated transport in both mammalian cells and yeast. *Nature* **339**, 355–359.

Yanagimachi, R. (1978). Sperm–egg association in mammals. *Curr. Top. Dev. Biol.* **12**, 83–105.

Zieseniss, E., and Plattner, H. (1985). Synchronous exocytosis in *Paramecium* cells involves very rapid (≤1 s), reversible dephosphorylation of a 65 kD phosphoprotein in exocytosis-competent strains. *J. Cell Biol.* **101**, 2028–2035.

Zimmerberg, J. (1987). Molecular mechanisms of membrane fusion: Steps during phospholipid and exocytotic membrane fusion. *Biosci. Rep.* **7**, 251–268.

Zimmerberg, J., Curran, M., Cohen, F. S., and Brodwick, M. (1987). Simultaneous electrical and optical measurements show that membrane fusion precedes secretory granule swelling during exocytosis of beige mouse mast cells. *Proc. Natl. Acad. Sci. USA* **84**, 1585–1589.

Zimmermann, U. (1991). Electrofusion and electropermeabilization in genetic engineering. *In* "Membrane Fusion" (J. Wilschut and D. Hoekstra, eds.), pp. 665–695. Marcel Dekker Inc., New York.

Zingsheim, H. P., and Plattner, H. (1976). Electron microscopic methods in membrane biology. *In* "Methods in Membrane Biology" (E. D. Korn, ed.), Vol. 7, pp. 1–146. Plenum Press, New York.

3

Osmotic Phenomena in Membrane Fusion

JACK A. LUCY

Department of Biochemistry and Chemistry
Royal Free Hospital School of Medicine
London, England

I. INTRODUCTION

A large number of observations have been made on the molecular mechanisms of membrane fusion in different systems. Does the existence of many differing findings on a variety of membranes indicate that there are numerous ways in which the fusion event can occur at the molecular level? Interest in osmotic phenomena in membrane fusion reflects recognition of the possibility that, despite the differences between fusion reactions occurring in individual systems, there may well be a common molecular mechanism that underlies most, or all, membrane fusion events. For a mechanism to be universal, it needs to be fundamental. Therein lies the attraction of a possible involvement of osmotic forces because the very existence of biomembranes composed of phospholipid bilayers that have a hydrophobic interior depends on their aque-

SIGNAL TRANSDUCTION DURING
BIOMEMBRANE FUSION

ous environment, and osmotically active solutes are present in all biological systems.

Osmotic phenomena may participate in membrane fusion reactions in at least four different ways. These possibilities are illustrated in Fig. 1. Before membranes can begin to fuse, they need to be brought closely together. The first stage in a membrane fusion reaction is the close approach of the two membranes (Stage 1 in Fig. 1). Contact between membranes, a prerequisite for fusion, must at some stage displace water from at least some part of their

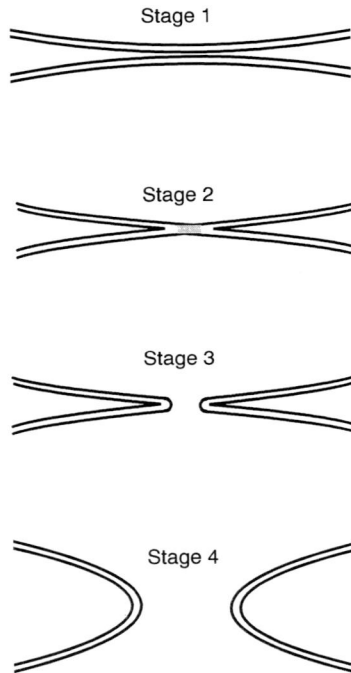

Fig. 1. A diagrammatic illustration of four stages in the fusion–fission reactions of biological membranes. Stage 1: The close apposition of two membranes. Stage 2: Hemifusion of the two membranes in which lipid mixing but not content mixing occurs. At this stage, the membranes may be linked at the fusion site (shown hatched in the diagram) by a nonbilayer phase, transient bilayer defects, or by a shared bilayer. Stage 3: Membrane fission occurs at the site of hemifusion, which allows contents mixing to occur via a narrow fusion pore. Stage 4: The fusion pore widens to permit extensive contents mixing.

surfaces, and it has been proposed that the primary work involved in bringing phospholipid bilayers into close apposition necessitates dehydration of their polar-head groups (Rand and Parsegian, 1989; Parsegian and Rand, 1991). By contrast, other researchers consider hydrophobic interactions more important in membrane adhesion and fusion (Ohki, 1988; Helm et al., 1992).

In a limited number of membrane systems it has been established that a state of hemifusion can occur in which the participating membranes fuse without the two aqueous compartments being joined, that is, a membranous entity still separates the aqueous compartments. Hemifusion has been well-documented in work with fluorescent probes on phospholipid vesicles under conditions that result in an intermixing of lipid probes without mixing the vesicular contents (Düzgüneş et al., 1987; Silvius et al., 1988). In Fig. 1, the hemifused state (Stage 2) is diagrammatically illustrated without commitment as to whether a nonbilayer phase, transient defects in bilayers, or a shared bilayer are involved. Osmotic forces may further contribute to membrane fusion reactions by pushing the contacting bilayers (Stage 1) into the hemifused state (Stage 2).

When membrane fusion proceeds via a transient hemifused state, osmotic forces may help to remove the intermediate membranous entity and establish continuity between the two aqueous phases; that is, they may participate in converting Stage 2 into Stage 3 (Fig. 1). Finally, osmotic phenomena appear to be primarily responsible for widening the fusion pore that is initially formed between the two aqueous phases, thus giving rise to Stage 4 in Fig. 1. Much current research on the molecular mechanisms of membrane fusion, particularly in relation to exocytosis, is concerned with determining whether osmotic forces contribute only to this final stage of the membrane fusion reaction (Stage 4), or whether they are also involved prior to the widening of the fusion pore.

The development of assays that unequivocally distinguish between the different stages of membrane fusion that are illustrated in Fig. 1 is not a simple matter. In particular, some lipid-mixing assays may not distinguish the aggregation of membranes (Stage 1) from their hemifusion (Stage 2). To investigate an overall membrane fusion reaction, it is also important to employ an assay for contents mixing of the two aqueous phases as well as a lipid-mixing assay in order to ascertain that the reaction proceeds beyond the stage of hemifusion, and to eliminate the possibility that lipid mixing and leakage of aqueous contents may have occurred without contents mixing (Düzgüneş et al., 1987). Difficulties experienced in this connection are illustrated by the fact that it has recently been concluded that, contrary to an earlier report, poly(ethylene glycol) causes lipid mixing but not fusion between large, unilamellar vesicles of phosphatidylcholine (Burgess *et al.*, 1991). In addition, in the electrofusion of human erythrocytes, it was unexpectedly observed that the number of contents-mixing events considerably exceeded the number of membrane-mixing events. This was shown to result from movement of fluorescein-labeled dextran, which had been released into the medium from an electropermeabilized ghost, into an adjacent ghost while its electro-induced pores were still open. After the pores closed, any fluorescence inside the ghosts was trapped, while that remaining in the released cloud outside the ghosts diffused away (Sowers, 1988).

The overall "membrane fusion" reaction is sometimes referred to, more precisely, as membrane fusion–fission in order to acknowledge that a stage of membrane fusion, however, transient, is followed by membrane fission (Palade, 1975). Data reported on "membrane fusion" phenomena nevertheless frequently relate only to the overall reaction rather than to any of the separate stages that are illustrated in Fig. 1. However, the observations on osmotic phenomena in membrane fusion that are reviewed here will be considered, so far as is possible, in relation to Stages 1–4 of Fig. 1.

II. CURRENT RESEARCH

A. Membrane Fusion in Model Phospholipid Systems

1. Phospholipid Vesicles and Planar Phospholipid Bilayers

Miller *et al.* (1976) showed that unilamellar proteoliposomes containing phosphatidylserine, which ceased to fuse in the presence of calcium ions after they had reached diameters of about 100 nm, could be induced to fuse further by an osmotic gradient (internal osmotic pressure higher than external) to yield single-walled liposomes with diameters that exceeded 1 μm, as well as forming multilayered vesicles. They suggested that the thermodynamic driving force for the Ca^{2+}-induced fusion was an excess surface energy which can be supplied by membrane curvature or transmembrane osmotic gradients. Vesicles of phosphatidylserine, which were aggregated by magnesium ions and then suspended in a hypotonic solution, were observed to fuse by Ohki (1984). The larger the diameter of the initial vesicles, the smaller was the osmotic pressure gradient (outside more hypotonic than inside) needed for fusion. For vesicles that were 1 μm in diameter, the threshold gradient required for fusion was 0.104 M sucrose. By contrast, for vesicles that were only 0.1 μm in diameter, there was no appreciable fusion with 0.138 M sucrose.

It was early reported by Miller and Racker (1976) that fusion of fragmented vesicles of the sarcoplasmic reticulum with planar phospholipid bilayers required the presence of at least 0.5 mM Ca^{2+}, an acidic phospholipid such as phosphatidylserine in the bilayer, and an osmotic gradient across the vesicle membranes with the internal osmolarity greater than the external. Much of the current interest in possible roles for osmotic forces in the fusion of biomembranes stems, however, from a series of experiments on the fusion of phospholipid vesicles with planar phospholipid bilayers that was undertaken subsequently as a model for exocytosis by Finkelstein et al. (1986). They showed that phospholipid vesicles could be induced to fuse with a planar phospholipid bilayer if there was an osmotic gradient across the planar membrane, with the cis side (the side containing the vesicles) hyperosmotic to the opposite (trans) side. A divalent cation was required on the cis side. Application of an osmotic gradient resulted in a discharge of the contents of multilamellar vesicles across (decane-containing) planar phospholipid membranes. Therefore, they proposed that fusion occurred by the osmotic swelling of vesicles in contact with the planar membrane, followed by rupture of the vesicular and planar membranes in the region of contact (Cohen et al., 1980, 1982; Zimmerberg et al., 1980a,b).

It was also found that unilamellar vesicles fuse with "hydrocarbon-free" planar bilayers under the same conditions. Furthermore, it was shown that vesicles made from uncharged lipids readily fused with planar membranes of phosphatidylethanolamine in the near absence of divalent cations (Cohen et al., 1984). Two experimentally distinguishable steps in the fusion of phospholipid vesicles with planar bilayers were demonstrated. Initially, the vesicles formed a stable, tightly bound, prefusion state with the planar membrane, for which divalent cations were necessary if either membrane contained negatively charged lipids. Osmotic swelling of the bound vesicle and its fusion with the planar bilayer then occurred in a second step (Akabas et al., 1984). Subsequently, it was demonstrated that the vesicular and planar membranes remained as individual phospholipid bilayers when Ca^{2+} ions promoted the adhesion of the vesicle to the bilayer (compare with Stage 1 in Fig. 1), and that the two membranes did not meld to form a shared bilayer at the point of contact (Stage 2, Fig. 1; Niles and Cohen, 1987). However, this observation would not have excluded the possibility of a transient, hemifusion intermediate participating in the ensuing fusion–fission reaction.

In more recent papers, Woodbury and Hall (1988) have considered the role of channels in the fusion of vesicles with a planar bilayer, while Cohen *et al.* (1989) have shown that fusion depends on the membrane permeability of the solute used to create the osmotic pressure. Niles *et al.* (1989) have employed irreversible thermodynamics to write expressions for the flows of water and solutes across the vesicular and planar membranes for all experimentally realized configurations and have concluded that the intravesicular pressures drive fusion.

The interaction and fusion of synaptic vesicles, in the presence of Ca^{2+} (100 μM to 15 mM) with planar phospholipid membranes composed of phosphatidylserine and phosphatidylethanolamine, has been investigated by Perin and MacDonald (1989) as a model system for the release by exocytosis of neurotransmitters. In more than half of the events recorded, the vesicle contents were released to the far side of the bilayer. The highest rate of release was obtained in potassium chloride (KCl) solutions and, although release was virtually eliminated in isotonic glucose, it was restored by replacing the glucose with KCl or urea. The action of urea indicated that the inhibitory effect of glucose was due to its low permeability through the vesicular membrane, and not to an absence of electrolyte. It was concluded that the fusion (and lysis) of synaptic vesicles adhering to the planar bilayer was the consequence of stress resulting from the entry into the vesicles of a permeable solute (KCl or urea) and accompanying water.

B. Experimentally Induced Cell Fusion

1. Chemically Induced Cell Fusion

a. Changes in Phospholipid Asymmetry. Hydration repulsion between anionic phospholipid bilayers is readily overcome by the addition of divalent cations to the aqueous phase (reviewed by Parsegian and Rand, 1991). Most of the water between the two membrane surfaces is displaced as the added cations bind to the polar groups of the anionic phospholipids, and the membranes then collapse together. The combination of Ca^{2+} ions and phosphatidylserine is particularly effective; no detectable water is left and the hydrocarbon chains of the phosphatidylserine molecules effectively freeze. It is for this reason that the model phospholipid systems described in Section II,A, which have been used to study membrane fusion, are based on the interactions of bilayers of phosphatidylserine in the presence of Ca^{2+} ions. In some mixtures of phosphatidylserine with phosphatidylcholine, the interaction between the anionic phospholipid and Ca^{2+} is strong enough to cause dehydration of the polar groups of the neutral phospholipid. With mixed lipid systems, segregation of different lipid species or the formation of nonlamellar lipid phases can occur, and such structures have received much attention as possible intermediates (compare Stage 2 in Fig. 1) in membrane fusion reactions (Ohki et al., 1988).

The addition of sufficient (>50%) phosphatidylserine to phospholipid vesicles composed of phosphatidylcholine leads to membrane fusion in the presence of Ca^{2+} (Düzgüneş et al., 1981). However, relatively little attention has been given to the fact that there is little or no phosphatidylserine in the outer monolayer of the plasma membranes of many cells, and that a change in phospholipid asymmetry appears to be required if a phosphatidylserine–Ca^{2+} complex is to participate in bringing plasma membranes together (Stage 1, Fig. 1) for cell fusion. The exposure of acidic phospholipids on the outside of cells yields procoagulant surfaces that facilitate the conversion of prothrombin into thrombin, and this can be used as a semiquantitative measure of the exposed acidic phospholipids. It has recently been observed that an increase in the procoagulant activity of human erythrocytes is associated with the fusion of these cells that occurs when they are induced to swell osmotically by the permeant molecule PEG 400 in the presence of Ca^{2+}. Cells that were incubated with ionophore A23187, subtilisin, and Ca^{2+} also developed procoagulant activity, and they fused on subsequent exposure to a hypotonic medium. From these experiments, it was concluded that a translocation of phosphatidylserine to the outer leaflet of the plasma membrane plays an important role in fusion protocols that involve cell swelling (Baldwin et al., 1990).

Human erythrocytes submitted to a "preswell lysis and resealing procedure" in the presence of Ca^{2+} also exhibited an increased susceptibility to fusion by high-molecular weight poly(ethylene glycol) (PEG 6000) (Tullius et al., 1989). Approximately 6% of the total amino group-containing phospholipids (phosphatidylserine and phosphatidylethanolamine) of human erythrocytes treated with 35% poly(ethylene glycol) were found to flip-flop from the inner monolayer to the outer monolayer 30 min after the PEG solution was diluted (Huang and Hui, 1990).

b. Poly(ethylene glycol). In relation to the fusogenic actions of high-molecular weight poly(ethylene glycol) on phospholipid vesicles and cells, Arnold et al. (1990) have concluded that there is no evidence for a direct interaction between membrane phospholipid molecules and the polymer.

Again, it appears that the aggregation of phospholipid vesicles and cells (Stage 1 in Fig. 1) is a consequence of membrane dehydration which, in this case, results from the high osmotic pressure of aqueous solutions of PEG, together with a reduced solubility of molecules of PEG in the water layer at the membrane surface.

In contrast to a previous report (Parente and Lentz, 1986), it has recently been observed that only lipid mixing, and not contents mixing that is indicative of fusion, can be detected with large unilamellar vesicles of pure synthetic phosphatidylcholines in the presence of up to 40% PEG 8000 (Burgess *et al.*, 1991). However, freeze-fraction electron microscopy has shown that the fusion of phospholipid vesicles, composed of phosphatidylcholine, phosphatidic acid, and cholesterol (or of egg phosphatidylcholine) occurs in concentrated solutions of poly(ethylene glycol), and that the subsequent dilution step that is important for cell fusion is not required (Boni *et al.*, 1981; Aldwinckle *et al.*, 1982). By contrast, Hui *et al.* (1985) found that osmotic swelling, achieved by using a hypotonic incubation medium, was necessary for the fusion of sealed erythrocyte ghosts after their exposure to poly(ethylene glycol). In subsequent work on the fusion of intact human erythrocytes induced by poly(ethylene glycol) 6000, carbocyanine and rhodamine membrane probes diffused from labeled to unlabeled cells when they were dehydrated by 40% poly(ethylene glycol) (Stage 2, Fig. 1), but there was no corresponding movement (Stage 3) of the cytoplasmic probe, 6-carboxyfluorescein, until the polymer solution was replaced by an isotonic buffer (Ahkong *et al.*, 1987). The rapid diffusion of carboxyfluorescein that then occurred (in some cases within 320 msec) was thought to be consistent with the rupture by osmotic stretching of an intermediate, shared bilayer, although the possible involvement of a transitory H_{II} hexagonal phase or other non-bilayer configuration could not be excluded. It was commented that, although some ultrastructural work has indicated that a single bilayer occurs at the site of membrane fusion in biological systems (compare Palade, 1975; Palade and Bruns, 1968), few such observations have been made. Whether comparable intermediate structures are absent from other systems, or whether they are extremely labile, is not clear. However, treatment of human erythrocytes with 40% PEG 6000 appears to provide a way of temporarily stabilizing an intermediate stage in cell fusion that may be only transient in many other systems.

As the surface concentration of Pr^{3+} and Mn^{2+} ions has been observed to increase about threefold when a 30% solution of poly(ethylene glycol) is added to vesicles of phosphatidylcholine, an increase in ion binding may be a further general feature of the action of poly(ethylene glycol) on neutral phospholipid bilayers (Gawrisch, 1986). Increased ion binding might result from a lower solubility of ions in the external phase, consequent on its decreased polarity in the presence of poly(ethylene glycol) (Arnold et al., 1988). This may be relevant to the fact that the ability of poly(ethylene glycol) to induce the hemifusion of human erythrocytes (compare Stage 2 of Fig. 1) apparently depends on the presence of unidentified metal ions in commercial preparations of the polymer. Thus, preparations of PEG that have been treated with Chelex beads are unable to induce the hemifusion of human erythrocytes. Their fusogenic properties can be restored by the addition of trace quantities of bivalent or trivalent metal ions (Ahkong and Lucy, 1990).

Observations made on osmotic swelling in human erythrocytes treated with PEG 6000 (Ahkong and Lucy, 1988) that may be relevant to the mechanisms of membrane fusion occurring in exocytosis are discussed in Section II,C,2).

c. Fusogenic Lipids. Hen erythrocytes that were treated with fusogenic lipids (e.g., unsaturated carboxylic acids and their glyceryl esters) were observed to exhibit colloid osmotic swelling in early work. The swollen cells lysed more readily than they fused when EDTA was present, but in the presence of Ca^{2+}, the cells aggregated and fused before they lysed (Ahkong et al., 1973a; Lucy, 1973). Every chemical agent that induced cell fusion also caused cell swelling, and it was therefore concluded that colloidal osmotic swelling plays an essential role in cell fusion (Ahkong et al., 1973b). Essentially similar events were observed by Kosower et al. (1977, 1978), and by Maggio et al. (1978), in the fusion of hen erythrocytes by 2-(2-methoxyethoxy)ethyl cis-8-(2-octylcyclopropyl)octanoate (described as a membrane-mobility agent), and by polysialogangliosides, respectively. Comparable relationships between cell swelling and cell fusion were also seen in experiments on the fusion of erythrocytes that was induced by heating to 48–50°C (Ahkong et al., 1973b), and when cultures of mouse fibroblasts were fused by treatment with oleylamine (Bruckdorfer et al., 1974).

In relation to the heat-induced fusion of erythrocytes, it is relevant that the swelling of these cells at osmolarities between 200 and 700 mOsm is normally restricted by the membrane skeleton. After heating to 50°C for 60 min, however, the swelling is approximately ideal. This is probably because the spectrin–actin network is disrupted at this temperature (Heubusch et al., 1985). Protein denaturation also appears to account for the early observation that heating human erythrocytes to 50°C caused their fragmentation into small hemoglobin-containing vesicles (Ponder, 1955). At this temperature, an unstable surface wave develops on the cell rim, from which externalized vesicles pinch off (Coakley and Deeley, 1980; Gallez and Coakley, 1986). Structural damage to the membrane skeleton of erythrocytes thus appears to be the feature that is common to heating erythrocytes and exposing them to lipid fusogens. This allows the cells to fuse in response to concomitant or subsequent osmotic swelling. Whereas heating erythrocytes denatures the structural proteins of the membrane skeleton, treatment with lipid fusogens can activate the degradation of skeletal proteins by endogenous proteinases (reviewed by Lucy and Ahkong, 1989).

2. Electrically Induced Cell Fusion

The electrofusion of cells is a valuable laboratory tool that can be usefully exploited in a variety of applications. Work on the molecular mechanisms involved in the electrofusion of membranes may also aid in understanding the fusion reactions of biomembranes in signaling processes. Several different approaches are used to bring cells into contact prior to electrofusion (reviewed by Lucy, 1993), but it is often done by dielectrophoresis. Dipoles induced on cells by an alternating electric field causes them to move in the direction of the greatest field strength when the field is nonuniform. As a consequence, cells in an electrofusion chamber migrate toward the electrodes. This effect (dielectrophoresis) also results in cells becoming attached to one another in so-called "pearl chains." Exposure of such pearl chains of cells to appropriate high-voltage (DC) pulses then leads to cell fusion (Arnold and Zimmermann, 1984).

By using the membrane probe ($DiIC_{16}$) and the cytoplasmic fluorophore (6-carboxyfluorescein) it has been shown that, as with human erythrocytes treated with poly(ethylene glycol) 6000 (Ahkong et al., 1987), hemifusion, that is, Stage 2 in Fig. 1, can occur under appropriate conditions when the cells are subjected to electrical breakdown (Song et al., 1991). There was an inverse relationship between conditions that permit complete fusion and those that favor hemifusion, with respect to breakdown pulse length, breakdown voltage, and, in particuar, osmolarity and temperature. Thus the incidence of hemifusion in 250 mM erythritol was twice that in 150 mM erythritol, and hemifusion was fivefold greater at 25°C than at 20°C. It seems possible that osmotic swelling may be responsible for converting a short-lived, hemifusion state to complete cell fusion and, if this were the case, resealing of electrically permeabilized erythrocytes would be expected to inhibit complete fusion. It is known that pores that are induced in cell membranes by an electrical breakdown pulse reseal increasingly rapidly between 10°C and 37°C. The marked increase in electrical hemifusion (and decrease in complete fusion) at 25°C, by comparison with that seen at 20°C, that was observed in this work may therefore have been due to rapid (although incomplete) resealing of the permeabilized cells at the higher temperature.

In other experiments on the effects of osmolarity on the efficiency of electrofusion in human erythrocytes labeled with carboxyfluorescein only, pearl chains of labeled and unlabeled cells were exposed to an electric field pulse in a hypotonic solution (150 M) of erythritol. Unlabeled erythrocytes that were adjacent to labeled cells immediately became fluorescent as the cells fused, and pearl chains containing up to 14 fluorescent cells were formed (Ahkong and Lucy, 1986). By contrast, the fluorophore transferred only to single cells that were immediately adjacent to fluorescent cells when the erythrocytes were suspended in 200 mM erythritol, and few fused cells were then formed. In 400 mM erythritol, the instantaneous transfer of fluorescence was negligible, and no fused cells were formed. Subsequently, NS1 mouse myeloma cells were found to behave in a similar manner, and it was suggested that hypotonic media may be of value in the fusion of myeloma cells with lymphocytes for the preparation of monoclonal antibodies (Brown *et al.*, 1986). In studies on the electro-acoustic fusion of human erythrocytes, Bardsley *et al.* (1989) found that the fusion yield in 170 mOsm solutions is much higher than the yield from cells in 272 mOsm solutions. With cells aligned by dielectrophoresis, Schmitt and Zimmermann (1989) also observed that the electrofusion of mammalian cells in strongly hypo-osmolar media containing sorbitol results in high yields of hybridoma cells. The efficacy of the hypo-osmolar electrofusion of these cells allowed the use of very few cells (about 10^5 lymphocytes per fusion chamber). Details of the application of hypo-osmolar fusion media in electrofusion protocols have been described by Zimmermann *et al.* (1989).

It was apparent from our work on the electrofusion of erythrocytes in erythritol solutions of differing osmolarity that the degree to which the cells were swollen affected the extent to which they fused. It was therefore proposed that the fusion process is governed by a combination of the electrical–compressive force (Zimmermann *et al.*, 1977, 1980) and the osmotic force applied prior to electrical breakdown (Ahkong and Lucy, 1986). Schmitt and Zimmermann (1989) have concluded, however, that neither pressure, nor osmotic gradients, nor the increase in volume of cells in hypo-osmolar solutions contribute directly to the mechanism of electrofusion because they found that treatment of myeloma cells with hypo-osmotic solutions enhanced the electrofusion yield even when the cells were returned to iso-osmolar conditions before being subjected to electrofusion. They suggested that the increase in fusion yield is due, instead, to the small increase in membrane permeability associated with the swelling process and to the dissolution of membrane skeletal proteins caused by osmotic stress. In experiments with Chinese hamster ovary cells, Rols and Teissié (1990) observed that the electrofusion index was affected by osmotic pressure in opposite ways under different conditions.

The fusion index increased with an increasing molarity of sucrose in experiments on cells that were growing in monolayers and were already in contact before exposure to the field pulses. By contrast, with cells in suspension that were brought into contact by centrifugation after exposure to electrical pulses, the fusion index decreased from 25 to 7% when the concentration of sucrose in the pulsing medium was increased from 100 to 400 mM sucrose. They concluded that the membrane fusion step of cells in suspension is not directly altered by osmotic pressure. Instead, it may modulate the extent of electropermeabilization by affecting the membrane sieving action of transient permeated structures and, as such, alter the fusogenic character of the permeabilized membrane.

Whether or not the electrofusion of cells is normally mediated via a transient, hemifused state, it is apparent that hemifused cells can give rise to completely fused cells because hemifused erythrocytes were occasionally found to fuse completely on heating to 50°C (Song *et al.*, 1991). The presence of Pronase may have contributed to this finding. However, a delayed cell fusion has also been observed (monitored by the diffusion of carboxyfluorescein) when electrically permeabilized erythrocytes are subsequently allowed to swell in 200 mM erythritol (Lucy, 1986).

Chernomordick and Sowers (1991) recently reported studies on factors that convert electrofused human erythrocyte ghosts, the cytoplasms of which are connected only by multiple, small fusion pores in their apposed plasma membranes, into giant cell poly-ghosts (compare Stage 4 in Fig. 1). They concluded that (1) the effects of heat and low-ionic strength, (2) the absence of demonstrable sources for osmotic pressure differences, and (3) the absence of a large difference in the rate of increase in the diameter of the fusion zone in planar diaphragms compared to open lumens reflect forces that are less likely to have an osmotic pressure origin than to result from a latent tension or stress in the membrane itself. Even so, forces due to osmotic pressure were apparent from the fact that placing the electrofused ghosts in a hypertonic sucrose medium caused a slowing of the "rounding-up" stage.

Recently, we have found that osmotic swelling, which results from the entry of small sugar molecules via electropores in the plasma membrane, is responsible for the rounding-up of human erythrocytes into giant cells, following their exposure to electrical breakdown pulses in low ionic strength media (Song, L. Y., Ahkong, Q. F., Baldwin, J. M., O'Reilly, R., and Lucy, J. A., unpublished observations). As with erythrocytes treated with (nonhaemolytic and nonleaky) Sendai virions (see Section II,C,1), osmotic swelling appears to be the driving force that results in cells, in which membrane fusion sites are present, being able to expand into poly-erythrocytes.

As in cell fusion protocols that involve cell swelling (Baldwin *et al.*, 1990), the plasma membranes of human erythrocytes exhibit a loss of asymmetry with respect to acidic phospholipids after treatment with electrical breakdown pulses of 20 μsec or longer (5 kV cm^{-1}) (Song *et al.*, 1992). Cell fusion and surface exposure of acidic phospholipids, as determined by the prothrombinase assay, increased approximately in parallel when the breakdown voltage was increased from 2 to 5 kV cm^{-1} (with breakdown pulses of 99 μsec). Furthermore, with 99 μsec pulses and a voltage of 3 kV cm^{-1}, a decrease in the osmolarity from 250 to 150 mM of the sucrose medium was accompanied by an increase in both the surface exposure of acidic phospholipids and the extent of cell fusion. It was suggested that a localized surface exposure of acidic phospholipids may contribute to the "long-lived fusogenic state" (Sowers, 1986) and the "transient permeant structures" (Teissié and Rols, 1986) that enable cell fusion to occur when contact between cells is established after they have been subjected to field pulses. The findings made were also thought to provide circumstantial support for the concept that changes in the phospholipid asymmetry of membranes may be important in physiologically occurring instances of biomembrane fusion.

C. Naturally Occurring Membrane Fusion

1. Virally Induced Membrane Fusion

Using an assay for lipid mixing (i.e., for Stage 2, Fig. 1), Blumenthal *et al.* (1987) investigated the dependence on pH of the fusion of vesicular stomatitis virions with Vero cells. Fusion at pH 7.4, via the endocytic pathway, was attenuated at hypo-osmotic solutions, but the rate of fusion of the virus with the plasma membrane at pH 5.9 was about the same in hypo-osmotic solutions (in which the cells were disrupted) as in iso- or hyperosmotic solutions. The latter finding contrasts with observations on the fusion of Sendai virus virions with human erythrocyte membranes treated with Pronase-neuraminidase and chromaffin granules (Citovsky and Loyter, 1985; Citovsky *et al.*, 1987) in which a lipid-mixing assay was also employed. In this work, fusion of Sendai virions with right-side-out erythrocyte vesicles, depleted of virus receptors, occurred only under hypotonic conditions. A threefold increase in virus-membrane fusion was also found on incubation of the virions with chromaffin granule membrane vesicles in a medium of low osmotic strength, but virus liposome fusion was not affected by the osmolarity of the medium. It was therefore suggested that the osmotic swelling of erythrocyte vesicles and chromaffin granule vesicles exposes their phospholipid bilayers, which can then interact with glycoproteins of the viral envelope in the absence of specific virus receptors. Herrmann *et al.* (1988) have reported findings that appear not to support this hypothesis because allowing resealed human erythrocyte ghosts to swell, before starting fusion induced by influenza virus, did not result in a higher percentage of ghost–ghost fusion as assayed by lipid mixing (Herrmann *et al.*, 1988). However, the parameters that govern the fusion of Sendai virion membranes with erythrocyte membranes that were studied by Loyter and his colleagues may not be the same as those governing the fusion of erythrocyte membranes with one another following exposure to the influenza virus.

Direct fusion between cells treated with hemolytic Sendai virus, rather than via a virion that fuses simultaneously with two cells and acts as a bridge, has been reported (Bächi *et al.*, 1973; Toister and Loyter, 1973). In this connection it has been pointed out that the fusion of erythrocytes by hemolytic Sendai virus has many similarities to the fusion of these cells by lipid-soluble fusogens, particularly in relation to the colloid-osmotic swelling and lysis that accompany cell fusion. It was therefore suggested that osmotic swelling may drive the fusion of erythrocytes with one another (Stages 2 and 3 in Fig. 1) after they have been permeabilized by hemolytic Sendai virus (Lucy and Ahkong, 1986, 1989). However, Herrmann *et al.* (1988) observed no effect of external osmotic pressure or of cell swelling on the fusion of human erythrocyte ghosts induced by the influenza virus. The virus also fused unsealed ghosts as effectively as resealed ghosts. Therefore, they concluded that (apart from Stage 4 in Fig. 1) neither osmotic forces nor the osmotic swelling of cells is necessary for virus-induced cell fusion.

The role of osmotically induced cell swelling in the formation of giant erythrocytes, or other polykaryons, by hemolytic Sendai virus particles (Stage 4) has been well documented. Fusion of Sendai virus particles with the plasma membrane of cells makes the membrane permeable to low molecular weight compounds and ions. As a consequence, K^+ ions leak out, Na^+ ions leak in (Poste and Pasternak, 1978), and the loss of cation asymmetry leads to an influx of water with resultant cell swelling (Knutton *et al.*, 1976). At least for fusion that involves a virion acting as a bridge between two cells, swelling is however responsible only for the rounding of already fused cells (Pasternak, 1984), that is, for Stage 4 in Fig. 1. This was demonstrated by the fact that the formation of giant cells was prevented when osmotic swelling was inhibited by hypertonic media (Knutton and Pasternak, 1979; Impraim *et al.*, 1980).

Secondly, when early-harvested (nonhemolytic and nonleaky) Sendai virions were used, giant cells were not formed (Knutton 1979; Wyke et al. 1980). That cell fusion nevertheless occurs under these circumstances was demonstrated by Knutton and Bächi (1980), who found that the treated erythrocytes enlarged into giant cells on subsequent exposure to a hypotonic medium. They therefore concluded that osmotic swelling appears to be the driving force that results in cells, in which membrane fusion sites are present, being able to expand into polyerythrocytes.

Similar conclusions were reached by Sekiguchi et al. (1981), who investigated the fusion by Sendai virus of human erythrocyte ghosts that were prepared with or without sequestered macromolecules, such as bovine serum albumin or dextran. In the absence of osmotic swelling, that is, with ghosts that were sealed without occluded macromolecules, no large poly-ghosts were observed following treatment with the virus. Similarly, poly-ghosts were not formed from ghosts that were loaded with bovine serum albumin but that were exposed to the virus in the presence of external macromolcules (e.g., 2% albumin). The occurrence of local cell fusion could nevertheless be demonstrated by the transfer of fluorescent-labeled albumin from one ghost to another, or by the fact that poly-ghosts were produced after osmotic swelling of the virus-treated cells in the cold. Colloidal osmotic swelling was thus seen to be responsible for the rounding up of virally fused ghosts.

2. Exocytosis

Isolated chromaffin granules release their contents under isotonic conditions when they are exposed to Mg^{2+}-ATP and large quantities of Cl^- or other permeant anions; this release is suppressed by raising the osmotic strength of the medium with salt or sucrose. These early observations indicated that the release reaction is the result of osmotic lysis, which permits the contents of the granules to escape. Several researchers have considered that this system might provide a general model for the membrane fission step in exocytosis, and a chemiosmotic model for exocytosis was suggested by Pollard et al. (1979). They proposed that close juxtaposition of the granule membrane to the cytoplasmic surface of the plasma membrane of a chromaffin cell is mediated by Ca^{2+} and synexin. (For more recent work on possible roles of synexin and other annexins in membrane fusion and exocytosis, see Pollard et al., 1990 and Meers et al., 1991.) In their model, it was envisaged that a region of shared bilayer (Palade, 1975; Palade and Bruns, 1968) is then formed by fusion of the two membranes, followed by an entry of anions and protons into the granule that increases the osmotic strength of the interior of the granule. This was considered to lead to osmotic rupture of the shared bilayer and release of the granule contents by exocytosis. Alternative early models, and other relevant observations, have been reviewed by Lucy and Ahkong (1989).

An elevation of osmotic strength with sucrose inhibits the secretion of catecholamines by exocytosis from chromaffin cells (Baker and Knight, 1981; Hampton and Holz, 1983; Pollard, et al., 1984). However, several observations showed that granule lysis and secretion are not identical even though they have strong qualitative similarities. In a critical review, Baker and Knight (1984) subsequently concluded that there was no clear evidence for an essential involvement of either anions or monovalent cations in the exocytotic release of catecholamines, and that the ATP-dependent proton pump in the membranes of secretory vesicles is very unlikely to play a part in exocytosis.

Many investigations have nevertheless indicated that osmotic forces play a role in a number of different exocytotic systems, including the following: mucocyst secretion in *Tetrahymena* (Satir et al., 1973); the secretion of serotonin from platelets (Pollard et al., 1977); the release of parathyroid hormone (Brown et al., 1978); endotoxin-induced degranulation in amebocyte blood cells of *Limulus polyphemus* (Ornberg and Reese, 1981); the discharge of nematocysts in sea anemones (Lubbock et al., 1981); exocytosis in the toad bladder induced by antidiuretic hormone (Kachadorian et al., 1981); the release of insulin from the pancreas (Hermans and Henquin, 1986; Orci and Malaisse, 1980); the cortical granule reaction in sea urchin eggs (Zimmerberg and Whitaker, 1985; Zimmerberg et al., 1985); and the exocytosis of prolactin from rat pituitary tumor cells *in vitro* (Day and Hinkle, 1988) and of myeloperoxidase from human polymorphonuclear leukocytes (Sato et al., 1990).

The pertinent question with regard to such observations on exocytosis is, however, at what stage in membrane fusion–fission do osmotic forces exert their effect? Are they involved only in widening the final fusion pore? Because the contents of intracellular storage granules are closely packed, rehydration of the stored material is important in exocytosis; a variety of polymers prevent exocytosis of the cortical granules in sea urchin eggs by inhibiting the dispersal of the granule contents after fusion of the granule and plasma membranes has occurred (Whitaker and Zimmerberg, 1987). However, in a rapid-freezing study of the release of blood-clotting proteins from amebocyte cells of horseshoe crabs, the earliest change observed was a separation between the granule core and its membrane that left a clear crescent beneath the granule membrane (Ornberg and Reese, 1981). Similar clear spaces in isolated chromaffin granules, aggregated by Ca^{2+} ions, were also attributed to an osmotically driven influx of water (Edwards et al., 1974). Such observations indicate that osmotic fusions are involved at early stages in membrane fusion–fission. In addition, Schmauder-Chock and Chock (1987) reported that the peri-granular membrane can enlarge prior to fusion with the plasma membrane, but the membrane enlargement was so extensive that it could not be explained by membrane stretching alone. Consequently, it was suggested that additional membrane was inserted during activation of secretory granules.

The behavior of erythrocytes in cell fusion induced by treatment with PEG 6000 is also of interest in relation to exocytosis because the hemoglobin of erythrocytes that are dehydrated by concentrated solutions of this polymer may be regarded as a model for the tightly packed, dehydrated contents of the granules of secretory cells. Under certain conditions of rehydration, the swelling of aqueous microdroplets that are located between the dehydrated hemo-globin and the plasma membrane is closely associated with the fusion of partially rehydrated but still shrunken, PEG-treated erythrocytes (Ahkong and Lucy, 1988). Therefore, it is apparent that osmotic forces, acting locally at the sites of aqueous microdroplets, can drive the fusion of membranes that encapsulate a dehydrated, concentrated protein, even though gross osmotic swelling at the level of the light microscope is absent (Fig. 2). This finding is consistent with the possibility that osmotic swelling may play a role in exocytotic membrane fusion if it is restricted to a small zone immediately under the granule membrane (Green, 1990).

The observations in the previous two paragraphs, taken together with those on membrane fusion reactions in planar phospholipid bilayers that are discussed in Section II,A,1 contribute interesting circumstantial evidence for a likely involvement of osmotic forces in exocytosis prior to enlargement of the fusion pore. However, measurements of membrane capacitation in mast cells of beige mice, designed to detect the moment at which the granule membrane and the plasma membrane become continuous in the exocytosis, have shown that the change in capacitance precedes granule swelling (Breckenridge and Almers, 1987a,b; Zimmerberg, 1987; Zimmberberg et al., 1987). In light of this work, it has been concluded that swelling cannot be the driving force for membrane fusion in this system. Nevertheless, the possibility that highly localized granule swelling may precede membrane fusion in exocytosis cannot be completely excluded. It has recently been proposed that the granule membrane is under tension, which is not due to secretory granule swelling, and that this tension causes a net transfer of membrane from the plasma membrane to the secretory granule while they are connected by the fusion pore (Monck et al., 1990, 1991). It is thought that the high membrane tension in the secretory granule may be the critical stress that is necessary for exocytotic fusion.

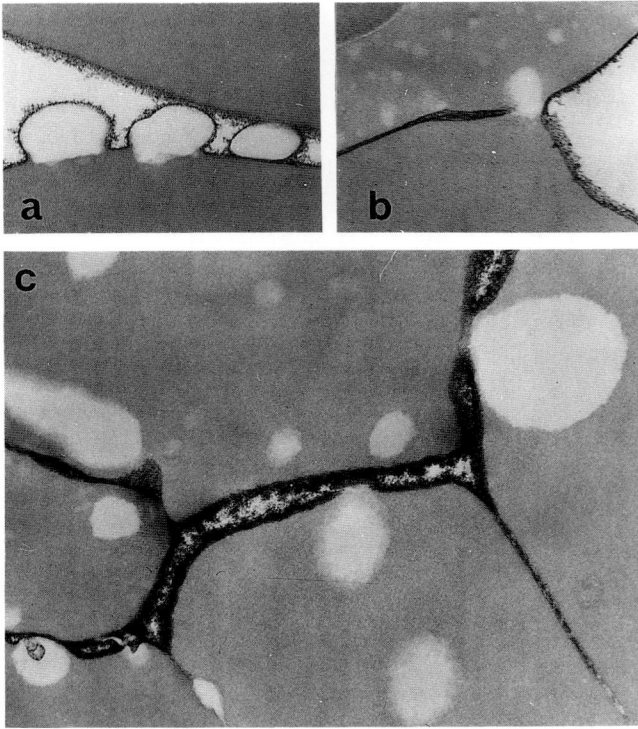

Fig. 2. Electron micrographs of thin sections of human erythrocytes during fusion by high-molecular weight poly(ethylene glycol) 6000 showing that the swelling of aqueous microdroplets between the dehydrated hemoglobin and the plasma membrane is closely associated with the fusion of partially rehydrated but still shrunken, poly(ethylene glycol)-treated cells. (a) Ballooning of the plasma membrane around the microdroplets. (b) Interactions of the droplets with the plasma membrane cause changes in membrane morphology that may precede fusion–fission. (c) An example of membrane fusion–fission (cell fusion) occurring at the site of an aqueous microdroplet. These phenomena were seen both when cells were treated with 50% poly(ethylene glycol) that was diluted to 35% before the addition of glutaraldehyde to fix the cells for electron microscopy, and when they were treated with 40–50% poly(ethylene glycol) that was then diluted more extensively (to 4.5%) with buffered saline-containing glutaraldehyde. Magnification × 30,000. (Reproduced with permission from Ahkong and Lucy, 1988.)

III. FINAL COMMENTS

Osmotic phenomena are undoubtedly of practical importance in membrane fusion processes and, as indicated above, they can be used to increase the yield of fused cells in electrofusion. The significance of osmotic phenomena in the mechanism of membrane fusion reactions in biological systems is, however, far from being satisfactorily resolved. Much work on the mechanisms of membrane fusion has been done with model systems (i.e., with planar phospholipid bilayers, phospholipid vesicles, and erythrocytes) on the implicit assumption that, if an event occurs in a model system, it can and probably does occur in biological systems. Unfortunately, this is not necessarily so. For example, two independent groups found that small phospholipid vesicles give rise to large vesicles, apparently by membrane fusion, in the presence of poly(ethylene glycol) (Boni et al., 1981; Aldwinckle et al., 1982). By contrast, cell fusion requires the treatment of cells with a concentrated solution of poly(ethylene glycol) to be followed by a dilution step or removal of the polymer. This difference has not been explained and, to complicate matters further, it was recently reported that poly(ethylene glycol) does not induce contents mixing with vesicles of pure synthetic phosphatidylcholine (Burgess et al., 1991).

One reason for the wealth of attention given to fusion phenomena in model systems (reviewed by Düzgüneş, 1985) is their relative simplicity. Such systems are nevertheless criticized by some as being too simple and not having sufficient relevance to membrane fusion occurring in vivo. It is hoped that the development of new techniques, for example, digital imaging in confocal microscopy, for the quantitative evaluation of events occurring at specific sites in single living cells will facilitate progress in the understanding of membrane fusion reactions in biological systems.

REFERENCES

Ahkong, Q. F., and Lucy, J. A. (1986). Osmotic forces in artificially induced cell fusion. *Biochim. Biophys. Acta* **858**, 206–216.

Ahkong, Q. F., and Lucy, J. A. (1988). Localized osmotic swelling and cell fusion in erythrocytes: Possible implications for exocytosis. *J. Cell Sci.* **91**, 597–601.

Ahkong, Q. F., and Lucy, J. A. (1990). Chelex-100 treated polyethylene glycol is nonfusogenic. *Proc. 10th Int. Congress of Biophysics.* [Abstract P3.5.10].

Ahkong, Q. F., Fisher, D., Tampion, W., and Lucy, J. A. (1973a). The fusion of erythrocytes by fatty acids, esters, retinol, and alpha-tocopherol. *Biochem. J.* **136**, 147–155.

Ahkong, Q. F., Cramp, F. C., Fisher, D., Howell, J. I., Tampion, W., Verrinder, M., and Lucy, J. A. (1973b). Chemically induced and thermally induced cell fusion. Lipid–lipid interactions. *Nature New Biol.* **242**, 215–217.

Ahkong, Q. F., Desmazes, J-P., Georgescauld, D., and Lucy, J. A. (1987). Movements of fluorescent probes in the mechanism of cell fusion induced by poly(ethylene glycol). *J. Cell Sci.* **88**, 389–398.

Akabas, M. H., Cohen, F. S., and Finkelstein, A. (1984). Separation of the osmotically driven fusion event from vesicle-planar membrane attachment in a model system for exocytosis. *J. Cell Biol.* **98**, 1063–1071.

Aldwinckle, T. J., Ahkong, Q. F., Bangham, A. D., Fisher, D., and Lucy, J. A. (1982). Effects of poly(ethylene glycol) on liposomes and erythrocyte permeability changes and fusion. *Biochim. Biophys. Acta* **689**, 548–560.

Arnold, K., Herrmann, A., Gawrisch, K., and Pratsch, L. (1988). Water-mediated effects of PEG on membrane properties and fusion. *In* "Molecular Mechanisms of Membrane Fusion" (S. Ohki, D. Doyle, T. D. Flanagan, S. Hui, and E. Mayhew, eds.), pp. 255–272. Plenum Press, New York.

Arnold, K., Zschoernig, O., Barthel, D., and Herold, W. (1990). Exclusion of poly(ethylene glycol) from liposome surfaces. *Biochim. Biophys. Acta* **1022**, 303–310.

Arnold, W. M., and Zimmermann, U. (1984). Electric-field-induced fusion and rotation of cells. *In* "Biological Membranes" (D. Chapman, ed.), Vol. 5, pp. 389–454. Academic Press, London.

Bächi, M., Aguet, M., and Howe, C. (1973). Fusion of erythrocytes by Sendai virus studied by immuno freeze-etching. *J. Virol.* **11**, 1004–1012.

Baker, P. F., and Knight, D. E. (1981). Calcium control of exocytosis and endocytosis in bovine adrenal medullary cells. *Philos. Trans. R. Soc. London (Biol.)* **296**, 83–103.

Baker, P. F., and Knight, D. E. (1984). Chemiosmotic hypotheses of exocytosis. A critique. *Biosci. Rep.* **4**, 285–298.

Baldwin, J. M., O'Reilly, R., Whitney, M., and Lucy, J. A. (1990). Surface exposure of phosphatidylserine is associated with the swelling and osmotically induced fusion of human erythrocytes in the presence of Ca^{2+}. *Biochim. Biophys. Acta* **1028**, 14–20.

Bardsley, D. W., Coakley, W. T., Jones, G., and Liddell, J. E. (1989). Electroacoustic fusion of millilitre volumes of cells in physiological media. *J. Biochem. Biophys. Meths.* **19**, 339–348.

Blumenthal, R., Bali-Puri, A., Walter, A., Covell, D., and Eidelman, O. (1987). pH-dependent fusion of vesicular stomatitis virus with Vero cells. *J. Biol. Chem.* **262**, 13614–13619.

Boni, L. T., Stewart, T. P., Alderfer, J. L., and Hui, S. W. (1981). Lipid-polyethylene glycol interactions. I. Induction of fusion between liposomes. *J. Membr. Biol.* **62**, 65–70.

Breckenridge, L. J., and Almers, W. (1987a). Currents through the fusion pore that forms during exocytosis of a secretory vesicle. *Nature (London)* **328**, 814–817.

Breckenridge, L. J., and Almers, W. (1987b). Final steps in exocytosis observed in a cell with giant secretory granules. *Proc. Natl. Acad. Sci. USA.* **84**, 1945–1949.

Brown, E. M., Pazoles, C. J., Creutz, C. E., Aurbach, G. D., and Pollard, H. B. (1978). Role of anions in parathyroid hormone release from dispersed bovine parathyroid cells. *Proc. Natl. Acad. Sci. USA.* **75**, 876–880.

Brown, S. M., Ahkong, Q. F., and Lucy, J. A. (1986). Osmotic pressure and the electrofusion of myeloma cells. *Biochem. Soc. Trans.* **14**, 1129–1130.

Bruckdorfer, K. R., Cramp, F. C., Goodall, A. H., Verrinder, M., and Lucy, J. A. (1974). Fusion of mouse fibroblasts with oleylamine. *J. Cell Sci.* **15**, 185–199.

Burgess, S. W., Massenburg, D., Yates, J., and Lentz, B. (1991). Poly(ethylene glycol)-induced lipid mixing, but not fusion, between synthetic phosphatidylcholine large unilamellar vesicles. *Biochemistry* **30**, 4193–4200.

Chernomordick, L. V., and Sowers, A. E. (1991). Evidence that the spectrin network and a nonosmotic force controls the fusion product morphology in electrofused erythrocyte ghosts. *Biophys. J.* **60**, 1026–1037.

Citovsky, V., and Loyter, A. (1985). Fusion of Sendai virions or reconstituted Sendai virus envelopes with liposomes or erythrocyte membranes lacking virus receptors. *J. Biol. Chem.* **260,** 12072–12077.

Citovsky, V., Laster, S., Schuldiner, S., and Loyter, A. (1987). Osmotic swelling allows fusion of Sendai virions with membranes of desialized erythrocytes and chromaffin granules. *Biochemistry* **26,** 3856–3864.

Coakley, W. T., and Deeley, J. O. T. (1980). Effects of ionic strength, serum protein, and surface charge on membrane movements and vesicle production in heated erythrocytes. *Biochim. Biophys. Acta* **602,** 355–375.

Cohen, F. S., Zimmerberg, J., and Finkelstein, A. (1980). Fusion of phospholipid vesicles with planar phospholipid bilayer membranes. II. Incorporation of a vesicular membrane marker into the planar membrane. *J. Gen. Physiol.* **75,** 251–270.

Cohen, F. S., Akabas, M. H., and Finkelstein, A. (1982). Osmotic swelling of phospholipid vesicles causes them to fuse with a planar phospholipid bilayer membrane. *Science* **217,** 458–460.

Cohen, F. S., Akabas, M. H., Zimmerberg, J., and Finkelstein, A. (1984). Parameters affecting the fusion of unilamellar phospholipid vesicles with planar bilayer membranes. *J. Cell Biol.* **98,** 1054–1062.

Cohen, F. S., Niles, W. D., and Akabas, M. H. (1989). Fusion of phospholipid vesicles with a planar membrane depends on the membrane permeability of the solute used to create the osmotic pressure. *J. Gen. Physiol.* **93,** 201–210.

Day, R. N., and Hinkle, P. M. (1988). Osmotic regulation of prolactin secretion: Possible role of chloride. *J. Biol. Chem.* **263,** 15915–15921.

Düzgüneş, N. (1985). Membrane fusion. *In* "Subcellular Biochemistry" (D. B. Roodyn, ed.), Vol. 11, pp. 195–286. Plenum Press, New York.

Düzgüneş, N., Nir, S., Wilschut, J., Bentz, J., Newton, C., Portis, A., and Papahadjopoulos, D. (1981). Calcium- and magnesium-induced fusion of mixed phosphatidylserine/phosphatidylcholine vesicles: Effect of ion binding. *J. Membr. Biol.* **59,** 115–125.

Düzgüneş, N., Allen, T. M., Fedor, J., and Papahadjopoulos, D. (1987). Lipid mixing during membrane aggregation and fusion: Why fusion assays disagree. *Biochemistry* **26,** 8435–8442.

Edwards, W., Phillips, J. H., and Morris, S. J. (1974). Structural changes in chromaffin granules induced by divalent cations. *Biochim. Biophys. Acta* **356,** 164–173.

Finkelstein, A., Zimmerberg, J., and Cohen, F. S. (1986). Osmotic swelling of vesicles: Its role in the fusion of vesicles with planar phosophlipid bilayer membranes and its possible role in exocytosis. *Annu. Rev. Physiol.* **48,** 163–174.

Gallez, D., and Coakley, W. T. (1986). Interfacial instability at cell membranes. *Prog. Biophys. Mol. Biol.* **48,** 155–199.

Gawrisch, K. (1986). "Molekulare mechanismen und membranveränderungen bei der durch polyethylenglykol induzierten zellfusion." Thesis B, Karl Marx Universität, Leipzig.

Green, D. P. L. (1990). Mechanical coupling of zymogen granule membrane with the granule core. *Biophys. J.* **58,** 1557–1558.

Hampton, R. Y., and Holz, R. W. (1983). Effects of changes in osmolarity on the stability and function of cultural chromaffin cells and the possible role of osmotic forces in exocytosis. *J. Cell Biol.* **96,** 1082–1088.

Helm, C. A., Israelachvili, J. N., and McGuiggan, P. M. (1992). Role of hydrophobic forces in bilayer adhesion and fusion. *Biochemistry,* **31,** 1794–1805.

Hermans, M. P., and Henquin, J. C. (1986). Is there a role for osmotic events in exocytotic release of insulin? *Endocrinology* **119,** 105–111.

Herrmann, A., Pritzen, C., Palesch, A., and Groth, T. (1988). The influenza virus induced fusion of erythrocyte ghosts does not depend on osmotic forces. *Biochim. Biophys. Acta* **943**, 411–418.

Heubusch, P., Jung, C. Y., and Green, F. A. (1985). The osmotic response of human erythrocytes and the membrane skeleton. *J. Cell Physiol.* **122**, 266–272.

Huang, S. K., and Hui, S. W. (1990). Fluorescence measurements of fusion between human erythrocytes induced by poly(ethylene glycol). *Biophys. J.* **58**, 1109–1117.

Hui, S. W., Isac, T., Boni, L. T., and Sen, A. (1985). Action of poly(ethylene glycol) on the fusion of human erythrocyte membranes. *J. Membr. Biol.* **85**, 137–146.

Impraim, C. C., Foster, K. A., Micklem, K. J., and Pasternak, C. A. (1980). Nature of virally mediated changes in membrane permeability to small molecules. *Biochem. J.* **186**, 847–860.

Kachadarian, W. A., Muller, J., and Finkelstein, A. (1981). Role of osmotic forces in exocytosis. Studies of ADH-induced fusion in toad urinary bladder. *J. Cell Biol.* **91**, 584–588.

Kosower, E. M., Kosower, N. S., and Wegman, P. (1977). Membrane mobility agents. IV. The mechanism of particle–cell and cell–cell fusion. *Biochim. Biophys. Acta* **471**, 311–329.

Kosower, N. S., Wegman, P. O., Neiman, T., and Kosower, E. M. (1978). Membrane mobility agents. V. Genetic variability in the fusibility of hen red cells. *Exp. Cell Res.* **116**, 454–456.

Knutton, S. (1979). Studies of membrane fusion. V. Fusion of erythrocytes with nonhemolytic Sendai virus. *J. Cell Sci.* **36**, 85–96.

Knutton, S., and Bächi, T. (1980). The role of cell swelling and haemolysis in Sendai virus-induced cell fusion and in the diffusion of incorporated viral antigens. *J. Cell Sci.* **42**, 153–167.

Knutton, S., and Pasternak, C. A. (1979). The mechanism of cell–cell fusion. *Trends. Biochem. Sci.* **4**, 220–223.

Knutton, S., Jackson, D., Graham, J. M., Micklem, K. J., and Pasternak, C. A. (1976). Microvilli and cell swelling. *Nature (London)* **262**, 52–54.

Lubbock, R., Gupta, B. L., and Hall, T. A. (1981). Novel role of calcium in exocytosis. Mechanism of nematocyst discharge as shown by X ray microanalysis. *Proc. Natl. Acad. Sci. USA* **78**, 3624–3628.

Jucy, J. A. (1973). The chemically induced fusion of cells. *In* "Membrane-mediated Information" (P. W. Kent, ed.), pp. 117–128. Medical and Technical Publishing, Lancaster, U.K.

Lucy, J. A. (1986). Salient features of artificially induced cell fusion. *Biochem. Soc. Trans.* **14**, 250–251.

Lucy, J. A. (1993). The electrofusion of animal cells. *In* "Electrical manipulation of cells" (P. T. Lynch and M. R. Davey, eds.). Chapman and Hall (in press).

Lucy, J. A., and Ahkong, Q. F. (1986). An osmotic model for the fusion of biological membranes. *FEBS Lett.* **199**, 1–11.

Lucy, J. A. and Ahkong, Q. F. (1989). Membrane fusion; fusogenic agents, and osmotic forces. *In* "Subcellular Biochemistry" (J. R. Harris and A.-H. Etémadi, eds.), Vol. 14, Artificial and Reconstituted Membrane Systems, pp. 189–228. Plenum Press, New York.

Maggio, B., Cumar, F. A., and Caputto, R. (1978). Induction of membrane fusion by polysialogangliosides. *FEBS Lett.* **90**, 149–152.

Meers, P., Daleke, D., Hong, K., and Papahadjopoulos, D. (1991). Interactions of annexins with membrane phospholipids. *Biochemistry* **30**, 2903–2908.

Miller, C., and Racker, E. (1976). Ca^{++}-induced fusion of fragmented sarcoplasmic reticulum with artificial planar bilayers. *J. Membr. Biol.* **30**, 283–300.

Miller, C., Arvan, P., Telford, J. N., and Racker, E. (1976). Ca^{2+}-induced fusion of proteoliposomes: Dependence on transmembrane osmotic gradient. *J. Membr. Biol.* **30**, 271–282.

Monck, J. R., Alvarez de Toledo, G., and Fernandez, J. M. (1990). Tension in secretory granule membranes causes extensive membrane transfer through the exocytotic fusion pore. *Proc. Natl. Acad. Sci. USA* **87**, 7804–7808.

Monck, J. R., Oberhauser, A. F., Alvarez de Toledo, G., and Fernandez, J. M. (1991). Is swelling of the secretory granule matrix the force that dilates the exocytotic fusion pore? *Biophys. J.* **59**, 39–47.

Niles, W. D., and Cohen, F. S. (1987). Video fluorescence microscopy studies of phospholipid vesicle fusion with a planar phospholipid membrane. *J. Gen. Physiol.* **90**, 703–735.

Niles, W. D., Cohen, F. S., and Finkelstein, A. (1989). Hydrostatic pressures developed by osmotically swelling vesicles bound to planar membranes. *J. Gen. Physiol.* **93**, 211–244.

Ohki, S. (1984). Effects of divalent cations, temperature, osmotic pressure gradient and vesicle curvature on phosphatidylserine vesicle fusion. *J. Membr. Biol.* **77**, 265–275.

Ohki, S. (1988). Surface tension, hydration energy, and membrane fusion. *In* "Molecular Mechanisms of Membrane Fusion" (S. Ohki, D. Doyle, T. D. Flanagan, S. W. Hui, and E. Mayhew, eds.) pp. 123–138, Plenum Press, New York.

Ohki, S., Doyle, D., Flanagan, T. D., Hui, S. W., and Mayhew, E. (1988). "Molecular Mechanisms of Membrane Fusion." p. 588. Plenum Press, New York.

Orci, L. and Malaise, W. (1980). Single and chain release of insulin secretory granules are related to anionic transport at exocytotic sites. *Diabetes* **29**, 943–944.

Ornberg, R. L., and Reese, T. S. (1981). Beginning of exocytosis captured by rapid-freezing of *Limulus* amebocytes. *J. Cell Biol.* **90**, 40–54.

Palade, G. E. (1975). Intracellular aspects of protein synthesis. *Science* **189**, 347–358.

Palade, G. E., and Bruns, R. R. (1968). Structural modifications of plasmalemmal vesicles. *J. Cell Biol.* **37**, 633–649.

Parente, R. A., and Lentz, B. R. (1986). Rate and extent of poly(ethylene glycol)-induced large vesicle fusion monitored by bilayer and internal contents mixing. *Biochemistry* **25**, 667–668.

Parsegian, V. A., and Rand, R. P. (1991). Forces governing lipid interaction and rearrangement. *In* "Membrane Fusion" (J. Wilschut and D. Hoekstra, eds.), pp. 65–85, Marcel Dekker, New York.

Pasternak, C. A. (1984). Virally mediated changes in cellular permeability. *In* "Membrane Processes" (G. Benga, H. Baum, and F. A. Kummerow, eds.), pp. 140–166. Springer-Verlag, New York.

Perrin, M. S., and MacDonald, R. C. (1989). Fusion of synaptic vesicle membranes with planar bilayer membranes. *Biophys. J.* **55**, 973–986.

Pollard, H. B., Tack-Goldman, K., Pazoles, C. J., Creutz, C. E., and Shulman, N. R. (1977). Evidence for control of serotonin secretion from human platelets by hydroxyl ion transport and osmotic lysis. *Proc. Natl. Acad. Sci. USA* **74**, 5295–5299.

Pollard, H. B., Pazoles, C. J., Creutz, C. E., and Zinder, O. (1979). The chromaffin granule and possible mechanisms of exocytosis. *Int. Rev. Cytol.* **58**, 159–197.

Pollard, H. B., Pazoles, C. J., Creutz, C. E., Scott, J. H., Zinder, O., and Hotchkiss, A. (1984). An osmotic mechanism for exocytosis from dissociated chromaffin cells. *J. Biol. Chem.* **259**, 1114–1121.

Pollard, H. B., Burns, A. L., and Rojas, E. (1990). Synexin (Annexin VII): A cytosolic calcium-binding protein that promotes membrane fusion and forms calcium channels in artificial bilayer and natural membranes. *J. Membr. Biol.* **117**, 101–112.

Ponder, E. (1955). "Red Cell Structure and Its Breakdown." Springer-Verlag, Vienna.

Poste, G., and Pasternak, C. A. (1978). Virus-induced cell fusion. *In* "Membrane Fusion" (G. Poste and G. L. Nicolson, eds.), pp. 305–367. Elsevier/North-Holland Biomedical Press, Amsterdam.

Rand, R. P., and Parsegian, V. A. (1989). Hydration forces between phospholipid bilayers. *Biochim. Biophys. Acta* **988**, 351–376.

Rols, M-P., and Teissié, J. (1990). Modulation of electrically induced permeabilization and fusion of Chinese hamster ovary cells by osmotic pressure. *Biochemistry* **29**, 4561–4567.

Sato, N., Wang, X, and Greer, M. A. (1990). Hyposmolarity stimulates myeloperoxidase exocytosis from human polymorphonuclear leukocytes. *Am. J. Med. Sci.* **299**, 309–312.

Satir, B., Shooley, C., and Satir, P. (1973). Membrane fusion in a model system: Mucocyst secretion in *Tetrahymena*. *J. Cell Biol.* **56**, 153–176.

Schmauder-Chock, E. A., and Chock, S. P. (1987). Mechanism of secretory granule exocytosis: Can granule enlargement precede pore formation? *Histochem. J.* **19**, 413–418.

Schmitt, J. J., and Zimmermann, U. (1989). Enhanced hybridoma production by electrofusion in strongly hypo-osmolar solutions. *Biochim. Biophys. Acta* **983**, 42–50.

Sekiguchi, K., Kuroda, K., Ohnishi, S-I., and Asano, A. (1981). Virus-induced fusion of human erythrocyte ghosts. I. Effect of macromolecules of the final stages of the fusion reaction. *Biochim. Biophys. Acta* **645**, 211–225.

Silvius, J. R., Leventis, R., and Brown, P. M. (1988). "Slow artifacts" in assay of lipid mixing between membranes. *In* "Molecular Mechanisms of Membrane Fusion" (S. Ohki, D. Doyle, T. D. Flanagan, S. Hui, and E. Mayhew, eds.), pp. 531–542. Plenum Press, New York.

Song, L., Ahkong, Q. F., Georgescauld, D., and Lucy, J. A. (1991). Membrane fusion without cytoplasmic fusion (hemifusion) in erythrocytes that are subjected to electrical breakdown. *Biochim. Biophys. Acta* **1065**, 54–62.

Song, L., Baldwin, J. M., O'Reilly, R., and Lucy, J. A. (1992). Relationships between the surface exposure of acidic phospholipids and cell fusion in erythrocytes subjected to electrical breakdown. *Biochim. Biophys. Acta* **1104**, 1–8.

Sowers, A. E. (1986). A long-lived fusogenic state is induced in erythrocyte ghosts by electric pulses. *J. Cell Biol.* **102**, 1358–1362.

Sowers, A. E. (1988). Fusion events and nonfusion contents-mixing events induced in erythrocyte ghosts by an electric pulse. *Biophys. J.* **54**, 619–626.

Teissié, J., and Rols, M. P. (1986). Fusion of mammalian cells in culture is obtained by creating the contact between cells after their electropermeabilization. *Biochem. Biophys. Res. Comm.* **140**, 258–266.

Toister, Z., and Loyter, A. (1973). The mechanism of cell fusion. II. Formation of chicken erythrocyte polykaryons. *J. Biol. Chem.* **248**, 422–432.

Tullius, E. K., Williamson, P., and Schlegel, R. A. (1989). Effect of transbilayer phospholipid distribution on erythrocyte fusion. *Biosci. Rep.* **9**, 623–633.

Whitaker, M., and Zimmerberg, J. (1987). Inhibition of secretory granule discharge during exocytosis in sea urchin eggs by polymer solutions. *J. Physiol.* **389**, 527–539.

Woodbury, D. J., and Hall, J. E. (1988). Role of channels in the fusion of vesicles with a planar bilayer. *Biophys. J.* **54**, 1053–1063.

Wyke, A. M., Impraim, C. C., Knutton, S., and Pasternak, C. A. (1980). Components involved in virally mediated membrane fusion and permeability changes. *Biochem. J.* **190**, 625–638.

Zimmerberg, J. (1987). Molecular mechanisms of membrane fusion. Steps during phospholipid and exocytotic membrane fusion. *Biosci. Rep.* **7**, 251–268.

Zimmerberg, J., and Whitaker, M. (1985). Irreversible swelling of secretory granules during exocytosis caused by calcium. *Nature (London)* **315**, 581–584.

Zimmerberg, J., Cohen, F. S., and Finkelstein, A. (1980a). Fusion of phospholipid vesicles with planar phospholipid bilayer membranes. I. Discharge of vesicular contents across the planar membrane. *J. Gen. Physiol.* **75**, 241–250.

Zimmerberg, J., Cohen, F. S., and Finkelstein, A. (1980b). Micromolar Ca^{2+} stimulates fusion of lipid vesicles with planar bilayers containing a calcium-binding protein. *Science* **210**, 906–908.

Zimmerberg, J., Sardet, C., and Epel, D. (1985). Exocytosis of sea urchin egg cortical vesicles *in vitro* is retarded by hyperosmotic sucrose: Kinetics of fusion monitored by quantitative light scattering. *J. Cell Biol.* **101**, 2398–2410.

Zimmerberg, J., Curran, M., Cohen, F. S., and Brodwick, M. (1987). Simultaneous electrical and optical measurements show that membrane fusion precedes secretory granule swelling during exocytosis of beige mast cells. *Proc. Natl. Acad. Sci. USA* **84**, 1585–1589.

Zimmermann, U., Beckers, F., and Coster, H. G. L. (1977). The effect of pressure on the electrical breakdown in the membranes of *Valonia utricularis*. *Biochim. Biophys. Acta* **464**, 399–416.

Zimmermann, U., Pilwat, G., Pequeux, A., and Gilles, R. (1980). Electromechanical properties of human erythrocyte membranes. The pressure-dependence of potassium permeability. *J. Membr. Biol.* **54**, 103–113.

Zimmermann, U., Gessner, P., Wander, M., and Foung, S. K. H. (1989). Electroinjection and electrofusion in hypo-osmolar solution. *In* "Electromanipulation in Hybridoma Technology" (C. A. K. Borrebaeck and I. Hagen, eds.), pp. 1–30. Stockton Press, New York.

II

Endocytosis, Exocytosis, and Vesicle Formation

4

Cell Signaling and Regulation of Exocytosis at Fertilization of the Egg

DOUGLAS KLINE

Department of Biological Sciences
Kent State University
Kent, Ohio

I. INTRODUCTION

Unfertilized eggs of many species contain membrane-bound secretory vesicles that lie just under the plasma membrane. Following fertilization, these vesicles, termed cortical granules, fuse with the egg membrane and release their contents into the extracellular matrix surrounding the egg, an event sometimes referred to as the cortical reaction. The cortical granule exudate transforms the extracellular matrix of the unfertilized egg; in frogs and sea urchins, the vitelline envelope is converted into the fertilization envelope, and in mice, two proteins comprising the *zona pellucida* are modified. The com-

75

SIGNAL TRANSDUCTION DURING
BIOMEMBRANE FUSION

position of cortical granules has been studied by examining isolated granules or the cortical granule exudate released from the egg following exocytosis. Among other substances, the cortical granules may contain structural elements of the fertilization envelope and various enzymes, including proteases, that alter the extracellular matrix. As a result of cortical granule exocytosis following fusion of the fertilizing sperm, the extracellular matrix becomes a hardened barrier to sperm entry and sperm receptors are inactivated, thereby preventing polyspermy (Grey *et al.*, 1974; Jaffe and Gould, 1985; Schuel, 1985; Wassarman, 1988).

Cortical granules are distinguished by their location, spherical shape, and size; they are 0.2–2.5 μm in diameter and are found in a 1–5 μm thick cortical layer of the egg adjacent to the egg plasma membrane (Fig. 1). This cortical region of the egg contains, in addition to cortical granules, cisternae

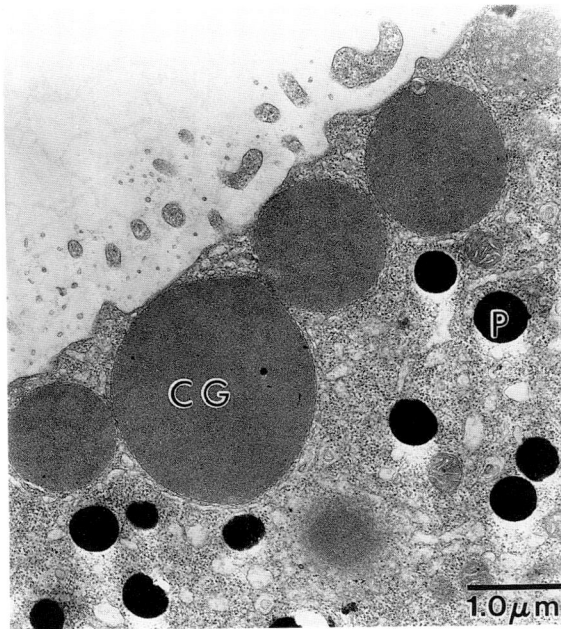

Fig. 1. Electron micrograph of the cortical region of an egg of the frog, *Xenopus laevis*. Cortical granules lie adjacent to the plasma membrane. CG, cortical granule; P, pigment granule. Magnification × 12,700; bar = 1.0 μm.

of endoplasmic reticulum, mitochondria, ribosomes, and cytoskeletal elements (Gulyas, 1980; Schuel, 1985; Charbonneau *et al.*, 1986; Longo, 1989). The attachment of cortical granules to the egg plasma membrane is sufficiently strong enough to allow the preparation of "cortical granule lawns" or "cortical complexes" from some eggs. Cortical granule lawns from sea urchin eggs are prepared by allowing the egg to adhere to a polylysine-coated surface and then rupturing the egg and removing the cytoplasm with a jet of solution (Vacquier, 1975). Cortical complexes, often called cell surface complexes or cortices, are composed of cortical granules, associated cytoplasm, the plasma membrane, and the overlaying vitelline envelope. Cortical complexes are made by homogenizing the eggs and separating the cortices from the cytoplasm by mild centrifugation (Detering *et al.*, 1977). These preparations have allowed the development of *in vitro* studies to examine the molecular mechanisms regulating exocytosis (see Section II,D). Exocytosis in eggs is also easily studied in the whole cell using microinjection and other manipulations of individual cells.

Current research is aimed to discover what signal passes from the fertilizing sperm to begin the sequence of events leading to exocytosis and to determine what biochemical and biophysical processes regulate cortical granule exocytosis. Interest in the signaling pathway is heightened because the pathway leading to exocytosis also appears to trigger activation of the egg to resume the cell cycle and begin embryonic development (egg activation). This review summarizes some of the information now available on signal transduction at fertilization and exocytosis in eggs, particularly in echinoderms, amphibians, and mammals.

II. CURRENT RESEARCH

A. Calcium and Exocytosis

Fertilization initiates an increase in calcium in eggs (reviewed in Jaffe, 1985; Whitaker and Steinhardt, 1985), which is a necessary and sufficient signal for cortical granule exocytosis. Exocytosis can be induced by injection of Ca^{2+} ions or Ca^{2+} buffers into cells or by the addition of a calcium ionophore. The evidence that Ca^{2+} is compulsory comes from experiments in which Ca^{2+} chelators are introduced into the egg before insemination; such experiments show that, although sperm fuse with the egg membrane and enter the egg, exocytosis does not occur (Zucker and Steinhardt, 1978; Hamaguchi and Hiramoto, 1981; Gilkey, 1983; Kline, 1988; Kline and Kline, 1992). Evidence that Ca^{2+} alone is sufficient to induce exocytosis has also been obtained from *in vitro* studies of cortical granule lawns and cell surface complexes (see Section II,D).

In fish and amphibians, the rise in calcium propagates over the egg surface as a wave that coincides with or just precedes a wave of cortical granule exocytosis (Gilkey *et al.*, 1978; Kubota *et al.*, 1987). The wave of increased intracellular Ca^{2+} also propagates across the sea urchin egg at about the same rate as the wave of exocytosis (Eisen *et al.*, 1984; Swann and Whitaker, 1986). In sea urchin, fish, and frogs, a single Ca^{2+} transient is produced at fertilization. The source of Ca^{2+} in all of these eggs is from an intracellular store (reviewed in Jaffe, 1985). In mammals (Miyazaki *et al.*, 1986; Kline and Kline, 1992) and ascidians (Speksnijder *et al.*, 1989), the initial Ca^{2+} transient is followed by periodic, transient increases in free Ca^{2+} that continue for over 1 hr in mammalian eggs and 25 min in ascidian eggs.

B. Phosphoinositide Metabolism and Second Messengers

Phosphoinositide messengers play a part in generating Ca^{2+} release in the egg (reviewed in Whitaker, 1989). The first evidence for the significance of phospholipid turnover in events leading to Ca^{2+} release and exocytosis in the egg came from measurements in sea urchin eggs. Turner and co-workers (1984) measured a 40% increase in the amount of phosphatidylinositol 4,5-bisphosphate (PIP_2) in eggs 15 sec after fertilization. This suggested the involvement of inositol 1,4,5-trisphosphate (IP_3) because PIP_2 is the membrane lipid from which IP_3 is formed on stimulation of phospholipase C (PIP_2 phosphodiesterase). Later, a measurable increase in intracellular IP_3 was detected within 10 sec of adding sperm to sea urchin eggs (Ciapa and Whitaker, 1986). Microinjection of IP_3 into sea urchin, frog, and hamster eggs triggers the release of Ca^{2+} from intracellular stores and initiates cortical granule exocytosis (Whitaker and Irvine, 1984; Busa *et al.*, 1985; Clapper and Lee, 1985; Swann and Whitaker, 1986; Miyazaki, 1988; Cran *et al.*, 1988; Larabell and Nuccitelli, 1992). IP_3 causes exocytosis by releasing intracellular Ca^{2+}; prior injection of a Ca^{2+} buffer to inhibit the rise in Ca^{2+} prevents any response to the IP_3 injection (Turner *et al.*, 1986; Swann and Whitaker, 1986). The IP_3 response in frog, sea urchin, and hamster eggs does not require external Ca^{2+} (Busa *et al.*, 1985; Crossley *et al.*, 1988; Miyazaki, 1988).

Although introduction of a small amount of IP_3 induces Ca^{2+} release, once a sufficient amount of Ca^{2+} is released, the rise in Ca^{2+} is regenerative and a Ca^{2+} wave is initiated. Two mechanisms may account for the regenerative release of Ca^{2+}. A rapid increase in Ca^{2+} may occur through cycles of Ca^{2+}-stimulated production of IP_3 (phospholipase C is activated by Ca^{2+}; Whitaker and Aitchison, 1985) and IP_3-stimulated Ca^{2+} release (Swann and Whitaker, 1986; reviewed in Meyer and Stryer, 1991). Alternatively, the regenerative rise in Ca^{2+} may occur through a process of Ca^{2+}-induced Ca^{2+} release similar to that which is known to occur in muscle and some other cells (reviewed in Fabiato, 1983; Endo, 1985; Fleischer and Inui, 1989; Petersen and Wakui, 1990). Ca^{2+}-induced Ca^{2+} release in skeletal muscle is mediated by the ryanodine receptor (Penner *et al.*, 1989).

Some eggs may contain the machinery for both IP_3-induced and Ca^{2+}-induced Ca^{2+} release. As indicated above, IP_3 initiates Ca^{2+} release from intracellular stores in sea urchin eggs. In addition, pharmacological agents thought to initiate or inhibit Ca^{2+}-induced Ca^{2+} release in sarcoplasmic reticulum of muscle cells have similar effects on sea urchin eggs. Potentiators of Ca^{2+} release from the sarcoplasmic reticulum include Ca^{2+}, caffeine, and ryanodine; inhibitors of Ca^{2+} release include procaine and ruthenium red (Fleischer and Inui, 1989). Galione *et al.* (1991) found that Ca^{2+}, caffeine, and ryanodine promote Ca^{2+} release or potentiate Ca^{2+}-induced Ca^{2+} release in sea urchin egg homogenates and that injection of ryanodine causes an increase in Ca^{2+} and exocytosis in the whole egg. Procaine and ruthenium red block Ca^{2+} release normally induced by caffeine in homogenates (Galione *et al.*, 1991) and inhibit cortical granule exocytosis in whole eggs (Fujiwara *et al.*, 1990). This data, together with the previous data on phosphoinositide lipid metabolism, suggest that both IP_3- and Ca^{2+}-induced Ca^{2+} release may be initiated by sperm at fertilization of the sea urchin egg. Rakow and Shen (1990) suggested that Ca^{2+} release in the sea urchin egg is from both IP_3-sensitive and IP_3-insensitive stores. The relative importance of these various Ca^{2+} release mechanisms in the signaling pathway at fertilization of sea urchin eggs awaits further study.

Release of intracellular Ca^{2+} in the hamster egg is mediated solely by IP_3-induced Ca^{2+} release. Injection of a function-blocking monoclonal antibody to the IP_3 receptor blocks Ca^{2+} release induced by IP_3 injection, as well as Ca^{2+} release induced by sperm (Miyazaki *et al.*, 1992a). Furthermore, there is no indication that a mechanism of Ca^{2+}-induced Ca^{2+} release mediated by the ryanodine receptor is utilized in the hamster egg. An apparent Ca^{2+}-induced Ca^{2+} release following injection of Ca^{2+} is blocked by the antibody to the IP_3 receptor (Miyazaki *et al.*, 1992a; Miyazaki *et al.*, 1992b). Thus, in contrast to the sea urchin, Ca^{2+}-induced Ca^{2+} release in the hamster egg is mediated by the IP_3 receptor. The generation of the Ca^{2+} wave in the hamster egg, which requires a regenerative or autocatalytic release of Ca^{2+} from intracellular stores, must involve some form of Ca^{2+} release initiated by Ca^{2+}. One possible mechanism for Ca^{2+} release triggered by Ca^{2+} is a positive feedback of IP_3-induced Ca^{2+} release and Ca^{2+} stimulated production of IP_3, presumably by activation of phospholipase C. This mechanism has also been proposed to contribute to the Ca^{2+} wave in the frog egg (Larabell and Nuccitelli, 1992). Another mechanism for Ca^{2+}-induced Ca^{2+} release involving the IP_3 receptor is the sensitization of the IP_3 receptor by Ca^{2+}, causing it to release Ca^{2+} from intracellular stores, even at low intracellular IP_3 concentrations (reviewed in Miyazaki *et al.*, 1992a). Data showing that the antibody to the IP_3 receptor blocks the sperm-induced Ca^{2+} transients in the hamster egg is, so far, the best evidence that sperm activate the egg through IP_3-induced Ca^{2+} release from intracellular stores.

The metabolism of PIP_2 at fertilization also produces the lipid, diacylglycerol (DAG; Ciapa and Whitaker, 1986). It has been suggested that DAG may promote cortical granule exocytosis by altering membrane properties and promoting the changes in the lipid phase necessary for the formation of the fusion pore (Whitaker, 1987). Alternatively, DAG may act with Ca^{2+} to activate protein kinase C (PKC), which may subsequently regulate exocytosis (Bement and Capco, 1990; Bement, 1992). The role of PKC in exocytosis is examined in Section II,D.

In addition to the phosphoinositide pathway, other metabolic pathways could influence Ca^{2+} release in the egg. One such pathway may be the production of cyclic ADP-ribose from nicotinamide adenine dinucleotide (NAD^+) by the enzyme ADP-ribosyl cyclase, which is present in sea urchin eggs and other cells (Lee and Aarhus, 1991). Cyclic ADP-ribose is as effective as IP_3 in causing Ca^{2+} release from homogenates of sea urchin eggs and, when injected into the sea urchin egg, causes a Ca^{2+} transient and cortical granule exocytosis (Dargie *et al.*, 1990). Cyclic ADP-ribose appears to act through the Ca^{2+}-induced Ca^{2+} release mechanism rather than the IP_3-induced Ca^{2+} release mechanism in the sea urchin egg. Inhibitors of Ca^{2+}-induced Ca^{2+} release (mediated by the ryanodine receptor channel) block the response to cyclic ADP-ribose (Galione *et al.*, 1991). Heparin, a competitive inhibitor of IP_3 binding, has no effect on cyclic ADP-ribose-induced Ca^{2+} release (Dargie *et al.*, 1990). The presence of cyclic ADP-ribose in the egg may serve to modulate Ca^{2+} release at fertilization. However, the role of cyclic ADP-ribose in eggs of the sea urchin and other species during fertilization is not known.

Ciapa and Epel (1991) suggest that another type of signal transduction mechanism could function during fertilization of the sea urchin egg; they investigated the possibility that sperm stimulate a protein kinase. Using an antiphosphotyrosine antibody, an early phosphorylation of tyrosine on two proteins was detected. Ciapa and Epel propose that this early phosphorylation may precede the rise in intracellular Ca^{2+} and could be involved in the early signaling pathway.

C. First Messengers at Fertilization

Whereas it is generally accepted that the rise in Ca^{2+} is necessary for cortical granule exocytosis and that Ca^{2+} release is probably triggered, at least in part, by IP_3, there is less agreement on what signal, or first message, must pass between the fertilizing sperm and the egg to initiate activation of phospholipase C and the rise in intracellular Ca^{2+}. Two hypotheses have been proposed: One proposes that the sperm introduces a soluble factor into the egg through the cytoplasmic bridge that forms following fusion of sperm and egg membranes (Fig. 2a). The other hypothesis suggests that the sperm interacts with a receptor on the cell surface that is coupled to phospholipase C by a guanine nucleotide-binding protein (G protein) (Fig. 2b).

1. Sperm Factors

Experiments in which sperm extracts are injected into eggs suggest that sperm might introduce a soluble factor into the egg (Dale et al., 1985; Stice and Robl, 1990; Swann, 1990). Sperm extracts, made in distilled water or buffer, can activate some eggs and, in some cases, cause exocytosis, but the identity of the factor that might accomplish this is unknown. If a sperm molecule acts as a cytoplasmic messenger to cause egg activation, it should trigger the entire sequence of events initiated by sperm, including the rise in intracellular Ca^{2+} and cortical granule exocytosis. It should also be found in sufficiently high concentrations in a single sperm. Once a sufficient amount of Ca^{2+} is released, the Ca^{2+} rise should be autocatalytic, either as the result of Ca^{2+}-stimulated production of IP_3 and IP_3-induced Ca^{2+} release or through Ca^{2+}-induced Ca^{2+} release. Any sperm factor should result in activation of phospholipase C, as it is known (at least for sea urchin) that IP_3 and

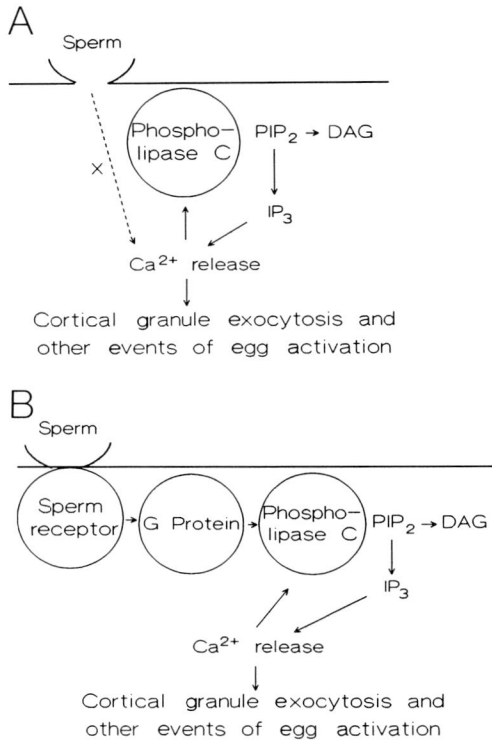

Fig. 2. Schematic drawings of two proposed mechanisms of signal transduction at fertilization in eggs. (A) A model based on the hypothesis that sperm introduce some factor (X) into the egg through the cytoplasmic bridge that forms following fusion of egg and sperm membranes (top left, drawing is not to scale). It is proposed that such a factor promotes Ca^{2+} release and that further Ca^{2+} release is initiated through the phosphoinositide–phospholipase C pathway (see text). (B) A model proposing that sperm activate a sperm receptor molecule in the egg membrane, which in turn activates a guanine nucleotide-binding protein (G protein), which then stimulates phospholipase C to produce IP_3 and DAG from PIP_2. In both models IP_3 is thought to promote Ca^{2+} release from intracellular stores. The regenerative increases in Ca^{2+} might result from repeated activation of phospholipase C by Ca^{2+} and/or by a process of Ca^{2+}-induced Ca^{2+} release (not shown in the schematic). G protein, guanine nucleotide-binding protein; PIP_2, phosphatidylinositol 4,5-bisphosphate; IP_3, inositol 1,4,5-trisphosphate; DAG, diacylglycerol.

diacylglycerol (DAG) are produced at fertilization. Stimulation of phospholipase C by a sperm factor might be direct, but the current hypothesis (Swann and Whitaker, 1990) is that a factor from sperm somehow causes Ca^{2+} release from an intracellular store and that phospholipase C is stimulated as a result of the initial Ca^{2+} rise (Fig. 2a).

One potential candidate for a cytoplasmic activator is Ca^{2+} itself. Jaffe (1980) proposed that Ca^{2+}, contained in sperm, might diffuse into the egg and trigger a wave of Ca^{2+} release within the egg. L. F. Jaffe (1990) has reformulated this hypothesis and suggests that Ca^{2+} entry into the egg may occur from a steady influx of Ca^{2+} through ion channels in the sperm membrane. This proposal for egg activation implies that extracellular Ca^{2+} is required for the initiation of the Ca^{2+} wave, but some experiments indicate that extracellular Ca^{2+} is not required for initiation of intracellular Ca^{2+} release (Schmidt et al., 1982). Further examination of this question is warranted.

Another proposed candidate for a cytoplasmic activator from sperm is IP_3, which, if present in sufficient amounts and if free to diffuse into the egg, could certainly cause Ca^{2+} release within the egg. Iwasa et al. (1990) have reported that sperm contain large amounts of IP_3, although the amount of IP_3 contained in a single sperm is just barely enough to activate an egg when compared to IP_3 injection experiments. Furthermore, it is unlikely that all of the IP_3 could be introduced rapidly to produce a high concentration in the egg.

Other potential cytoplasmic activators derived from sperm have been proposed. cGMP produced Ca^{2+} transients when injected into sea urchin eggs, but at far greater concentrations than required for IP_3, and it is unlikely that sperm contain a sufficiently large amount of cGMP (see Whitaker and Crossley, 1990). Because cyclic ADP-ribose can induce Ca^{2+} release in sea urchin eggs (see Section II,B), it is also a possible candidate for an activating factor from sperm. However, it is not known if cyclic ADP-ribose is present in sperm. Swann (1990) injected cytosolic extracts from mammalian sperm into hamster eggs. These extracts produced the characteristic periodic Ca^{2+} transients normally associated with fertilization and, although exocytosis was not examined, resumption of meiosis occurred. Swann suggests that the sperm factor might be a high-molecular weight protein.

The difficulty in interpreting these injection experiments is that many substances may initiate Ca^{2+} release. The finding that sperm extracts cause exocytosis might be an artifact caused by introduction of some factor that would not normally be introduced during sperm–egg fusion. It will be necessary to isolate, purify, and identify any potential egg-activating factor from sperm, demonstrate that a single sperm contains enough of the factor to activate the egg, and show that the normal fertilization response is abolished by inhibiting the activity of the specific factor.

Attempts have been made to examine the time of sperm–egg fusion with respect to the earliest changes in the egg associated with fertilization. A very early response of the sea urchin egg to fertilization is an increase in ion conductance resulting in a change in membrane potential. The sperm-induced electrical changes that accompany fertilization have been examined extensively by voltage-clamp experiments in which the membrane potential is held constant and sperm-induced ion currents are recorded. The currents induced by sperm consist of an inward current of abrupt onset which increases slowly, and that is followed by a large inward Na^+ current, coinciding with the large increase in intracellular Ca^{2+} (Lynn et al., 1988). Sperm–egg fusion, as indicated by electron microscopy (Longo et al., 1986) or by transfer of a DNA dye from the egg to the sperm through the fusion pore (Hinkley et al., 1986), is not seen until 5–10 sec after the beginning of the first sperm-induced ion current. This suggests that the process of egg activation begins just before sperm fusion, perhaps by a receptor-mediated mechanism. However, these detection techniques are limited by the time it takes to fix the egg and the time it takes for the dye to reach a detectable level within the sperm.

The hypothesis that sperm might activate the egg by introduction of some factor after sperm–egg fusion has received some support from other electrophysiological experiments with sea urchin eggs. The initial fertilization current in the sea urchin egg appears to be localized near the site of sperm–egg interaction and may be due to the local activation of ion channels in the egg membrane or to the introduction of sperm channels following sperm–egg fusion. McCulloh and Chambers (1992) report that the fusion of egg and sperm is coincident with the initial sperm-induced electrical response. An increase in ion conductance occurs at the same time as an increase in the membrane capacitance of the egg (a measure of membrane area) measured at the site of sperm–egg interaction. The increase in membrane capacitance may reflect the addition of sperm membrane to the egg membrane at the fusion event. Chambers and his associates (Chambers, 1989; McCulloh and Chambers, 1992) have proposed that the ion currents may be due to the introduction of ion channels from the sperm membrane into the egg at the time of fusion. The initial ion currents and the increase in egg membrane capacitance appear to precede detectable increases in intracellular messengers such as IP_3, DAG, and Ca^{2+} by several seconds. This is consistent with the model that sperm may introduce an activating factor into the egg following fusion, but these observations do not rule out the possibility that a receptor-mediated mechanism is involved in triggering the release of intracellular Ca^{2+}. Further discussion of the possible signaling mechanism involving introduction of a sperm factor into the sea urchin egg is contained in a number of reviews (Chambers, 1989; Whitaker *et al.*, 1989; Swann and Whitaker, 1990).

2. Activation of Guanine Nucleotide-Binding Proteins

The first steps in egg activation may depend on stimulation of a guanine nucleotide-binding protein (G protein) in the egg membrane. G proteins are present in eggs of sea urchins, frogs, and mammals, and may play a role in fertilization (Garty *et al.*, 1988; Jones and Schultz, 1990; Kline *et al.*, 1991; reviewed in Turner and Jaffe, 1989). In other cells, specific G proteins couple receptor and effector molecules such as phospholipase C. It has been proposed that the fertilizing sperm activates a receptor in the egg membrane, which in turn activates a G protein, thus stimulating phospholipase C (Fig. 2b).

Primary evidence for the participation of a G protein comes from experiments in which the G protein is artificially activated by injection of the hydrolysis-resistant analog of GTP, GTPγS, which constitutively activates G proteins. In the sea urchin, activation of the egg's G protein by GTPγS causes an increase in intracellular Ca^{2+} and cortical granule exocytosis (Turner *et al.*, 1986; Swann *et al.*, 1987). In the frog egg, injection of GTPγS causes a change in membrane potential that mimics the Ca^{2+}-dependent fertilization potential (Kline *et al.*, 1991). Similarly, in the hamster egg, GTPγS causes a rise in Ca^{2+} (Miyazaki, 1988). Cortical granule exocytosis in the sea urchin egg can be blocked by injection of a Ca^{2+} chelator prior to injection of GTPγS. Therefore, cortical granule exocytosis is not regulated directly by a G protein, but is initiated by the subsequent rise in Ca^{2+}.

Additional evidence for the role of G proteins in fertilization comes from experiments using cholera toxin, which activates certain G proteins; injection of cholera toxin into sea urchin eggs causes exocytosis (Turner *et al.*, 1987). However, injection of cholera toxin into frog or hamster eggs does not promote egg activation (Miyazaki, 1988; Kline *et al.*, 1991), indicating that if a G protein is involved in eggs of these species, it must be insensitive to cholera toxin. Adenosine diphosphate (ADP) ribosylation experiments demonstrated no cholera toxin-sensitive substrates in the egg membrane or cortices of the frog (Kline *et al.*, 1991).

G proteins in eggs can be activated by neurotransmitters following injection of mRNA for neurotransmitter receptors known to function through G proteins. Exogenous neurotransmitter receptors that act by way of a G protein to stimulate phospholipase C have been introduced into *Xenopus*, starfish, and mouse eggs. Application of acetylcholine or serotonin after expression of muscarinic type m1 or serotonin type 1c receptor in the frog egg caused cortical granule exocytosis as well as other Ca^{2+}-dependent events (Kline et al., 1988, 1991). Cortical granule exocytosis was also stimulated in the starfish egg by serotonin after introduction of the serotonin receptors into the egg (Shilling et al., 1990). Application of acetylcholine after introduction of exogenous type m1 muscarinic receptors in the mouse egg caused zona pellucida modifications indicative of cortical granule exocytosis (Williams et al., 1992). These results indicate that receptor activation of a G protein in the egg membrane will produce Ca^{2+}-dependent cortical granule exocytosis and other fertilization responses.

Mature frog and starfish eggs do not contain endogenous serotonin receptors; however, the hamster egg contains serotonin receptors (apparently type 2). Application of 10 μM serotonin to the hamster egg produced 4–6 Ca^{2+} transients similar to those produced by sperm. External application of serotonin also potentiated the response to injection of GTPγS (Miyazaki et al., 1990). The egg of at least one strain of mouse may contain endogenous acetylcholine receptors since application of acetylcholine causes Ca^{2+} transients (Kline and Kline, unpublished results). The role of these endogenous neurotransmitter receptors, if any, is not known, but further investigations may give additional insights about receptor- and G protein-mediated Ca^{2+} release in eggs.

It is clear that activating a G protein in the egg, either by injection of GTPγS or stimulation of an endogenous or exogenous neurotransmitter receptor, causes Ca^{2+} release and exocytosis. If a G protein functions in the pathway linking sperm–egg interaction to Ca^{2+} release, it should be possible to inhibit the response to sperm by inhibiting the G protein. The stable analog of GDP, GDPβS, which can inhibit G protein function, has been injected into eggs in attempts to demonstrate G protein involvement in the signaling pathway.

The initial experiments with sea urchin eggs suggested that GDPβS inhibited exocytosis by inhibiting the egg's G protein (Turner et al., 1986). However, further examination revealed that inhibition of exocytosis in the sea urchin egg by GDPβS appears to occur at a step following the sperm-induced rise in Ca^{2+} (L. A. Jaffe, 1990; Crossley et al., 1991). It is not known why GDPβS inhibits exocytosis, but it might alter the GTP-dependent polymerization of cytoskeletal tubulin and modify the egg cortex, thereby interfering with exocytosis (L. A. Jaffe, 1990). Crossley et al. (1991) observed that GDPβS does not inhibit the Ca^{2+} rise in the sea urchin egg, but the amount of GDPβS injected may not have inhibited all G proteins in the egg. Similar questions about the interpretation of GDPβS injections arose in experiments in which GDPβS was injected into frog eggs (Kline et al., 1990). However,

GDPβS did inhibit the sperm-induced Ca^{2+} increase in the hamster egg (Miyazaki, 1988), suggesting that GDPβS does inhibit G protein function in this species and that the signaling pathway may involve a G protein. It would be valuable to try other methods to inhibit G proteins.

One possible way to inhibit G protein function is to treat cells with pertussis toxin. Pertussis toxin catalyzes the adenosine diphosphate (ADP) ribosylation of some, but not all, G proteins. Pertussis toxin-catalyzed ADP ribosylation of G proteins prevents their normal interaction with receptors, thus inhibiting receptor-mediated signal transduction (reviewed in Gilman, 1987). Pertussis toxin can also identify G proteins if the ADP ribosylation is carried out in the presence of ^{32}P-NAD.

The *Xenopus* egg membrane contains a pertussis toxin-sensitive G protein with an α subunit of M_r 40,000; however this G protein does not appear to be coupled to the Ca^{2+} release mechanism that functions at fertilization. Sperm or serotonin, after introduction of the serotonin 1c receptor, cause nearly identical activation responses, including cortical granule exocytosis. Under conditions in which the pertussis-sensitive G protein was inhibited (at least 80–90% of the pertussis toxin substrate was ADP-ribosylated), both sperm and serotonin initiated the normal activation responses (Kline *et al.,* 1991). These results suggest that both sperm and serotonin, acting through an introduced receptor, activate the *Xenopus* egg by way of a pertussis toxin-insensitive G protein.

The identity of the G protein that may mediate the response in eggs to sperm and serotonin is not known; however, there are signaling pathways that activate phospholipases that are resistant to pertussis toxin. These pathways appear to rely on G proteins containing Gα subunits that lack the pertussis toxin-sensitive ADP ribosylation site. One particular Gα subunit likely to be involved in activation of phospholipase C is $Gα_q$. $Gα_q$ is pertussis toxin-resistant, it has been partially purified, and has been shown to activate phospholipase C in reconstitution experiments (reviewed in Simon *et al.,* 1991). A G protein containing the $Gα_q$ subunit, or one of the other pertussis toxin-insensitive Gα subunits, may couple sperm–egg interaction to stimulation of phospholipase C. This is supported by the observations that antibodies to the $Gα_q$ subunit recognize proteins in mouse (Williams *et al.,* 1992), starfish, sea urchin, and clam egg membranes (C. Gallo and L. A. Jaffe, personal communication).

3. A Sperm Receptor in the Egg Membrane?

The results from experiments in sea urchin, frog, and hamster using GTPγS and endogenous and introduced neurotransmitter receptors support the hypothesis that egg activation and the subsequent exocytosis of cortical granules result from activation of a G protein in the egg membrane. Activation of the G protein may occur through direct stimulation of the G protein in a manner suggested by the wasp venom peptide, mastoparan (Higashijima, 1988) or, in some cases, by the neurotransmitter, substance P (Mousli *et al.*, 1990). Alternatively, G protein activation may occur through interaction of the sperm with a specific receptor molecule in the egg membrane.

Presently, there is no evidence for a G protein-coupled sperm receptor in the egg membrane. If such a receptor is present, it could activate the egg's G protein much like the introduced or endogenous neurotransmitter receptors. It need not completely resemble a neurotransmitter receptor, since receptors of quite different overall structures can activate the same G protein and the amino acid differences between different classes of G protein-coupled receptors may be high (Birnbaumer *et al.*, 1990). Nevertheless, it is likely that a sperm receptor would belong to the large family of G protein-coupled receptors, which consists of proteins that span the plasma membrane seven times and have an extracellular amino terminus and an intracellular carboxyl terminus.

A plasma membrane receptor appears to be involved in the activation of the egg of the marine worm, *Urechis caupo*. In this species, an acrosomal protein contacts the egg membrane upon sperm–egg binding and elicits activation responses even under conditions in which sperm–egg fusion is prevented (Gould and Stephano, 1987). Furthermore, an active peptide prepared from the protein retains the ability to activate the egg even when coupled to bovine serum albumin, preventing it from crossing the egg plasma membrane (Gould and Stephano, 1991). Therefore, in this species, activation of the egg does not appear to require the introduction of an activating molecule into the egg, but rather suggests that an interaction at the plasma membrane, perhaps with a specific receptor, is the initial event leading to activation of the egg.

The mechanisms involved in the activation of protostome species such as *Urechis* are not completely known. The activating Ca^{2+} in these species comes primarily from entry of Ca^{2+} across the plasma membrane (Jaffe, 1985) and the role of phosphoinositide metabolism and G protein activation have not been examined in detail. Nevertheless, the experiments of Gould and Stephano indicate that a receptor-mediated mechanism for egg activation is likely. It would be useful to apply similar approaches to the study of sperm–egg interactions in the more thoroughly studied deuterostomes like echinoderms, amphibians, and mammals. Bindin, the protein in sea urchin sperm analogous to the acrosomal protein isolated from *Urechis* sperm, does not activate sea urchin eggs (Vacquier and Moy, 1977), but this should be examined in other species.

D. The Calcium Target

The rise in Ca^{2+} at fertilization appears to be necessary for exocytosis, because inhibition of the Ca^{2+} rise prevents cortical granule exocytosis (see Section II,A). Calcium is clearly the trigger, but what is the target of this Ca^{2+} rise? A distinct target has yet to be identified, partly because the fundamental mechanism of membrane fusion and its initiation remain unknown. It is likely that specific fusion proteins are involved in exocytosis, and that Ca^{2+} is a regulator of some catalytic step in exocytosis (reviewed in Augustine *et al.*, 1987; Plattner, 1989; Almers, 1990). Calcium may act directly on a fusion protein involved in exocytosis or on some regulatory protein. The mechanism of exocytosis in eggs may be simpler than in some other cells, because Ca^{2+} appears to be the sole trigger, rather than a cofactor with guanine nucleotides, as is the case in mast cells and some other cells (reviewed in Gomperts, 1990). The ability of guanine nucleotides to induce exocytosis in eggs appears to be due to mobilizing Ca^{2+} and not due to any effect on the exocytotic apparatus (Turner *et al.*, 1986; Whitaker, 1987). Presently there is no evidence to suggest the involvement of a guanine-nucleotide-binding protein in exocytosis other than the involvement of a G protein in activation of phospholipase C. A second G protein (G_E) may regulate exocytosis in mast cells and other cells at a later step (Gomperts, 1990).

1. Calmodulin

The Ca^{2+}-dependent regulatory protein, calmodulin, may be necessary in the sequence of events coupling the rise in Ca^{2+} to exocytosis. Steinhardt and Alderton (1982) inhibited cortical granule exocytosis in cortical granule lawns prepared from sea urchin eggs by applying a high concentration of an antibody to calmodulin. The egg cortices lost their sensitivity to Ca^{2+} when the antibody was applied. Exocytosis was triggered by Ca^{2+} in controls isolated with medium alone, preimmune serum, or the calmodulin antibody preincubated with excess calmodulin. This experiment suggested that a Ca^{2+}–calmodulin-dependent protein kinase or phosphatase might be involved in the steps leading to exocytosis.

However, Picard and Doree (1982) found that cortical granule exocytosis was not inhibited by intracellular injection of a variety of anticalmodulin drugs into sea urchin eggs. Furthermore, Jackson and Crabb (1988) reported that several anticalmodulin drugs did not inhibit exocytosis in cell surface complexes and that several samples of calmodulin antibody failed to inhibit exocytosis. Treatment of cortical granule lawns or cell surface complexes with the phenothiazine drug, trifluoperazine, a calmodulin antagonist, does inhibit exocytosis, although it is more likely that trifluoperazine, a charged hydrophobic drug, is acting by altering the cell membrane. Cations that alter the surface potential of lipid bilayers inhibit exocytosis by increasing the concentration of Ca^{2+} required for exocytosis (Crabb and Jackson, 1986; McLaughlin and Whitaker, 1988). Therefore, experimental data on whether calmodulin is specifically required for exocytosis in sea urchin eggs is equivocal. Calmodulin may be associated with exocytosis in other cells (see Section II,D,2; reviewed in Burgoyne, 1990); it would be worthwhile to examine the role of calmodulin in eggs of other species.

2. Calcium-Dependent Dephosphorylation?

Studies of exocytosis in *Paramecium* suggest that protein dephosphorylation may be a step in the exocytosis of trichocysts. A phosphoprotein of M_r 63,000 is rapidly and reversibly dephosphorylated at exocytosis and it may be present in a wide variety of secretory cells (Satir *et al.*, 1989). The phosphoprotein, calmodulin, and a protein phosphatase resembling calcineurin (a protein phosphatase regulating Ca^{2+} channel function in neurons) are found in the cortex of *Paramecium*. Exocytosis is inhibited by antibodies to calmodulin and calcineurin (reviewed by Plattner, 1989; Plattner *et al.*, 1991). The finding that sea urchin eggs contain a Ca^{2+}–calmodulin-dependent phosphatase similar to calcineurin (Iwasa and Ishiguro, 1986) suggests that a calcineurinlike phosphatase may regulate exocytosis in sea urchin eggs. Despite the conflicting evidence on the participation of calmodulin in exocytosis, further examination of protein phosphatases in eggs would be informative.

The potential function of phosphoproteins in exocytosis has been examined in sea urchin eggs. Walley *et al.* (1991) have used ATPγS (adenosine 5'-*O*-3-thiotriphosphate), a substrate for certain protein kinases, which results in thiophosphoproteins that are not readily dephosphorylated by phosphoprotein phosphatases, and okadaic acid, which is an inhibitor of protein phosphatases. Microinjection of ATPγS or okadaic acid inhibited exocytosis at a step following the increase in intracellular Ca^{2+}. However, ATPγS or okadaic acid did not prevent exocytosis in *in vitro* experiments with egg cortices, so it remains to be clarified how phosphatase activity might regulate cortical granule exocytosis.

3. Protein Kinase C

The participation of protein kinase C (PKC) in exocytosis has been suggested for a number of cell types including eggs (reviewed in Baker, 1988; Plattner, 1989; Bement, 1992). Protein kinase C is activated by Ca^{2+} and DAG (Nishizuka, 1986), and both Ca^{2+} and DAG concentrations increase at fertilization. The potential role of protein kinase C in exocytosis of cortical granules in eggs has been examined primarily by using agents that activate protein kinase C; these include diacylglycerols and phorbol esters. More recent experiments have employed inhibitors of protein kinase C.

Some indication that artificial activation of protein kinase C might cause exocytosis in mouse eggs was obtained by Endo et al. (1987). Application of a phorbol ester or the diacylglycerol, sn-1,2-dioctanoyl glycerol, partially modifies the proteins in the zona pellucida of eggs, a change thought to be due to exocytosis of cortical granules. Colonna et al. (1989) also reported that a phorbol ester or the DAG analog, 1-oleoyl-2-acetylglycerol (OAG) hardened the zona pellucida, another indication of cortical granule exocytosis. Although these experiments suggest that activation of protein kinase C causes exocytosis, exocytosis itself was not examined.

Artificial activation of protein kinase C may initiate Ca^{2+} release in some cells. Therefore, it is important to determine if protein kinase C is involved in exocytosis, or whether artificial activation of protein kinase C merely elevates Ca^{2+} which, by itself, triggers exocytosis through some other Ca^{2+}-dependent step. It was reported that stimulation of protein kinase C by a phorbol ester caused oscillations in intracellular Ca^{2+} in mouse eggs somewhat like those initiated by sperm (Cuthbertson and Cobbold, 1985); however, Colonna et al. (1989) found no change in Ca^{2+} over a 20- to 30-min application of phorbol ester in the mouse egg and Miyazaki et al. (1990) found no increase in Ca^{2+} in hamster eggs following application of a phorbol ester. Although stimulation of protein kinase C causes Ca^{2+} release in a few cells, it usually acts instead to inhibit agonist-induced Ca^{2+} oscillations (reviewed in Berridge, 1990) and indeed, such a role for protein kinase C is suggested for the desensitization of the serotonin-induced Ca^{2+} increase in the hamster egg (Miyazaki et al., 1990). Artificial activation of protein kinase C in mammalian eggs may cause exocytosis without an increase in intracellular Ca^{2+}, but further study is needed.

The role of protein kinase C in exocytosis has been examined in the egg of the frog, Xenopus laevis. It was found that activation of protein kinase C by a phorbol ester initiated exocytosis without an increase in intracellular Ca^{2+}, and that such treatment initiated exocytosis even when a calcium chelator was introduced into the egg prior to application of phorbol ester (Bement and Capco, 1990). Treatment of *Xenopus* eggs with the phorbol ester, phorbol 12-myristate 13-acetate (PMA), triggered cortical granule exocytosis in the absence of any detectable Ca^{2+} rise (Bement and Capco, 1990; Grandin and Charbonneau, 1991).

Whereas artificial activation of protein kinase C induces exocytosis, it must be shown that activation of protein kinase C is a step in the pathway linking sperm–egg interaction to exocytosis. This would best be demonstrated by use of a specific protein kinase C inhibitor during fertilization. This experiment has not yet been done in the frog egg; however, application of the protein kinase C inhibitors, sphingosine (100 μM) or 5-(*tert*-isoquinolinylsulfonyl)-2-methylpiperazine (H-7; 10–100 μM), each fairly specific for protein kinase C, at least partially inhibited cortical granule exocytosis in *Xenopus* eggs following application of phorbol ester or a Ca^{2+} ionophore in a Ca^{2+}-free medium (Bement and Capco, 1990). Grandin and Charbonneau (1991) report quite different findings for protein kinase C inhibitors; 100 μM sphingosine or 20 μM staurosporine by themselves actually caused an increase in intracellular calcium and cortical granule exocytosis in a calcium-containing medium. Thus, the use of these protein kinase C inhibitors has given ambiguous results, and more specific inhibitors of protein kinase C (such as the synthetic peptide of the pseudosubstrate domain of the kinase) will need to be used.

Two models for the potential roles of Ca^{2+} and protein kinase C are illustrated in Fig. 3. It remains to be shown whether the normal pathway linking fertilization to exocytosis requires activation of protein kinase C alone, or if protein kinase C serves only to modulate some aspect of exocytosis, such as the sensitivity to Ca^{2+} (Fig. 3a). Bement and Capco (1990) proposed that the rise in Ca^{2+} results in protein kinase C activation and that protein kinase C stimulates exocytosis (Fig. 3b). The substrate for protein kinase C in the egg would need to be identified. It may be that protein kinase C is involved in the initiation of exocytosis in the *Xenopus* egg and perhaps the mouse egg; however, treatment with phorbol ester does not induce exocytosis of cortical granules in the sea urchin egg. In fact, prolonged treatment with phorbol ester partially inhibits exocytosis normally triggered by fertilization by displacing cortical granules from the inner face of the plasma membrane (Ciapa *et al.*, 1988). Thus, protein kinase C activity might initiate or modulate exocytosis in some eggs but not in others.

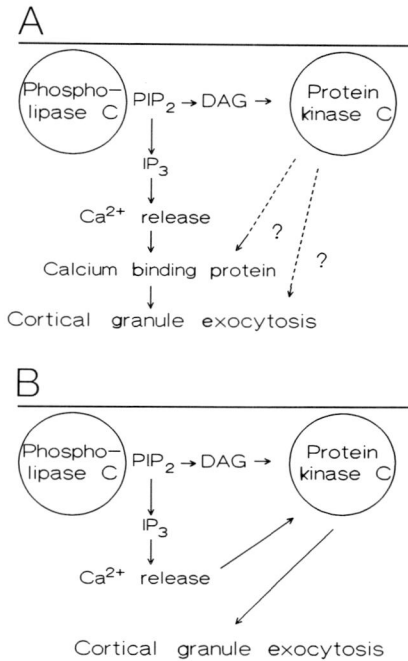

Fig. 3. Schematic drawings indicating the potential roles of Ca^{2+} and protein kinase C in cortical granule exocytosis. (A) A model in which Ca^{2+} acts specifically on some Ca^{2+}-dependent step leading to exocytosis. Protein kinase C, if it has any role, may possibly modulate Ca^{2+} release, one of the Ca^{2+}-dependent steps, or the exocytosis event. (B) A model in which protein kinase C plays a more direct role in promoting exocytosis and Ca^{2+} serves primarily to activate protein kinase C. See text for details. PIP_2, phosphatidylinositol 4,5-bisphosphate; IP_3, inositol 1,4,5-trisphosphate; DAG, diacylglycerol.

4. Additional Observations from in Vitro Studies

Development of cortical granule lawns and cell surface complexes from sea urchin eggs as *in vitro* models of exocytosis (see Section I; reviewed in Jackson and Crabb, 1988) has generated additional information that could not be obtained from studies in whole eggs. It is possible to remove cortical granules from cortical lawn preparations and obtain plasma membrane complexes without cortical granules and then reconstitute exocytotically competent cortical granule lawns by reintroducing purified cortical granules obtained from cortical granule lawns or cell surface complexes (Crabb and Jackson, 1985). The reconstituted cortical granule lawns are somewhat less sensitive to Ca^{2+} (25–36 μM Ca^{2+} for half maximal stimulation compared to approximately 6 μM for intact preparations; Jackson and Modern, 1990b; Whalley and Whitaker, 1988), indicating that the reconstituted system does not fully mimic the intact cortical granule–plasma membrane preparation; nevertheless, this technique has made it possible to examine some of the factors involved in exocytosis. The reconstituted cortical granule lawn allows independent manipulation of the cortical granules and plasma membrane and may provide important information about the molecules responsible for binding the cortical granules to the plasma membrane and those necessary for exocytosis.

The binding of cortical granules to the plasma membrane appears to be a specific protein-mediated event. Jackson and Modern (1990b) found that binding purified cortical granules to plasma membrane lawns is prevented by chymotrypsin treatment of the plasma membrane fragments. Cortical granules of one species bound to egg plasma membranes prepared from several different species of sea urchin, but did not bind to plasma membrane lawns prepared from human red blood cells, suggesting that the reassociation is mediated by a specific membrane protein.

Pretreatment of cortical granules or the plasma membrane with high concentrations of Ca^{2+} did not inhibit or alter exocytosis when the system was reconstituted (Walley and Whitaker, 1988). This indicates that if a Ca^{2+}-dependent enzymatic step is important, that step occurs only when the cortical granule and plasma membranes are in contact, or that the Ca^{2+}-dependent event is reversible; otherwise, one might expect a Ca^{2+}-dependent loss of function by prior exposure to high Ca^{2+}.

The participation of a specific protein in the fusion process is suggested by experiments in which cell surface complexes are treated with proteases or with sulfhydryl-modifying reagents such as N-ethylamaleimide (NEM). Proteolytic treatment increases the Ca^{2+} concentration necessary for exocytosis, and treatment with NEM prevents exocytosis in cell surface complexes (Jackson *et al.*, 1985). These data suggest that a sulfhydryl-containing protein may be part of the exocytotic apparatus. Furthermore, NEM-sensitive proteins appear to be present on both the cortical granule and plasma membrane surfaces. Interestingly, exocytosis can occur as long as one component of the reconstituted system is not treated with NEM; it does not matter if the untreated component is the cortical granule preparation or the plasma membrane. This suggests that both cortical granules and the plasma membrane contain the same or similar sulfhydryl-containing proteins that function in exocytosis

(Jackson and Modern, 1990a). The dissociation of cortical granules from the membrane may alter the localization of these proteins, so it remains to be determined if this protein or proteins are on both surfaces in the intact egg.

The exocytosis activity of NEM-treated cell surface complexes can be restored by mild proteolytic treatment (Jackson *et al.*, 1985), indicating that the NEM-sensitive protein is not absolutely necessary for fusion, but that the NEM-sensitive protein must be closely associated with the putative fusion protein. The NEM-sensitive protein in sea urchin eggs may be related to an NEM-sensitive factor (NSF) required for the membrane fusion step in the secretory pathway in other cells (Wilson *et al.*, 1989). Vogel and Zimmerberg (1992) provide additional evidence that cortical granules may have a protein that mediates fusion on their surface. Additional study is needed to determine what specific protein or proteins may be involved in exocytosis of cortical granules in eggs, and to discover how they may be related to putative proteins thought to have a role in vesicle fusion in other cells (see Almers, 1990; Burgoyne, 1990).

III. FINAL COMMENTS

The fertilizing sperm provokes an increase in intracellular calcium in the egg that initiates the beginning of embryonic development and causes exocytosis of cortical granules. Progress has been made in learning what signal may pass between sperm and egg to initiate the rise in calcium, but many questions remain. How do sperm initiate the hydrolysis of membrane polyphosphoinositides to produce IP_3 and DAG? Do sperm introduce a Ca^{2+}-releasing factor following sperm–egg fusion or is there a specific receptor coupled to G protein-stimulated activation of phospholipase C? If a cytosolic factor from sperm is the first messenger in this signal transduction pathway, the process may be unique to fertilization since pathways for activation of phospholipase C generally use receptor-mediated activation of a G protein. Can it be demonstrated that G proteins are in fact responsible for the increase in IP_3 at fertilization? Will the signaling pathway in eggs be similar for all species or will there be variation? Might there be parallel pathways, including, for example, the introduction of a cytosolic sperm factor together with the receptor–G protein pathway to ensure that the fertilizing sperm activates the egg in this once-in-a-lifetime event of development?

Clearly, much remains to be learned about how calcium regulates cortical granule exocytosis in eggs. What is the calcium target? Does protein kinase C play a critical part in exocytosis in some species? Is there a role for protein dephosphorylation in exocytosis in eggs? Is exoctyosis in eggs a good model for exocytosis in other cells? Further use of *in vitro* preparations from sea urchin eggs and eggs of other species will help answer these questions. The next few years should provide some interesting research on fertilization and the regulation of exocytosis in eggs.

ACKNOWLEDGMENT

I thank Joanne Kline for her assistance in preparing this chapter. This work was supported in part by a grant from the National Institute of Child Health and Human Development (HD 26032).

REFERENCES

Almers, W. (1990). Exocytosis. *Annu. Rev. Physiol.* **52**, 607–624.

Augustine, G. J., Charlton, M. P., and Smith, S. J. (1987). Calcium action in synaptic transmitter release. *Annu. Rev. Neurosci.* **10**, 633–693.

Baker, P. F. (1988). Exocytosis in electropermeabilized cells: Clues to mechanism and physiological control. *In* "Current Topics in Membranes and Transport" (N. Düzgüneş and F. Bronner, eds.), Vol. 32, pp. 115–138. Academic Press, San Diego.

Bement, W. M. (1992). Signal transduction by calcium and protein kinase C during egg activation. *J. Exp. Zool.* **263**, 382–397.

Bement, W. M., and Capco, D. G. (1990). Protein kinase C acts downstream of calcium at entry into the first mitotic interphase of *Xenopus laevis*. *Cell Regul.* **1**, 315–326.

Berridge, M. J. (1990). Calcium oscillations. *J. Biol. Chem.* **265**, 9583–9586.

Birnbaumer, L., Abramowitz, J., and Brown, A. M. (1990). Receptor–effector coupling by G proteins. *Biochim. Biophys. Acta* **1031**, 163–224.

Burgoyne, R. D. (1990). Secretory vesicle-associated proteins and their role in exocytosis. *Annu. Rev. Physiol.* **52**, 647–659.

Busa, W. B., Ferguson, J. E., Joseph, S. K., Williamson, J. R., and Nuccitelli, R. (1985). Activation of frog (*Xenopus laevis*) eggs by inositol trisphosphate. I. Characterization of Ca^{2+} release from intracellular stores. *J. Cell Biol.* **101**, 677–682.

Chambers, E. L. (1989). Fertilization in voltage-clamped sea urchin eggs. *In* "Mechanisms of Egg Activation" (R. Nuccitelli, G. N. Cherr, and W. H. Clark, Jr., eds.), pp. 1–18. Plenum Press, New York.

Charbonneau, M., Grey, R. D., Baskin, R. J., and Thomas, D. (1986). A freeze-fracture study of the cortex of *Xenopus laevis* eggs. *Dev. Growth Differ.* **28**, 75–84.

Ciapa, B., and Epel, D. (1991). A rapid change in phosphorylation on tyrosine accompanies fertilization of sea urchin eggs. *FEBS Lett.* **295**, 167–170.

Ciapa, B., and Whitaker, M. (1986). Two phases of inositol polyphosphate and diacylglycerol production at fertilisation. *FEBS Lett.* **195**, 347–351.

Ciapa, B., Crossley, I., and De Renzis, G. (1988). Structural modifications induced by TPA (12-O-tetradecanoyl phorbol-13-acetate) in sea urchin eggs. *Dev. Biol.* **128**, 142–149.

Clapper, D. L., and Lee, H. C. (1985). Inositol trisphosphate induces calcium release from nonmitochondrial stores in sea urchin egg homogenates. *J. Biol. Chem.* **260**, 13947–13954.

Colonna, R., Tatone, C., Malgaroli, A., Eusebi, F., and Mangia, F. (1989). Effects of protein kinase C stimulation and free Ca^{2+} rise in mammalian egg activation. *Gamete Res.* **24**, 171–183.

Crabb, J. H., and Jackson, R. C. (1985). *In vitro* reconstitution of exocytosis from plasma membrane and isolated secretory vesicles. *J. Cell Biol.* **101**, 2263–2273.

Crabb, J. H., and Jackson, R. C. (1986). Polycation inhibition of exocytosis from sea urchin egg cortex. *J. Membr. Biol.* **91**, 85–96.

Cran, D. G., Moor, R. M., and Irvine, R. F. (1988). Initiation of the cortical reaction in hamster and sheep oocytes in response to inositol trisphosphate. *J. Cell Sci.* **91**, 139–144.

Crossley, I., Swann, K., Chambers, E., and Whitaker, M. (1988). Activation of sea urchin eggs by inositol phosphates is independent of external calcium. *Biochem. J.* **252**, 257–262.

Crossley, I., Whalley, T., and Whitaker, M. (1991). Guanosine 5'-thiotriphosphate may stimulate phosphoinositide messenger production in sea urchin eggs by a different route than the fertilizing sperm. *Cell Regul.* **2**, 121–133.

Cuthbertson, K. S. R., and Cobbold, P. H. (1985). Phorbol ester and sperm activate mouse oocytes by inducing sustained oscillations in cell Ca^{2+}. *Nature (London)* **316**, 541–542.

Dale, B., DeFelice, L. J., and Ehrenstein, G. (1985). Injection of a soluble sperm fraction into sea-urchin eggs triggers the cortical reaction. *Experientia* **41**, 1068–1070.

Dargie, P. J., Agre, M. C., and Lee, H. C. (1990). Comparison of Ca^{2+} mobilizing activities of cyclic ADP-ribose and inositol trisphosphate. *Cell Regul.* **1**, 279–290.

Detering, N. K., Decker, G. L., Schmell, E. D., and Lennarz, W. J. (1977). Isolation and characterization of plasma membrane-associated cortical granules from sea urchin eggs. *J. Cell Biol.* **75**, 899–914.

Eisen, A., Kiehart, D. P., Wieland, S. J., and Reynolds, G. T. (1984). Temporal sequence and spatial distribution of early events of fertilization in single sea urchin eggs. *J. Cell Biol.* **99**, 1647–1654.

Endo, M. (1985). Calcium release from sarcoplasmic reticulum. *In* "Current Topics in Membranes and Transport" (F. Bronner and A. E. Shamoo, eds.), Vol. 25, pp. 181–230. Academic Press, Orlando.

Endo, Y., Schultz, R. M., and Kopf, G. S. (1987). Effects of phorbol esters and a diacylglycerol on mouse eggs: Inhibition of fertilization and modification of the *zona pellucida. Dev. Biol.* **119**, 199–209.

Fabiato, A. (1983). Calcium-induced release of calcium from the cardiac sarcoplasmic reticulum. *Am. J. Physiol.* **245**, C1–C14.

Fleischer, S., and Inui, M. (1989). Biochemistry and biophysics of excitation-contraction coupling. *Annu. Rev. Biophys. Biophys. Chem.* **18**, 333–364.

Fujiwara, A., Taguchi, K., and Yasumasu, I. (1990). Fertilization membrane formation in sea urchin eggs induced by drugs known to cause Ca^{2+} release from isolated sarcoplasmic reticulum. *Dev. Growth Differ.* **32**, 303–314.

Galione, A., Lee, H. C., and Busa, W. B. (1991). Ca^{2+}-induced Ca^{2+} release in sea urchin egg homogenates and its modulation by cyclic ADP-ribose. *Science* **253**, 1143–1145.

Garty, N. B., Galiani, D., Aharonheim, A., Ho, Y., Phillips, D. M., Dekel, N., and Salomon, Y. (1988). G-proteins in mammalian gametes" An immunocytochemical study. *J. Cell Sci.* **91**, 21–31.

Gilkey, J. C. (1983). Roles of calcium and pH activation of eggs of the medaka fish, *Oryzias latipes*. *J. Cell Biol.* **97**, 669–678.

Gilkey, J. C., Jaffe, L. F., Ridgway, E. B., and Reynolds, G. T. (1978). A free calcium wave traverses the activating egg of the medaka, *Oryzias latipes*. *J. Cell Biol.* **76**, 448–466.

Gilman, A. G. (1987). G proteins: Transducers of receptor-generated signals. *Annu. Rev. Biochem.* **56**, 615–649.

Gomperts, B. D. (1990). G_E: A GTP-binding protein mediating exocytosis. *Annu. Rev. Physiol.* **52**, 591–606.

Gould, M., and Stephano, J. L. (1987). Electrical responses of eggs to acrosomal protein similar to those induced by sperm. *Science* **235**, 1654–1656.

Gould, M. C., and Stephano, J. L. (1991). Peptides from sperm acrosomal protein that initiate egg development. *Dev. Biol.* **146**, 509–518.

Grandin, N., and Charbonneau, M. (1991). Intracellular pH and intracellular free calcium responses to protein kinase C activators and inhibitors in *Xenopus* eggs. *Development* **112**, 461–470.

Grey, R. D., Wolf, D. P., and Hedrick, J. L. (1974). Formation and structure of the fertilization envelope in *Xenopus laevis*. *Dev. Biol.* **36**, 44–61.

Gulyas, B. J. (1980). Cortical granules of mammalian eggs. *Int. Rev. Cytol.* **63**, 357–392.

Hamaguchi, Y., and Hiramoto, Y. (1981). Activation of sea urchin eggs by microinjection of calcium buffers. *Exp. Cell Res.* **134**, 171–179.

Higashijima, T., Uzu, S., Nakajima, T., and Ross, E. M. (1988). Mastoparan, a peptide toxin from wasp venom, mimics receptors by activating GTP-binding regulatory proteins (G proteins). *J. Biol. Chem.* **263**, 6491–6494.

Hinkley, R. E., Wright, B. D., and Lynn, J. W. (1986). Rapid visual detection of sperm–egg fusion using the DNA-specific fluorochrome Hoechst 33342. *Dev. Biol.* **118**, 148–154.

Iwasa, F., and Ishiguro, K. (1986). Calmodulin-binding protein (55K + 17K) of sea urchin eggs has a Ca^{2+}- and calmodulin-dependent phosphoprotein phosphatase activity. *J. Biochem.* **99**, 1353–1358.

Iwasa, K. H., Ehrenstein, G., DeFelice, L. J., and Russell, J. T. (1990). High concentration of inositol 1,4,5-trisphosphate in sea urchin sperm. *Biochem. Biophys Res. Commun.* **172**, 932–938.

Jackson, R. C., and Crabb, J. H. (1988). Cortical exocytosis in the sea urchin egg. *In* "Current Topics in Membranes and Transport" (N. Düzgüneş and F. Bronner, eds.), Vol. 32, pp. 45–85. Academic Press, San Diego.

Jackson, R. C., and Modern, P. A. (1990a). *N*-ethylmaleimide-sensitive protein(s) involved in cortical exocytosis in the sea urchin egg: Localization to both cortical vesicles and plasma membrane. *J. Cell Sci.* **96**, 313–321.

Jackson, R. C., and Modern, P. A. (1990b). Reassociation of cortical secretory vesicles with sea urchin egg plasma membrane: Assessment of binding specificity. *J. Membr. Biol.* **115**, 83–93.

Jackson, R. C., Ward, K. K., and Haggerty, J. G. (1985). Mild proteolytic digestion restores exocytotic activity to *N*-ethylmaleimide-inactivated cell surface complex from sea urchin eggs. *J. Cell Biol.* **101**, 6–11.

Jaffe, L. A. (1990). First messengers at fertilization. *J. Reprod. Fert. Suppl.* **42**, 107–116.

Jaffe, L. A., and Gould, M. (1985). Polyspermy-preventing mechanisms. *In* "Biology of Fertilization" (C. B. Metz and A. Monroy, eds.), Vol. 3, pp. 223–250. Academic Press, Orlando.

Jaffe, L. F. (1980). Calcium explosions as triggers of development. *Ann. N.Y. Acad. Sci.* **339**, 86–101.

Jaffe, L. F. (1985). The role of calcium explosions, waves, and pulses in activating eggs. *In*

"Biology of Fertilization" (C. B. Metz and A. Monroy, eds.), Vol. 3, pp. 127–165. Academic Press, Orlando, Florida.

Jaffe, L. F. (1990). The roles of intermembrane calcium in polarizing and activating eggs. *In* "Mechanism of Fertilization: Plants to Humans" (B. Dale, ed.), NATO ASI Series, Vol. H45, pp. 389–417. Springer-Verlag, Berlin.

Jones, J., and Schultz, R. M. (1990). Pertussis toxin-catalyzed ADP-ribosylation of a G protein in mouse oocytes, eggs, and preimplantation embryos: Developmental changes and possible functional roles. *Dev. Biol.* **139**, 250–262.

Kline, D. (1988). Calcium-dependent events at fertilization of the frog egg: Injection of a calcium buffer blocks ion channel opening, exocytosis, and formation of pronuclei. *Dev. Biol.* **126**, 346–361.

Kline, D., and Kline, J. T. (1992). Repetitive calcium transients and the role of calcium in exocytosis and cell cycle activation in the mouse egg. *Dev. Biol.* **149**, 80–89.

Kline, D., Simoncini, L., Mandel, G., Maue, R. A., Kado, R. T., and Jaffe, L. A. (1988). Fertilization events induced by neurotransmitters after injection of mRNA in *Xenopus* eggs. *Science* **241**, 464–467.

Kline, D., Kado, R. T., Kopf, G. S., and Jaffe, J. A. (1990). Receptors, G proteins, and activation of the amphibian egg. *In* "Mechanism of Fertilization: Plants to Humans" (B. Dale, ed.), NATO ASI Series, Vol. H45, 529–541.

Kline, D., Kopf, G. S., Muncy, L. F., and Jaffe, L. A. (1991). Evidence for the involvement of a pertussis toxin-insensitive G-protein in egg activation of the frog, *Xenopus laevis*. *Dev. Biol.* **143**, 218–229.

Kubota, H. Y., Yoshimoto, Y., Yoneda, M., and Hiramoto, Y. (1987). Free calcium wave upon activation in *Xenopus* eggs. *Dev. Biol.* **119**, 129–136.

Larabell, C., and Nuccitelli, R. (1992). Inositol lipid hydrolysis contributes to the Ca^{2+} wave in the activating egg of *Xenopus laevis*. *Dev. Biol.* **153**, 347–355.

Lee, H. C., and Aarhus, R. (1991). ADP-ribosyl cyclase: An enzyme that cyclizes NAD^+ into a calcium-mobilizing metabolite. *Cell Regul.* **2**, 203–209.

Longo, F. J. (1989). Egg cortical architecture. *In* "The Cell Biology of Fertilization" (H. Schatten and G. Schatten, eds.), pp. 105–138. Academic Press, Orlando.

Longo, F. J., Lynn, J. W., McCulloh, D. H., and Chambers, E. L. (1986). Correlative ultrastructural and electrophysiological studies of sperm–egg interactions of the sea urchin, *Lytechinus variegatus*. *Dev. Biol.* **118**, 155–166.

Lynn, J. W., McCulloh, D. H., and Chambers, E. L. (1988). Voltage clamp studies of fertilization in sea urchin eggs. II. Current patterns in relation to sperm entry, nonentry, and activation. *Dev. Biol.* **128**, 305–323.

McCulloh, D. H., and Chambers, E. L. (1992). Fusion of membranes during fertilization. *J. Gen. Physiol.* **99**, 137–175.

McLaughlin, S., and Whitaker, M. (1988). Cations that alter surface potentials of lipid bilayers increase the calcium requirement for exocytosis in sea urchin eggs. *J. Physiol. (London)* **396**, 189–204.

Meyer, T., and Stryer, L. (1991). Calcium spiking. *Annu. Rev. Biophys. Biophys. Chem.* **20**, 153–174.

Miyazaki, S. (1988). Inositol 1,4,5-trisphosphate-induced calcium release and guanine nucleotide-binding protein-mediated periodic calcium rises in golden hamster eggs. *J. Cell Biol.* **106**, 345–353.

Miyazaki, S., Hashimoto, N., Yoshimoto, Y., Kishimoto, T., Igusa, Y., and Hiramoto, Y. (1986). Temporal and spatial dynamics of the periodic increase in intracellular free calcium at fertilization of golden hamster eggs. *Dev. Biol.* **118**, 259–267.

Miyazaki, S., Katayama, Y., and Swann, K. (1990). Synergistic activation by serotonin and GTP analogue and inhibition by phorbol ester of cyclic Ca^{2+} rises in hamster eggs. *J. Physiol. (London)* **426,** 209–227.

Miyazaki, S., Yuzaki, M., Nakada, K., Shirakawa, H., Nakanishi, S., Nakade, S., and Mikoshiba, K. (1992a). Block of Ca^{2+} wave and Ca^{2+} oscillation by antibody to the inositol 1,4,5-trisphosphate receptor in fertilized hamster eggs. *Science* **257,** 251–255.

Miyazaki, S., Shirakawa, H., Nakada, K., Honda, Y., Yuzaki, M., Nakade, S., and Mikoshiba, K. (1992b). Antibody to the inositol trisphosphate receptor blocks thimerosal-enhanced Ca^{2+}-induced Ca^{2+} release and Ca^{2+} oscillations in hamster eggs. *FEBS Lett.* **309,** 180–184.

Mousli, M., Bronner, C., Landry, Y., Bockaert, J., and Rouot, B. (1990). Direct activation of GTP-binding regulatory proteins (G-proteins) by substance P and compound 48/80. *FEBS Lett.* **259,** 260–262.

Nishizuka, Y. (1986). Studies and perspectives of protein kinase C. *Science* **233,** 305–312.

Penner, R., Neher, E., Takeshima, H., Nishimura, S., and Numa, S. (1989). Functional expression of the calcium release channel from skeletal muscle ryanodine receptor cDNA. *FEBS Lett.* **259,** 217–221.

Petersen, O. H., and Wakui, M. (1990). Oscillating intracellular Ca^{2+} signals evoked by activation of receptors linked to inositol lipid hydrolysis: Mechanism of generation. *J. Membr. Biol.* **118,** 93–105.

Picard, A., and Doree, M. (1982). Intracellular microinjection of anticalmodulin drugs does not inhibit the cortical reaction induced by fertilization, ionophore A23187 or injection of calcium buffers in sea urchin eggs. *Dev. Growth Differ.* **24,** 155–162.

Plattner, H. (1989). Regulation of membrane fusion during exocytosis. *Int. Rev. Cytol.* **119,** 197–286.

Plattner, H., Lumpert, C. J., Knoll, G., Kissmehl, R., Höhne, B., Momayezi, M., and Glas-Albrecht, R. (1991). Stimulus-secretion coupling in *Paramecium* cells. *Eur. J. Cell Biol.* **55,** 3–16.

Rakow, T. L., and Shen, S. S. (1990). Multiple stores of calcium are released in the sea urchin egg during fertilization. *Proc. Natl. Acad. Sci. USA* **87,** 9285–9289.

Satir, B. H., Hamasaki, T., Reichman, M., and Murtaugh, T. J. (1989). Species distribution of a phosphoprotein (parafusin) involved in exocytosis. *Proc. Natl. Acad. Sci. USA* **86,** 930–932.

Schmidt, T., Patton, C., and Epel, D. (1982). Is there a role for the Ca^{2+} influx during fertilization of the sea urchin egg? Dev. Biol. **90,** 284–290.

Schuel, H. (1985). Functions of egg cortical granules. *In* "Biology of Fertilization" (C. B. Metz and A. Monroy, eds.), Vol. 3, pp. 1–43. Academic Press, Orlando, Florida.

Shilling, F., Mandel, G., and Jaffe, L. A. (1990). Activation by serotonin of starfish eggs expressing the rat serotonin 1c receptor. *Cell Regul.* **1,** 465–469.

Simon, M. I., Strathmann, M. P., and Gautam, N. (1991). Diversity of G proteins in signal transduction. *Science* **252,** 802–808.

Speksnijder, J. E., Corson, D. W., Sardet, C., and Jaffe, L. F. (1989). Free calcium pulses following fertilization in the ascidian egg. *Dev. Biol.* **135,** 182–190.

Steinhardt, R. A., and Alderton, J. M. (1982). Calmodulin confers calcium sensitivity on secretory exocytosis. *Nature (London)* **295,** 154–155.

Stice, S. L., and Robl, J. M. (1990). Activation of mammalian oocytes by a factor obtained from rabbit sperm. *Mol. Reprod. Dev.* **25,** 272–280.

Swann, K. (1990). A cytosolic sperm factor stimulates repetitive calcium increases and mimics fertilization in hamster eggs. *Development* **110,** 1295–1302.

Swann, K., and Whitaker, M. (1986). The part played by inositol trisphosphate and calcium in the propagation of the fertilization wave in sea urchin eggs. *J. Cell Biol.* **103**, 2333–2342.

Swann, K., and Whitaker, M. J. (1990). Second messengers at fertilization in sea urchin eggs. *J. Reprod. Fert. Suppl.* **42**, 141–153.

Swann, K., Ciapa, B., and Whitaker, M. (1987). Cellular messengers and sea urchin egg activation. *In* "Molecular Biology of Invertebrate Development" (D. O'Connor, ed.), pp. 45–69. A. R. Liss, New York.

Turner, P. R., and Jaffe, L. A. (1989). G-proteins and the regulation of oocyte maturation and fertilization. *In* "The Cell Biology of Fertilization" (H. Schatten and G. Schatten, eds.) pp. 297–318. Academic Press, Orlando, Florida.

Turner, P. R., Sheetz, M. P., and Jaffe, L. A. (1984). Fertilization increases the polyphosphoinositide content of sea urchin eggs. *Nature (London)* **310**, 414–415.

Turner, P. R., Jaffe, L. A., and Fein, A. (1986). Regulation of cortical vesicle exocytosis in sea urchin eggs by inositol 1,4,5-trisphosphate and GTP-binding protein. *J. Cell Biol.* **102**, 70–76.

Turner, P. R., Jaffe, L. A., and Primakoff, P. (1987). A cholera toxin-sensitive G-protein stimulates exocytosis in sea urchin eggs. *Dev. Biol.* **120**, 577–583.

Vacquier, V. D. (1975). The isolation of intact cortical granules from sea urchin eggs: Calcium ions trigger granule discharge. *Dev. Biol.* **73**, 62–74.

Vacquier, V., and Moy, G. (1977). Isolation of bindin: The protein responsible for adhesion of sperm to sea urchin eggs. *Proc. Natl. Acad. Sci. USA* **74**, 2456–2460.

Vogel, S. S., and Zimmerberg, J. (1992). Proteins on exocytic vesicles mediate calcium-triggered fusion. *Proc. Natl. Acad. Sci. USA* **89**, 4749–4753.

Wassarman, P. M. (1988). Zona pellucida glycoproteins. *Annu. Rev. Biochem.* **57**, 415–442.

Whalley, T., and Whitaker, M. (1988). Exocytosis reconstituted from the sea urchin egg is unaffected by calcium pretreatment of granules and plasma membrane. *Biosci. Rep.* **8**, 335–343.

Whalley, T., Crossley, I., and Whitaker, M. (1991). Phosphoprotein inhibition of calcium-stimulated exocytosis in sea urchin eggs. *J. Cell Biol.* **113**, 769–778.

Whitaker, M. (1987). How calcium may cause exocytosis in sea urchin eggs. *Biosci. Rep.* **7**, 383–397.

Whitaker, M. (1989). Phosphoinositide second messengers in eggs and oocytes. *In* "Inositol Lipids in Cell Signalling" (R. H. Michell, A. H. Drummond, and C. P. Downes, eds.), pp. 459–483. Academic Press, London.

Whitaker, M., and Aitchison, M. (1985). Calcium-dependent polyphosphoinositide hydrolysis is associated with exocytosis *in vitro*. *FEBS Lett.* **182**, 119–124.

Whitaker, M., and Crossley, I. (1990). How does a sperm activate a sea urchin egg? *In* "Mechanism of Fertilization: Plants to Humans" (B. Dale, ed.), NATO ASI Series, Vol. H45, pp. 433–443.

Whitaker, M., and Irvine, R. F. (1984). Inositol 1,4,5-trisphosphate microinjection activates sea urchin eggs. *Nature (London)* **312**, 636–639.

Whitaker, M. J., and Steinhardt, R. A. (1985). Ionic signaling in the sea urchin egg at fertilization. *In* "Biology of Fertilization" (C. B. Metz and A. Monroy, eds.), Vol. 3, pp. 167–221. Academic Press, Orlando, Florida.

Whitaker, M., Swann, K., and Crossley, I. (1989). What happens during the latent period at fertilization. *In* "Mechanisms of Egg Activation" (R. Nuccitelli, G. N. Cherr, and W. H. Clark, Jr., eds.), pp. 157–171. Plenum Press, New York.

Williams, C. J., Schultz, R. M., and Kopf, G. S. (1992). Role of G proteins in mouse egg activation: Stimulatory effects of acetylcholine on the ZP2 to ZP2$_f$ conversion and pronuclear formation in eggs expressing a functional m1 muscarinic receptor. *Dev. Biol.* **151,** 288–296.

Wilson, D. W., Wilcox, C. A., Flynn, G. C., Chen, E., Kuang, W., Henzel, W. J., Block, M. R., Ullrich, A., and Rothman, J. E. (1989). A fusion protein required for vesicle-mediated transport in both mammalian cells and yeast. *Nature (London)* **339,** 355–359.

Zucker, R. S., and Steinhardt, R. A. (1978). Prevention of the cortical reaction in fertilized sea urchin eggs by injection of calcium-chelating ligands. *Biochim. Biophys. Acta* **541,** 459–466.

5

Signal Transduction during Exocytosis in Mast Cells

WITTE KOOPMANN*

Department of Biochemistry
Dartmouth Medical School
Hanover, New Hampshire

I. INTRODUCTION

A specialized set of intracellular signal transduction pathways couples the stimulation of plasma membrane receptors to the process of exocytosis. Such pathways are present in both excitable and nonexcitable cells, and regulate functions such as hormonal release and neurotransmission. Despite the importance of this type of signaling, termed stimulus–secretion coupling (Douglas and Rubin, 1963), very little is known regarding its precise biochemical

Present address: Department of Immunology, Duke University Medical Center, Durham, North Carolina

103

SIGNAL TRANSDUCTION DURING
BIOMEMBRANE FUSION

nature. It is well established that elevated levels of intracellular calcium are associated with exocytosis in many cell types. In excitable cells, receptor stimulation leads to changes in membrane potential which, in turn, results in the opening of voltage-gated calcium channels and the influx of extracellular calcium. In nonexcitable cells, stimulation of cell surface receptors leads to both the mobilization of intracellular calcium stores and the influx of extracellular calcium. However, the exact role of calcium in exocytosis is unclear; in fact, in some cell types, exocytosis has been demonstrated to occur under conditions in which there is no detectable rise in intracellular calcium concentrations. Other messengers, such as cyclic AMP, diacylglycerols, and nucleotides are known to modulate exocytosis, as are various kinases and phosphatases, in complex interactions. This chapter focuses on recent experiments regarding signal transduction during exocytosis in mast cells; recent reviews, comprising more comprehensive reviews of the literature, are available (Almers, 1990; Gomperts, 1990).

Exocytosis may be defined as the fusion of secretory vesicles with the cytoplasmic face of the plasma membrane, resulting in the release of the vesicle contents to the extracellular space. Exocytosis thus represents the terminal step in vesicular protein transport, the process by which proteins destined for secretion are carried to their final destinations. The basic features of this pathway have been known for some time (Palade, 1975); secreted proteins are synthesized in the rough endoplasmic reticulum, transported via vesicles to and through the Golgi stack, and carried to the plasma membrane via secretory vesicles. The secretory pathway may be divided into constitutive and regulated pathways (Kelly, 1985). In the regulated pathway, the vesicles are stored in the cytosol until the proper signal (or set of signals) is received by membrane-bound receptors. Following receptor stimulation, the vesicles undergo exocytosis. This is in contrast to the constitutive secretory pathway, in which exocytosis occurs continually, in the absence of a defined stimulus. While the mechanisms underlying the final steps in constitutive and regulated secretion (i.e., membrane fusion) may be the same, it is clear that control mechanisms have evolved to fine-tune the stimulation of exocytosis in the regulated pathway.

Mast cells have become a favorite model system for researchers studying exocytosis for a variety of reasons. Large quantities of mast cells can be quickly and easily purified from rat peritoneal fluid; more important, the cells are highly adapted to the process of regulated secretion. An individual mast cell may contain more than 1000 secretory vesicles in the cytosol; the vesicles contain a variety of mediators including histamine, proteases, and interleukins. On stimulation of the cells by cross-linking of high-affinity receptors for immunoglobulin E at the cell surface (Ishizaka and Ishizaka, 1984), a rapid and dramatic exocytotic event ensues in which almost all of the vesicles are released. In the past several years, mast cell exocytosis has been studied using whole-cell, permeabilized-cell, and electrophysiological methods. The goal of all of these lines of research has been an understanding of the biochemical mechanisms by which mast cell secretion is controlled.

II. SIGNAL TRANSDUCTION DURING EXOCYTOSIS IN MAST CELLS

A. The Calcium Signal

The first control mechanism implicated in the regulation of exocytosis was an elevation in the intracellular concentration of free calcium ions (Ca^{2+}_i). Early experiments by Katz and his colleagues suggested that neurotransmission was dependent on the influx of extracellular calcium (Ca^{2+}_o) into neurons (Katz, 1969). The idea that calcium was required for mast cell exocytosis was suggested from early experiments by Mongar and Schild (1958) in which antigen-stimulated exocytosis in lung mast cells was determined to be dependent on the presence of Ca^{2+}_o. The calcium target molecules important for stimulus–secretion coupling are not yet known, but calpactin (Ali *et al.*, 1989) and other calcium-binding proteins (Momayezi *et al.*, 1987) have been suggested to play roles in exocytosis.

Support for the involvement of calcium in exocytosis has come from the development of methods for monitoring Ca^{2+}_i in secretory cells. Membrane-permeant dyes such as Quin 2 and Fura 2, developed by R. Y. Tsien, undergo changes in fluorescence when bound to calcium. These changes allow the spatial and temporal monitoring of changes in Ca^{2+}_i at both population and single-cell levels (Tsien, 1980; Grynkiewicz *et al.*, 1985). Experiments using such indicators revealed that stimulation of receptors in a wide variety of secretory cell types resulted in the elevation of Ca^{2+}_i from a resting level of approximately 100 nM to the micromolar range (Beaven *et al.*, 1984; Knight and Kesteven, 1983; Rink *et al.*, 1982). This elevation of Ca^{2+}_i in response to receptor stimulation is known as the calcium signal.

The calcium signal in mast cells is complex. Cross-linking of IgE receptors results in a biphasic calcium signal (Neher and Almers, 1986). The initial phase is a rapid calcium transient, or "spike," which peaks at 1.0–1.5 μM Ca^{2+}_i. This fast calcium transient is followed by a prolonged plateau phase, during which Ca^{2+}_i reaches approximately 1 μM. As mentioned, extracellular calcium is required for receptor-stimulated exocytosis in mast cells, although the nature of the channel by which extracellular calcium traverses the plasma membrane remains controversial (Penner *et al.*, 1988). Recent experiments have demonstrated that generation of the fast calcium transient does not require extracellular calcium, but results from the mobilization of intracellular calcium stores (Neher and Almers, 1986).

The nature of the calcium transient and the mechanism of its generation has received much attention in recent years. The picture that has emerged is that stimulation of a class of cell surface receptors, termed calcium-mobilizing receptors, leads to the activation of a plasma membrane-associated phosphatidylinositol-specific phospholipase C (PI-PLC) (Putney, 1987); the high-affinity IgE receptor belongs to the functional category of calcium-mobilizing receptors (Metzger *et al.*, 1986). The activation of PI-PLC is coupled to receptor stimulation by an as yet unidentified GTP-binding regulatory protein, or G protein, termed G_P (Cockroft, 1987). Activated PI-PLC cleaves a minor species of plasma membrane phospholipid, phosphatidylinositol 4,5-bisphosphate (PIP_2) to generate two messengers: inositol 1,4,5-trisphosphate (IP_3) and diacylglycerol (DAG). IP_3 diffuses from its site of generation at the plasma membrane to bind to a specific receptor (Spat *et al.*, 1986). Binding of IP_3 to this receptor results in the liberation of calcium from a storage organelle, originally thought to be the endoplasmic reticulum and now referred to as the calciosome (Volpe *et al.*, 1988). Calcium can be released by exogenous IP_3 as well as by endogenous IP_3; perfusion of patch-dialyzed mast cells with IP_3 has been shown to result in generation of a fast calcium transient (Penner *et al.*, 1988).

The receptor-stimulated calcium transient alone is insufficient to trigger exocytosis in mast cells, despite the fact that Ca^{2+}_i reaches the same stimulatory levels seen in the plateau phase (Neher and Almers, 1986). In addition, the calcium transient is temporally inconsistent with exocytosis in receptor-stimulated mast cells. Secretion begins at the same time as Ca^{2+}_i reaches its plateau, some 20–40 sec after receptor stimulation (Neher and Almers, 1986). Recent work from Beaven and his collaborators (Lo *et al.*, 1987; Beaven *et al.*, 1987) support the contention that calcium is not the only signal for mast cell exocytosis. Both reports show that elevation of Ca^{2+}_i alone with ionophore is insufficient to drive exocytosis in RBL cells. The additional observation by Lo *et al.* (1987) that exocytosis requires the use of concentrations of ionophore sufficient to stimulate PIP_2 hydrolysis (200 nM-1 μM) suggests that the products of this reaction, DAG and inositol phosphates, may be involved in stimulation. Taken together, the data indicate that elevation of Ca^{2+}_i alone is not sufficient for the stimulation of secretion in mast cells, and, therefore, additional signals must be required.

B. Permeabilized Cell Systems

This discovery has led to a search for additional control mechanisms. Whereas Ca^{2+}_i may be monitored in intact cells and manipulated with ionophores, and enzymes such as protein kinase C may be stimulated using membrane-permanent reagents such as phorbol esters, biochemical analysis of other components in stimulus–secretion coupling are not easily studied in intact cells. The development of permeabilized cell systems has greatly advanced the molecular dissection of these pathways. In such systems, the plasma membranes of the cells are rendered leaky, such that the composition of the cytosol may be manipulated while leaving the secretory function intact. A variety of methods have been devised for the permeabilization of cells; these include patch dialysis (Hamill *et al.*, 1981), exposure to electrical fields (Baker and Knight, 1978), incubation with bacterial toxins (Bader *et al.*, 1986; Howell and Gomperts, 1987) or plant glycosides (Wilson and Kirshner, 1983; Dunn and Holz, 1983), and exposure to enveloped viruses (Impraim *et al.*, 1980). The characteristics of permeabilization vary according to the method used; for example, plasma membrane lesions generated by streptolysin *O* (SLO; Howell and Gomperts, 1987) or digitonin (Wilson and Kirshner, 1983; Dunn and Holz, 1983) allow the efflux of relatively large molecules, whereas lesions generated by α toxin (Bader *et al.*, 1986) are much smaller. Initial experiments (Baker and Knight, 1978; Wilson and Kirshner, 1983; Dunn and Holz, 1983) using permeabilized adrenal chromaffin cells suggested that an elevation of intracellular calcium was sufficient to stimulate exocytosis so long as cellular ATP levels were maintained; elevated Ca^{2+}_i was achieved simply by buffering the free calcium concentration of the permeabilization media into the micromolar range.

In order to study the regulation of mast cell exocytosis, conditions were established for the permeabilization of rat peritoneal mast cells with digitonin. In initial experiments, the conditions for selective permeabilization of mast cell plasma membranes were determined (Fig. 1). Mast cells were purified from rat peritoneal fluid and treated with various concentrations of digitonin in buffers containing either 2 mM EGTA or the combination of 10 μM free calcium and 40 μM GTPγS, a stable GTP analog. Following a 5-min incubation, the cell supernatants were analyzed for the release of the secretory product, histamine (Fig. 1a), and the high-molecular-weight marker enzyme, lactate dehydrogenase (LDH; Fig. 1b). Treatment of mast cells with low concentrations of digitonin results in permeabilization of the cells as evidenced by LDH release, both in the presence and absence of $Ca^{2+}/GTPγS$.

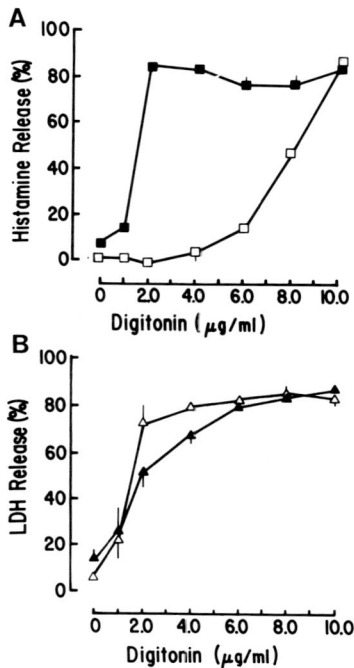

Fig. 1. Release of histamine and lactate dehydrogenase (LDH) from digitonin-permeabilized rat mast cells. Purified mast cells (2.5×10^4/tube) were incubated for 5 min at 37°C in mast cell buffer containing the indicated concentration of digitonin as well as 10 μM Ca^{2+} and 40 μM GTPγS (■,▲) or 2 mM EGTA (□,△). (A) At the end of the incubation, release of histamine was determined using *O*-phthaldialdehyde. (B) In a separate experiment, release of LDH was determined. Both LDH and histamine release are expressed as a percentage of the total. From Koopmann, W. R., and Jackson, R. C. (1990). Calcium- and Guanine Nucleotide-dependent Exocytosis in Digitonin-permeabilized Mast Cells: Modulation by Protein Kinase C. *Biochemical Journal* **265**, 365–373.

Histamine release was stimulated by the combination of Ca^{2+}/GTPγS at low concentrations of digitonin, whereas release of histamine in the presence of EGTA required higher digitonin concentrations. Next, the concentrations of Ca^{2+} and GTPγS required to stimulate exocytosis in the digitonin-permeabilized system were established (Fig. 2). The dual requirement for Ca^{2+} and GTPγS, as well as for the concentrations of these substances needed for half-maximal histamine release, are in agreement with results obtained with SLO-permeabilized cells (Howell and Gomperts, 1987). The close correspondence between the values obtained with these two permeabilizing agents

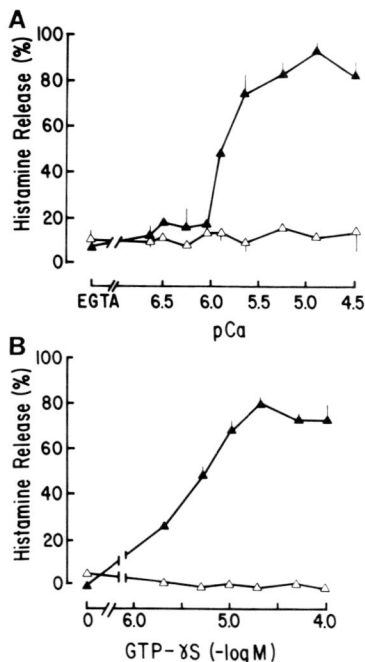

Fig. 2. Histamine release from digitonin-permeabilized rat mast cells as a function of calcium and GTPγS concentrations. (A) Purified mast cells were incubated for 5 min at 37°C in mast cell buffer containing 4 μg/ml digitonin and the indicated concentration of free calcium in the presence (▲) or absence (△) of 40 μM GTPγS. (B) Purified mast cells were incubated for 5 min at 37°C in mast cell buffer containing 4 μg/ml digitonin and either 10 μM Ca^{2+} (▲) or 2 mM EGTA (△), as well as the indicated concentration of GTPγS. From Koopmann, W. R. and Jackson, R. C. (1990). Calcium- and Guanine Nucleotide-dependent Exocytosis in Digitonin-permeabilized Mast Cells: Modulation by Protein Kinase C. *Biochemical Journal* **265**, 365–373.

strongly suggests that the results are not related to the specific membrane perturbant used for permeabilization, but rather are a reliable representation of the exocytotic process.

C. Guanine Nucleotides and Exocytosis

The observation of a requirement for guanine nucleotide in exocytosis in the permeabilized cell system raises the question of the site of the effect. One known function of guanine nucleotides in mast cell stimulus–secretion coupling is the stimulation of PI-PLC by G_P (Cockroft, 1987). Therefore, in permeabilized cells, it is possible that the requirement for both calcium and GTPγS reflects a requirement for G_P activity. The liberation of intracellular Ca^{2+} stores by IP_3 could not be the sole function of G_P activation in such a scheme because the cells are provided with micromolar calcium. Instead, it is more likely that the other product of PIP_2 hydrolysis, DAG, might provide the necessary stimulus.

An obligatory role for G_P-stimulated PI-PLC activity in exocytosis in permeabilized mast cells was made unlikely, however, by a recent study using the aminoglycoside antibiotic neomycin in the SLO-permeabilized mast cell system (Cockroft *et al.*, 1987). Neomycin inhibits PI-PLC activity by binding to the polar head groups of PIP_2 (Schacht, 1978). When mast cells were treated with neomycin prior to permeabilization, GTPγS-stimulated generation of inositol phosphates (a measure of PI-PLC activity) was abolished, yet exocytosis stimulated by the combination of Ca^{2+}/GTPγS remained intact. The authors interpreted this result to mean that there is a second GTPγS-reactive site (in addition to G_P), presumably a GTP-binding regulatory protein, or G protein, required for the stimulation of exocytosis in mast cells.

The identity of this second G protein (designated G_E, for exocytosis) is currently unknown. Several possible explanations have been proposed for the GTP requirement. Kennerly and his colleagues have provided evidence that a phospholipase D may be involved (Gruchalla *et al.*, 1990; Kennerly, 1987). The authors reported that IgE receptor cross-linking resulted in the activation of a phospholipase D (PLD) in mast cells, that mast cells contain a phosphatidic acid-phosphohydrolase (PA-PHase) activity capable of converting PA, the product of PLD, to DAG, and that the primary source of DAG in receptor-stimulated mast cells is not PIP_2, but phosphatidylcholine (PC). Thus, if IgE receptor cross-linking activates formation of DAG from PC via a G protein, either by sequential PLD and PA-PHase activity or by PC-specific PLC, then the guanine nucleotide requirement for exocytosis from permeabilized mast cells might still reflect a requirement for DAG even though activation of PI-PLC is not required. An indication that this might be the case comes from studies in which a plasma-membrane associated PLD was shown to be regulated by guanine nucleotides (Bocckino *et al.*, 1987)

Another possible explanation for the GTP requirement is a requirement for the activation of a phospholipase A_2 (PLA_2). IgE receptor cross-linking is known to stimulate PLA_2 activity in mast cells (as measured by the release of [$_3$H]arachidonic acid) (Yamada *et al.*, 1987), and PLA_2 activity may be stimulated by guanine nucleotides in some cell types (Burch *et al.*, 1986; Silk *et al.*, 1989). However, recent evidence suggests that there are conditions under which exocytosis from SLO-permeabilized mast cells is stimulated by the combination of Ca^{2+}/GTPγS, but PLA$_2$ is not activated (Churcher et al., 1990). This finding makes the possibility that PLA$_2$ activation is the second GTP-dependent site an unlikely one.

An alternative explanation for the GTP requirement for mast cell exocytosis is based on possible analogies with earlier steps in the vesicular protein transport pathway. GTP-binding proteins have recently been implicated in vesicular transport in both yeast and mammalian cell systems. Mutations in two separate yeast genes that block constitutive secretion have been determined to be caused by alterations in two small (24 kDa) GTP-binding proteins, at least one of which is associated with secretory granules and the plasma membrane (Salminen and Novick, 1987; Segev et al., 1988). In mammalian cell-free systems, GTPγS was found to inhibit vesicular transport at several steps in the secretory pathway (Melancon et al., 1987; Beckers and Balch, 1988; Baker et al., 1990). Based on these results, it has been suggested (Bourne, 1988) that the function of GTP-binding and hydrolysis in the early steps of protein transport is to ensure the directionality of transport. It is possible that a similar mechanism accounts for the GTP requirement for mast cell exocytosis, and thus that GTP is involved in transport events all along the secretory pathway. However, this interpretation appears to present a paradox, since GTPγS inhibits transport in cell-free systems and stimulates exocytosis in permeabilized mast cells. The difference, as pointed out by Bourne (1988), may be due to the differential need to recycle G proteins. Because GTPγS is nonhydrolyzable, it blocks G protein recycling after only one round. In the cell-free vesicular transport systems, a single round of transport may be missed, whereas a single round of vesicle fusion is all that is required for the exocytotic release of the entire complement of vesicles from a mast cell.

The precise role of guanine nucleotides in mast cell exocytosis is still a matter of controversy. It is of interest that adrenal chromaffin cells permeabilized with digitonin do not require guanine nucleotides for exocytosis; the addition of nonhydrolyzable GTP analogs to permeabilized chromaffin cells has only modest effects on secretion (Bittner et al., 1986). This observation indicates that there may be mechanistic differences in exocytosis between excitable and nonexcitable cells.

D. Protein Kinase C

The role of DAG generated by PLC-mediated PIP$_2$ hydrolysis in the stimulation of exocytosis was originally described in terms of the activation of the calcium and phospholipid-dependent protein kinase, protein kinase C (PKC). This enzyme becomes translocated from the cytosol to the plasma membrane following stimulation of calcium-mobilizing receptors (see Nishizuka, 1986, for a review). At the plasma membrane, PKC becomes activated by the combination of calcium, phosphatidylserine, and DAG (Kishimoto *et al.*, 1980). It was suggested that PKC activation was an integral part of the stimulus–secretion coupling pathway and acted synergistically with Ca^{2+} for the stimulation of exocytosis. This hypothesis was based in part on the discovery that the tumor-promoting phorbol esters (for example, 12-*O*-tetradecanoyl-13-acetate or TPA), which activate PKC directly (Castagna *et al.*, 1982), act synergistically with calcium ionophore in the stimulation of exocytosis in many cell types (Zawalich *et al.*, 1983; Zurgil and Zisapel, 1985; Sagi-Eisenberg *et al.*, 1985; Kaibuchi *et al.*, 1983), including mast cells (Katakami *et al.*, 1984). The idea that activation of PKC stimulates exocytosis is consistent with the observation that IgE receptor cross-linking results in PKC activation, as detected by the translocation of PKC to the plasma membrane in mast cells and RBL cells (White and Metzger, 1988; White *et al.*, 1985). Additionally, TPA pretreatment lowers the concentration of Ca^{2+} required to elicit exocytosis in several permeabilized cell systems (Ahnert-Hilger *et al.*, 1987; Lee and Holz, 1986; Vallar *et al.*, 1987). Thus, activation of PKC by DAG could conceivably provide the "second signal" (in addition to the elevation of $Ca^{2+}{}_i$) required for exocytosis.

To test the effects of PKC activation on mast cell exocytosis, cells were treated with increasing concentrations of TPA prior to permeabilization in the presence of either $Ca^{2+}/GTP\gamma S$, Ca^{2+} alone, GTPγS alone, or EGTA. The results (Fig. 3) indicate that the stimulation of mast cell exocytosis by either Ca^{2+} alone or GTPγS alone is dramatically enhanced by phorbol ester pretreatment. If this enhancement reflects a sensitization of the exocytotic machinery by PKC, it is possible that the effect of PKC activation is to lower the concentrations of Ca^{2+} and GTPγS required to elicit exocytosis. Figure 4 illustrates the results of an experiment designed to test the effects of TPA pretreatment on the Ca^{2+} requirement for exocytosis in the permeabilized cell system. The results indicate that TPA treatment reduces the Ca^{2+} requirement both in the presence and absence of guanine nucleotide. Parallel experiments designed to test the effects of TPA on the GTPγS requirement indicate that TPA also reduces the concentration of GTPγS required to elicit exocytosis (Fig. 5). In the digitonin-permeabilized cell system, then, activation of PKC has dual effects, reducing both the Ca^{2+} and GTPγS requirements for exocytosis.

Recent studies, however, have provided evidence that activation of PKC is

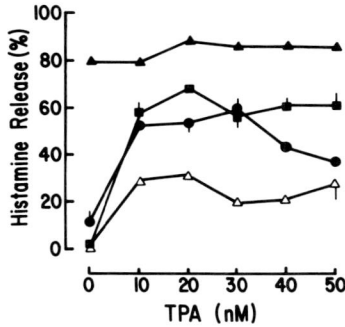

Fig. 3. Effect of TPA on exocytosis from digitonin-permeabilized rat mast cells. Purified mast cells were preincubated for 5 min at 37°C with vehicle (0.1% DMSO) or with the indicated concentration of TPA. The cells were then treated for an additional 5 min at 37°C with mast cell buffer containing 4 μg/ml digitonin and 2 mM EGTA (△), 10 μM Ca^{2+} + 40 μM GTPγS (▲), 10 μM Ca^{2+} alone (●), or 40 μM GTPγS alone (■). From Koopmann, W. R. and Jackson, R. C. (1990). Calcium- and Guanine Nucleotide-dependent Exocytosis in Digitonin-permeabilized Mast Cells: Modulation by Protein Kinase C. *Biochemical Journal* **265**, 365–373.

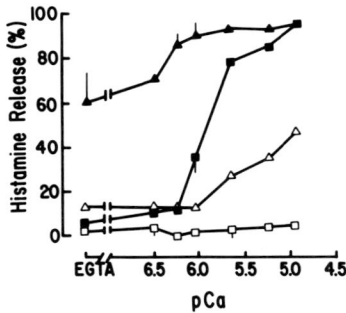

Fig. 4. Effect of TPA pretreatment on calcium dependence of exocytosis from permeabilized mast cells. Purified mast cells were incubated for 5 min at 37°C either with 25 nM TPA (▲,△) or with vehicle (■,□). Cells were then treated for an additional 5 min at 37°C with mast cell buffer containing 4 μg/ml digitonin and the indicated concentrations of free calcium in the presence (▲,■) or absence (△,□) of 40 μM GTPγS. From Koopmann, W. R. and Jackson, R. C. 1990). Calcium- and Guanine Nucleotide-dependent Exocytosis in Digitonin-permeabilized Mast Cells: Modulation by Protein Kinase C. *Biochemical Journal* **265**, 365–373.

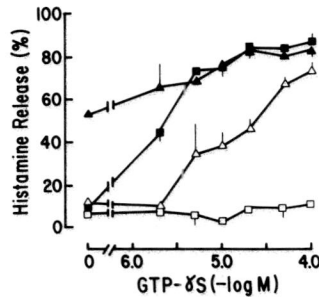

Fig. 5.   Effect of TPA pretreatment on GTPγS dependence of exocytosis from permeabilized mast cells. Purified mast cells were incubated for 5 min at 37°C either with 25 nM TPA (▲,△) or with vehicle (■,□). Cells were then treated with mast cell buffer containing 4 μg/ml digitonin and the indicated concentrations of GTPγS, as well as 10 μM free calcium (▲,■) or 2 mM EGTA (△,□). From Koopmann, W. R. and Jackson, R. C. (1990). Calcium- and Guanine Nucleotide-dependent Exocytosis in Digitonin-permeabilized Mast Cells: Modulation by Protein Kinase C. *Biochemical Journal* **265**, 365–373.

probably not required for the exocytotic event. White and Metzger (1988) demonstrated that antigen was able to elicit a small but significant exocytotic response in RBL cells in which PKC had been down-regulated. Insulin-secreting pancreatic islet cells are also capable of undergoing exocytosis following PKC-depletion (Hii *et al.*, 1987). The question of PKC involvement in exocytosis was addressed using the kinase inhibitor staurosporine. PKC is inhibited by this compound both *in vitro* and *in vivo* (Tamaoki *et al.*, 1986; Watson *et al.*, 1988; Vegesna *et al.*, 1988; Rogers *et al.*, 1988). Treatment of mast cells with 1 μM staurosporine abolished the effects of TPA on exocytosis in response to subsequent permeabilization in the presence of both Ca^{2+} and GTPγS (Fig. 6), showing that the effects of TPA on exocytosis in the permeabilized mast cell system are due to the activation of PKC. More important, the stimulation of exocytosis by the combination of Ca^{2+} and GTPγS was not affected by staurosporine, indicating that in permeabilized cells the exocytotic event is insensitive to the inhibitor and hence not dependent on PKC. Taken together, the data are consistent with a modulatory, but not an obligatory, role for PKC in the control of mast cell exocytosis.

The observation that activation of PKC is not an obligatory event in the stimulus–secretion coupling pathway raises the question of whether DAG may be playing other roles in the pathway, apart from the stimulation of PKC. A series of recent experiments have provided evidence that DAG may provide

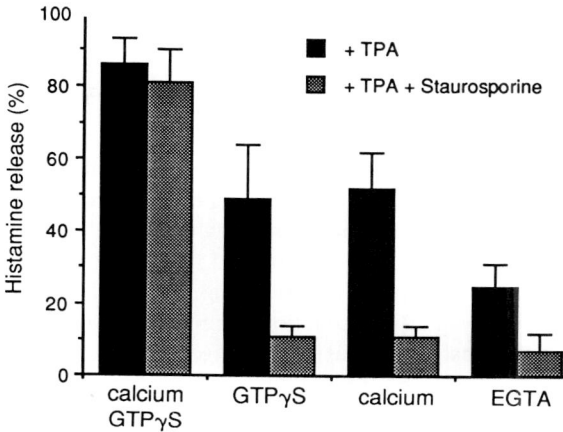

Fig. 6. Effect of staurosporine on exocytosis from permeabilized mast cells. Purified mast cells were incubated for 5 min at 37°C either with 1 μM staurosporine (hatched bars) or with vehicle (0.1% DMSO, solid bars). Mast cell buffer containing sufficient TPA to achieve a final concentration of 25 nM was added to each sample, and the cells were incubated for an additional 5 min at 37°C. The exocytotic capability of the cells was assessed by adding mast cell buffer containing sufficient digitonin, Ca^{2+}, GTPγS, and EGTA to achieve final concentrations of 4 μg/ml digitonin, 10 μM calcium, 40 μM GTPγS, and 2 mM EGTA, as indicated. From Koopmann, W. R. and Jackson, R. C. (1990). Calcium- and Guanine Nucleotide-dependent Exocytosis in Digitonin-permeabilized Mast Cells: Modulation by Protein Kinase C. *Biochemical Journal* **265**, 365–373.

a second stimulatory signal for exocytosis (Siegel *et al.,* 1989; Ellens *et al.,* 1989). The authors show that DAG is fusogenic to liposomes at low concentrations (2 mol%). Because the plasma membranes of cells stimulated through calcium-mobilizing receptors may contain 1 mol% DAG, it is possible that DAG may be exerting a stimulatory effect through the direct promotion of membrane fusion, in addition to stimulation of PKC. If this interpretation is correct, then generation of DAG may constitute a required stimulus for mast cell exocytosis, even though activation of PKC is not.

III. CONCLUSIONS

The use of permeabilized cell systems has allowed analysis of the messenger molecules involved in the signal transduction pathways that regulate exocytosis. In mast cells permeabilized with either digitonin or SLO, there is a requirement for the combination of calcium and guanine nucleotide; this dual requirement may be bypassed in digitonin-permeabilized cells by activating PKC directly with phorbol ester. However, the results of the staurosporine experiments provide evidence that PKC is not required for exocytosis. The role of PKC *in vivo* is likely to be that of a molecular signal amplifier; that is, activation of PKC resulting from cross-linking the IgE receptor sensitizes the exocytotic machinery, both to endogenous guanine nucleotides and to calcium. The nature of the PKC modulation and the identity of the phosphorylated proteins that are important in modulating exocytosis are not known. A full understanding of signal transduction during exocytosis depends on answering these questions, as well as identifying the relevant target molecules for both calcium and guanine nucleotides.

ACKNOWLEDGMENTS

I thank Bob Jackson, in whose laboratory the original studies described in this chapter were conducted, for advice and guidance; Paul Modern, for excellent technical assistance, and Annika Sanfridson, for support and encouragement. This work was supported by NIH Grant GM 26763.

REFERENCES

Ahnert-Hilger, G., Brautigam, M., and Gratzl, M. (1987). Ca^{2+}-stimulated catecholamine release from α-toxin-permeabilized PC12 cells: Biochemical evidence for exocytosis and its modulation by protein kinase C and G proteins. *Biochemistry* **26,** 7842–7848.

Ali, S. M., Geisow, M. J., and Burgoyne, R. D. (1989). A role for calpactin in calcium-dependent exocytosis in adrenal chromaffin cells. *Nature* **340,** 313–315.

Almers, W. (1990). Exocytosis. *Annu. Rev. Physiol.* **52,** 607–624.

Bader, M.-F., Thierse, D., Aunis, D., Ahnert-Hilger, G., and Gratzl, M. (1986). Characterization of hormone and protein release from α-toxin-permeabilized chromaffin cells in primary culture. *J. Biol. Chem.* **261,** 5777–5783.

Baker, D., Wuestehube, L., Schekman, R., Botstein, D., and Segev, N. (1990). GTP-binding YPT1 protein and Ca^{2+} function independently in a cell-free protein transport reaction. *Proc. Natl. Acad. Sci. USA* **87,** 355–359.

Baker, P. F., and Knight, D. E. (1978). Calcium-dependent exocytosis in bovine adrenal medullary cells with leaky plasma membranes. *Nature* **276,** 620–622.

Beaven, M. A., Guthrie, D. F., Moore, J. P., Smith, G. A., Hesketh, T. R., and Metcalfe, J. C. (1987). Synergistic signals in the mechanism of antigen-induced exocytosis in 2H3 cells: Evidence for an unidentified signal required for histamine release. *J. Cell Biol.* **105,** 1129–1136.

Beckers, C. J. M., and Balch, W. E. (1989). Calcium and GTP: Essential components in vesicular trafficking between the endoplasmic reticulum and Golgi apparatus. *J. Cell Biol.* **108,** 1245–1256.

Bittner, M. A., Holz, R. W., and Neubig, R. R. (1986). Guanine nucleotide effects on catecholamine secretion from digitonin-permeabilized adrenal chromaffin cells. *J. Biol. Chem.* **261,** 10182–10188.

Bocckino, S. B., Blackmore, P. F., Wilson, P. B., and Exton, J. H. (1987). Phosphatidate accumulation in hormone-treated hepatocytes via a phospholipase D mechanism. *J. Biol. Chem.* **262,** 15309–15315.

Bourne, H. R. (1988). Do GTPases direct membrane traffic in secretion? *Cell* **53,** 669–671.

Burch, R. M., Luini, A., and Axlerod, J. (1986). Phospholipase A_2 and phospholipase C are activated by distinct GTP-binding proteins in response to α1-adrenergic stimulation in FRTL5 thyroid cells. *Proc. Natl. Acad. Sci. USA* **83,** 7201–7205.

Castagna, M., Takai, Y., Kaibuchi, K., Sano, K., Kikkawa, U., and Nishizuka, Y. (1982). Direct activation of calcium-activated, phospholipid-dependent protein kinase by tumor-promoting phorbol esters. *J. Biol Chem.* **257,** 7847–7851.

Churcher, Y., Allan, D., and Gomperts, B. D. (1990). Relationship between arachidonate generation and exocytosis in permeabilized mast cells. *Biochem. J.* **266,** 157–163.

Cockroft, S. (1987). Polyphosphoinositide phosphodiesterase: Regulation by a novel guanine nucleotide-binding protein, G_P. *Trends Biochem. Sci.* **12,** 75–78.

Cockroft, S., Howell, T. W., and Gomperts, B. D. (1987). Two G proteins act in series to control stimulus-secretion coupling in mast cells: Use of neomycin to distinguish between G proteins controlling polyphosphoinositide phosphodiesterase and exocytosis. *J. Cell Biol.* **105,** 2745–2750.

Douglas, W. W., and Rubin, R. P. (1963). The mechanism of catecholamine release from the adrenal medulla and the role of calcium in stimulus–secretion coupling. *J. Physiol.* **167,** 288–310.

Dunn, L. A., and Holz, R. W. (1983). Catecholamine secretion from digitonin-treated adrenal medullary chromaffin cells. *J. Biol. Chem.* **258,** 4989–4993.

Gomperts, B. D. (1990). G_E: A GTP-binding protein mediating exocytosis. *Annu. Rev. Physiol.* **52,** 591–606.

Gruchalla, R. S., Dinh, T. T., and Kennerly, D. A. (1990). An indirect pathway of receptor mediated 1,2-diacylglycerol formation in mast cells. I. IgE receptor-mediated activation of phospholipase D. *J. Immunol.* **144,** 2334–2342.

Grynkiewicz, G., Poenie, M., and Tsien, R. Y. (1985). A new generation of Ca^{2+} indicators with greatly improved fluorescence properties. *J. Biol. Chem.* **260,** 3440–3450.

Hamill, O., Marty, A., Neher, E., Sakmann, B., and Sigworth, F. (1981). Improved patch-clamp techniques for high-resolution current recordings from cells and cell-free membrane patches. *Eur. J. Physiol.* **391,** 85–100.

Hii, C. S. T., Jones, P. M., Persaud, S. J., and Howell, S. L. (1987). A reassesment of the role of protein kinase C in glucose-stimulated insulin secretion. *Biochem. J.* **246,** 489–493.

Howell, T. W., and Gomperts, B. D. (1987). Rat mast cells permeabilized with streptolysin-O secrete histamine in response to Ca^{2+} at concentrations buffered in the micromolar range. *Biochim. Biophys. Acta* **927,** 177–183.

impraim, C. C., Foster, K. A., Micklem, K. J., and Pasternak, C. A. (1980). Nature of virally mediated changes in membrane permeability to small molecules. *Biochem. J.* **186,** 847–860.

Ishizaka, T., and Ishizaka, K. (1984). Activation of mast cells for mediator release through IgE receptors. *Prog. Allergy* **34,** 188–235.

Kaibuchi, K., Takai, Y., Sawamura, M., Hoshijima, M., Fujikura, T., and Nishizuka, Y. (1983).

Synergistic functions of protein phosphorylation and calcium mobilization in platelet activation. *J. Biol. Chem.* **258**, 6701–6704.

Katakami, Y., Kaibuchi, K., Sawamura, M., Takai, Y., and Nishizuka, Y. (1984). Synergistic action of protein kinase C and calcium for histamine release from rat peritoneal mast cells. *Biochem. Biophys. Res. Commun.* **121**, 573–578.

Katz, B. (1969). "The Release of Neural Transmitter Substances." Charles Thomas, Springfield, Illinois.

Kelly, R. B. (1985). Pathways of protein secretion in eukaryotes. *Science* **230**, 25–32.

Kennerly, D. A. (1987). Diacylglycerol metabolism in mast cells. Analysis of lipid metabolic pathways using molecular species analysis of intermediates. *J. Biol. Chem.* **262**, 16,305–16,313.

Kishimoto, A., Takai, Y. Mori, T., Kikkawa, U., and Nishizuka, Y. (1980). Activation of calcium- and phospholipid-dependent protein kinase by diacylglycerol: Its possible relation to phosphatidylinositol turnover. *J. Biol. Chem.* **255**, 2273–2276.

Knight, D. E., and Kesteven, N. T. (1983). Evoked transient intracellular free Ca^{2+} changes and secretion in isolated bovine adrenal medullary cells. *Proc. Royal Soc. London (B)* **218**, 177–199.

Koopmann, W. R., Jr., and Jackson, R. C. (1990). Calcium and guanine nucleotide-dependent exocytosis in permeabilized rat mast cells: Modulation by protein kinase C. *Biochem. J.* **265**, 365–373.

Lee, S. A., and Holz, R. W. (1986). Protein phosphorylation and secretion in digitonin-permeabilized adrenal chromaffin cells. *J. Biol. Chem.* **261**, 17089–17098.

Lo, T. N., Saul, W., and Beaven, M. A. (1987). The actions of Ca^{2+} ionophores on rat basophilic (2H3) cells are dependent on cellular ATP and hydrolysis of inositol phospholipids: A comparison with antigen stimulation. *J. Biol. Chem.* **262**, 4141–4145.

Melancon, P., Glick, B. S., Malhotra, V., Weidman, P. J., Sarafini, T., Gleason, M. L., Orci, L., and Rothman, J. E. (1987). Involvement of GTP-binding "G" proteins in transport through Golgi stack. *Cell* **51**, 1053–1062.

Metzger, H., Alcaraz, G., Hohman, R., Kinet, J.-P., Pribluda, V., and Quarto, R. (1986). The receptor with high affinity for immunoglobulin E. *Annu. Rev. Immunol.* **4**, 419–470.

Momayezi, M., Lumpert, C. J., Kersken, H., Gras, U., Plattner, H., Krinks, M. H., and Klee, C. B. (1987). Exocytosis induction in *Paramecium tetraurelia* cells by exogenous phosphoprotein phosphatase *in vivo* and *in vitro:* Possible involvement of calcineurin in exocytotic membrane fusion. *J. Cell Biol.* **105**, 181–189.

Mongar, J. L., and Schild, H. O. (1958). The effect of calcium and pH on the anaphylactic reaction. *J. Physiol.* **140**, 272–284.

Neher, E., and Almers, W. (1986). Fast calcium transients in rat peritoneal mast cells are not sufficient to trigger exocytosis. *EMBO J.* **5**, 51–53.

Nishizuka, Y. (1986). Studies and perspectives of protein kinase C. *Science* **233**, 305–312.

Palade, G. (1975). Intracellular aspects of the process of protein synthesis. *Science* **189**, 347–358.

Penner, R., Matthews, G., and Neher, E. (1988). Regulation of calcium influx by second messengers in rat mast cells. *Nature* **334**, 499–504.

Putney, J. W., Jr. (1987). Calcium-mobilizing receptors. *Trends Pharmacol. Sci.* **8**, 481–486.

Rink, T. J., Smith, S. W., and Tsien, R. Y. (1982). Cytopasmic free Ca^{2+} in human platelets: Ca^{2+} thresholds and Ca^{2+}-independent activation for shape change and secretion. *FEBS Lett.* **148**, 21–26.

Rogers, J., Hughes, R. G., and Matthews, E. K. (1988). Cyclic GMP inhibits protein kinase C-mediated secretion in rat pancreatic acini. *J. Biol. Chem.* **263**, 3713–3719.

Sagi-Eisenberg, R., Lieman, H., and Pecht, I. (1985). Protein kinase C regulation of the receptor-coupled calcium signal in histamine-secreting rat basophilic leukemia cells. *Nature* **313**, 59–60.

Salminen, A., and Novick, P. (1987). A ras-like protein is required for a post-Golgi event in yeast secretion. *Cell* **49**, 527–538.

Schacht, J. (1978). Purification of polyphosphoinositides by chromatography on immobilized neomycin. *J. Lipid Res.* **19**, 1063–1067.

Segev, N., Mulholland, J., and Botstein, D. (1988). The yeast GTP-binding YPT1 protein and a mammalian counterpart are associated with the secretion machinery. *Cell* **52**, 915–924.

Siegel, D. P., Banschbach, J., Alford, D., Ellens, H., Lis, L. J., Quinn, P. J., Yeagle, P. L., and Bentz, J. (1989). Physiological levels of diacylglycerols in phospholipid membranes induce membrane fusion and stabilize inverted phases. *Biochemistry* **28**, 3703–3709.

Silk, S. T., Clejan, S., and Witkom, K. (1989). Evidence of GTP-binding protein regulation of phospholipase A_2 activity in isolated human platelet membranes. *J. Biol. Chem.* **264**, 21,466–21,469.

Spat, A., Bradford, P. G., McKinney, J. S., Rubin, R. P., and Putney, J. W. (1986). A saturable receptor for ^{32}P-inositol-1,4,5-trisphosphate in hepatocytes and neutrophils. *Nature* **319**, 514–516.

Tamaoki, T., Nomoto, H., Takahashi, I., Kato, Y., Morimoto, M., and Tomita, F. (1986). Staurosporine, a potent inhibitor of phospholipid/Ca^{2+}-dependent protein kinase. *Biochem. Biophys. Res. Commun.* **135**, 397–402.

Tsien, R. Y. (1980). New calcium indicators and buffers with high selectivity against magnesium and protons: Design, synthesis, and properties of prototype structures. *Biochemistry* **19**, 2396–2404.

Vallar, L., Biden, T. J., and Wollheim, C. B. (1987). Guanine nucleotides induce Ca^{2+}-independent insulin secretion from permeabilized RINm5F cells. *J. Biol. Chem.* **262**, 5049–5056.

Vegesna, R. V. K., Wu, H.-L., Mong, S., and Crooke, S. T. (1988). Staurosporine inhibits protein kinase C and prevents phorbol ester-mediated leukotriene D_4 receptor desensitization in RBL-1 cells. *Mol. Pharm.* **33**, 537–542.

Volpe, P., Krause, K.-H., Hashimoto, S., Zorzato, F., Pozzan, T., Meldolesi, J., and Lew, D. P. (1988). "Calciosome," a cytoplamic organelle: The inositol 1,4,5-trisphosphate-sensitive Ca^{2+} store of nonmuscle cells? *Proc. Natl. Acad. Sci. USA* **85**, 1091–1095.

Watson, S. P., McNally, J., Shipman, L. J., and Godfrey, P. P. (1988). The action of the protein kinase C inhibitor, staurosporine, on human platelets. *Biochem. J.* **249**, 345–350.

White, K. N., and Metzger, H. (1988). Translocation of protein kinase C in rat basophilic leukemic cells induced by phorbol ester or by aggregation of IgE receptors. *J. Immunol.* **141**, 942–947.

White, J. R., Pluznik, D. H., Ishizaka, K., and Ishizaka, T. (1985). Antigen-induced increase in protein kinase C activity in plasma membrane of mast cells. *Proc. Natl. Acad. Sci. USA* **82**, 8193–8197.

Wilson, S. P., and Kirshner, N. (1983). Calcium-evoked secretion from digitonin-permeabilized adrenal medullary chromaffin cells. *J. Biol. Chem.* **258**, 4994–5000.

Yamada, K., Okano, Y., Miura, K., and Nozawa, Y. (1987). A major role for phospholipase A_2 in antigen-induced arachidonic acid release in rat mast cells. *Biochem. J.* **247**, 95–99.

Zawalich, W., Brown, C., and Rasmussen, H. (1983). Insulin secretion: Combined effects of phorbol ester and A23187. *Biochem. Biophys. Res. Commun.* **117**, 448–455.

Zurgil, N., and Zisapel, N. (1985). Phorbol ester and calcium act synergistically to enhance neurotransmitter release by brain neurons in culture. *FEBS Lett.* **185**, 257–261.

6

Protein Kinase C and Granule Membrane Fusion in Platelets

**JON M. GERRARD, ARCHIBALD McNICOL,
AND SATYA P. SAXENA**

Manitoba Institute of Cell Biology
University of Manitoba
Winnipeg, Manitoba, Canada

I. INTRODUCTION

The secretion or exocytosis of granules is of fundamental importance to the function of platelets. Adenosine diphosphate (ADP) and, to a lesser extent,

121

SIGNAL TRANSDUCTION DURING
BIOMEMBRANE FUSION

serotonin and calcium secreted from dense granules can promote or amplify platelet aggregation. Thrombospondin and fibrinogen secreted from platelet-α granules are extracellular adhesive proteins that form part of the "glue" binding aggregated platelets together. The secretion of the contents of platelet granules appears to depend critically on two principal processes.

1. Platelet granules must be moved to a site at which they can fuse with other granules, with channels of the surface connected canalicular system (SCCS), or with the plasma membrane before exocytosis.
2. The membranes of the granules must fuse with the membranes of other granules or with the membranes of the canalicular system or the plasma membrane in order for their contents to be secreted or exocytosed.

The role of protein kinase C (PKC) in secretion will be discussed in the context of these two processes. The action of PKC in platelets has been linked to the stimulation of the latter process involving membrane fusion. Understanding the relationship of PKC activation to membrane fusion in platelets and the role this plays in cell function is facilitated by a general understanding of platelet anatomy and function.

A. Platelet Anatomy

Platelets are small discoid cells (Fig. 1). Beneath the plasma membrane in the plane of largest diameter there is an encircling band of microtubules (Figs. 1 and 2). On the platelet exterior there are openings of the SCCS, an invagination of the plasma membrane. The lumen of the SCCS connects to the surrounding plasma and serves as the conduit through which the contents of granules can pass during secretion. In the platelet interior, associated with the SCCS, is another membrane system, the dense tubular system. The dense tubular system is a smooth endoplasmic reticulum membrane that has the capacity to sequester calcium and to release it on cell activation. The dense tubular system is also involved in some aspects of lipid metabolism, in particular, the synthesis of prostaglandin endoperoxides and thromboxane A_2 from arachidonic acid. Four types of granules are present in the platelet cytoplasm. The most numerous of these, the α granule, contains a variety of polypeptides including fibrinogen, thrombospondin, fibronectin, platelet factor 4, platelet-derived growth factor, β-thromboglobulin, albumin, platelet factor V, von Willebrand factor, and platelet factor VIII. The second most numerous gran-

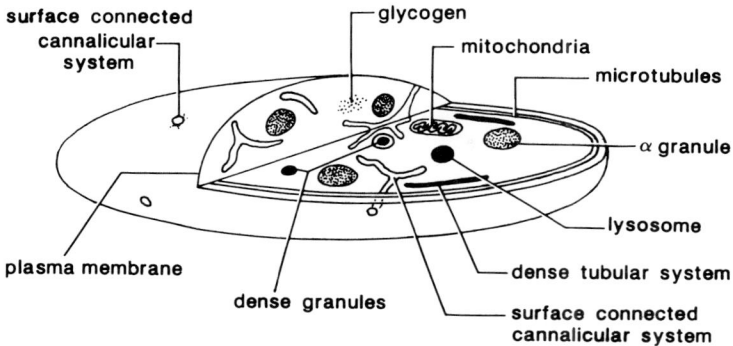

surface connected cannalicular system

glycogen

mitochondria

microtubules

α granule

lysosome

plasma membrane

dense granules

dense tubular system

surface connected cannalicular system

Fig. 1. Diagram showing the anatomy of a resting platelet. In the equatorial plane, a band of microtubules encircles the cell just beneath the plasma membrane. Granules including α granules, dense granules, lysosomes, and peroxisomes (not shown) are distributed randomly in the interior of the platelet. The surface connected canalicular system (SCCS), an invagination of the plasma membrane, has an important role in secretion. The dense tubular system is a smooth endoplasmic reticulum membrane with a critical role in the regulation of calcium transport and the synthesis of thromboxane A_2.

ule is the dense granule which contains ADP, ATP, serotonin, calcium, and pyrophosphate. Lysosomes containing hydrolytic enzymes and peroxisomes containing catalase are the other granular forms.

B. Historical Developments in Knowledge of Platelet Secretion

The first demonstration that serotonin could be secreted from washed platelets was by Zucker and Borelli in 1955. However, it was not until the late 1960s and early 1970s, when rapid advances in the biochemistry and morphology of secretion occurred, that an integrated picture of platelet secretion emerged. From the biochemical perspective, Holmsen, in particular, working with Day (Holmsen and Day, 1968, 1970), characterized the platelet secretory process, which Grette (1962) termed the release reaction because it occurred with great rapidity. In brief, when platelets are stimulated by a variety of agonists, they secrete their granule contents; the extent of secretion is proportionate to the concentration and nature of the stimulus. Weak agonists secrete primarily the contents of α granules and dense granules, whereas stronger agonists, particularly thrombin, cause the additional secretion of lysosomal

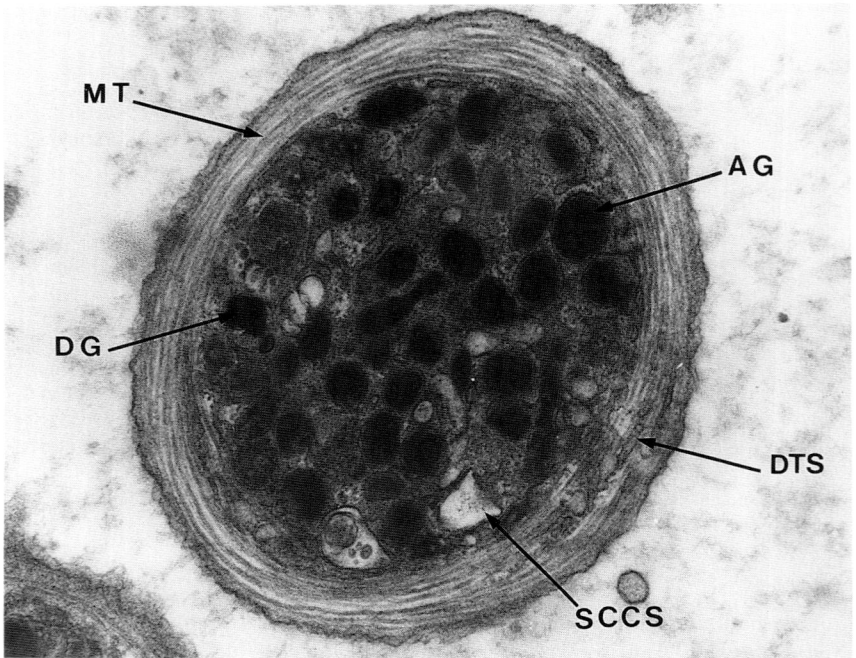

Fig. 2. A transmission electron micrograph showing a platelet sectioned through the equatorial plane. AG, α granules; DG, dense granules; MT, microtubules; SCCS, surface connected canalicular system; DTS, dense tubular system are visible. Magnification × 35,000.

enzymes. Secretion can be monitored by measuring the appearance of one of the granule contents in the solution surrounding the platelets. ATP and serotonin found in dense granules have proved to be particularly useful for this purpose.

At about the same time, White was studying the morphological changes that occur when platelets are stimulated by agonists. In 1968, he provided a lucid description of the movement of granules to the center of individual platelets when they were stimulated by ADP (White, 1968a). Furthermore, White recognized the process to be contractile in nature and suggested that it might involve the contractile actin and myosin proteins initially described in platelets by Bettex-Galland and Luscher (1959). Puzzled by the fact that secretion in other cells usually occurs with movement of granules to the periphery of the cell, White used electron-dense tracers and cytochemical studies to stabilize platelet proteins to further study the secretory pathway. The results of these investigations clearly showed that contents of platelet granules are dumped into a SCCS, and then proceed through the channels of this system to the exterior of the platelet (White, 1968b, 1970, 1972, 1974; White and Estensen, 1972).

The concepts proposed by White have been expanded, clarified, and modified in subsequent years. Substantial additional evidence now supports the contention that actin–myosin contraction drives the process that centralizes platelet granules (Gerrard and White 1976; Gerrard et al., 1987). The work of Lebowitz and Cook (1978), Dabrowska and Hartshorne (1978), and Hathaway and Adelstein (1979) has showed that platelet actin and myosin can generate tension, and that stimulation (in the presence of calmodulin) of

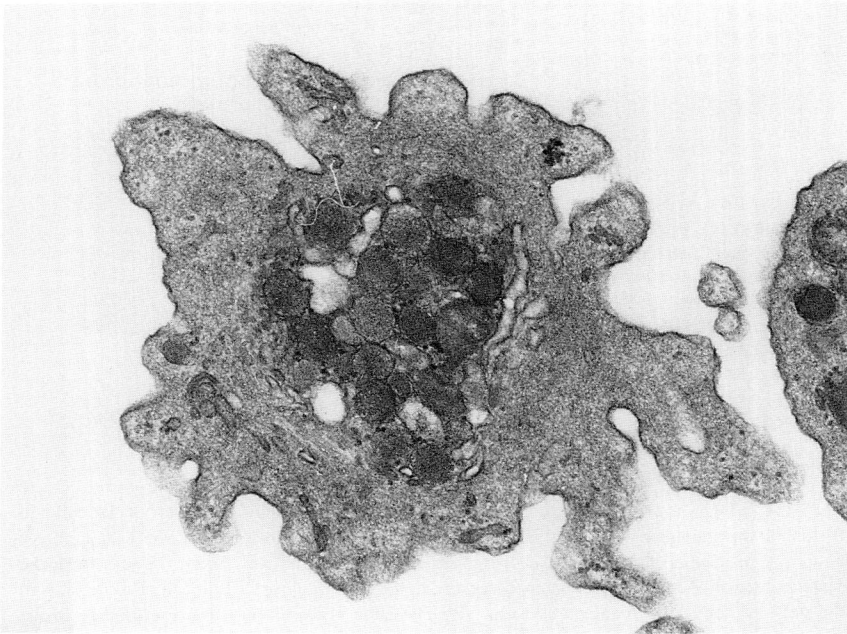

Fig. 3. A platelet exposed to the calcium ionophore A23187 (1 μM) for 5 min. The platelet has extended pseudopods and granules have been moved to the center of the platelet. Magnification × 35,000.

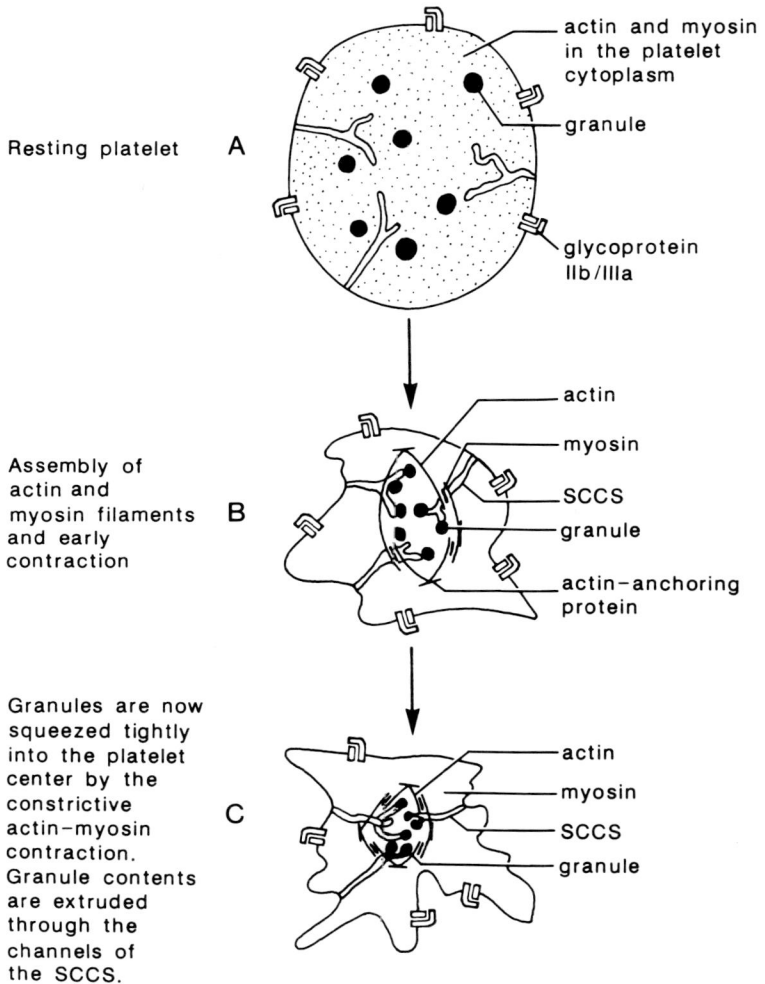

Resting platelet   **A**

actin and myosin
in the platelet
cytoplasm

granule

glycoprotein
IIb/IIIa

Assembly of
actin and
myosin filaments   **B**
and early
contraction

actin

myosin

SCCS

granule

actin–anchoring
protein

Granules are now
squeezed tightly
into the platelet
center by the
constrictive   **C**
actin–myosin
contraction.
Granule contents
are extruded
through the
channels of
the SCCS.

actin

myosin

SCCS

granule

Fig. 4 Diagram showing stages in movement of granules toward the center of the cell and fusion of the granules with the channels of the SCCS. (A) A resting platelet with granules randomly distributed throughout the cytoplasm and with channels of the SCCS present in the platelet interior. Actin and myosin are dispersed in the platelet cytoplasm; much of the actin is in the gel and not in the filamentous form. (B) The actin converts in greater percentage to the filamentous form and begins to interact with myosin, constricting the granules and moving them toward the center of the cell. (C) The granules are shown fused with the channels of the SCCS, and the continued squeezing and contraction of the actin–myosin ring now drives the extrusion of granule contents to the outside through the channels of the SCCS.

myosin light chain (MLC) kinase by calcium regulates the phosphorylation of myosin light chain to initiate myosin contraction. Immunofluorescent studies by Painter and Ginsberg (1984) showed the centripetal movement of myosin around the granules as the platelets undergo granule centralization. Studies by Carroll and co-workers (1982) identified the presence of two separate cytoskeletons in activated platelets and showed that the assembly of the actin–myosin cytoskeleton is closely correlated with the centralization of platelet granules (Carroll *et al.*, 1982). The calcium ionophore A23187, which increases cytoplasmic calcium levels, is a potent stimulus for granule centralization, consistent with the concept that raised cytoplasmic calcium initiates the actin–myosin contraction responsible for this process (Gerrard *et al.*, 1974). This process is shown in Figs. 3 and 4.

The concept that secretion occurs solely by fusion of granules with membranes of the SCCS has been challenged and modified. Ginsberg and co-workers (1980) suggested that a significant proportion of secretion occurred when granules fused with each other to form large multigranular vesicles. Ginsberg further suggested that such vesicles, by virtue of their increased distensibility, could escape through the encircling and contracting actin–myosin web to move toward and fuse with the plasma membrane in order to release their contents directly to the exterior of the platelet.

Tannic acid, used to stain both the channels of the SCCS and those granules that have already fused with it so they are functionally connected to the extracellular space (i.e., the lumen is in continuity with the extracellular space), has provided further clarification of the secretory process (Stenberg *et al.*, 1984; Stenberg and Bainton, 1986). These studies emphasized that granules in the platelet center can become connected to the extracellular space soon after thrombin exposure, presumably through fusion with the channels of the SCCS. Furthermore, the large vesicular structures described by Ginsberg *et al.* (1980) were shown to be connected to the platelet exterior and were not simply the result of fusing with other granules. Careful studies of resting platelets showed that the channel connecting the SCCS to the platelet exterior have narrow necks (Fig. 5). After thrombin stimulation, the neck connecting the SCCS to the outside widens, facilitating exocytosis.

From these studies has emerged the current picture of exocytosis, in which the granules are driven toward the cell center by the constricting actin–myosin contraction. Concomitantly, the granule membranes fuse with the membranes of the SCCS and probably with the membranes of other granules. As secretion continues, the neck (Fig. 5) connecting the SCCS to the exterior at the plasma membrane widens. Continued compression of the granules by the contracting actin–myosin web results in rapid extrusion of the granule contents.

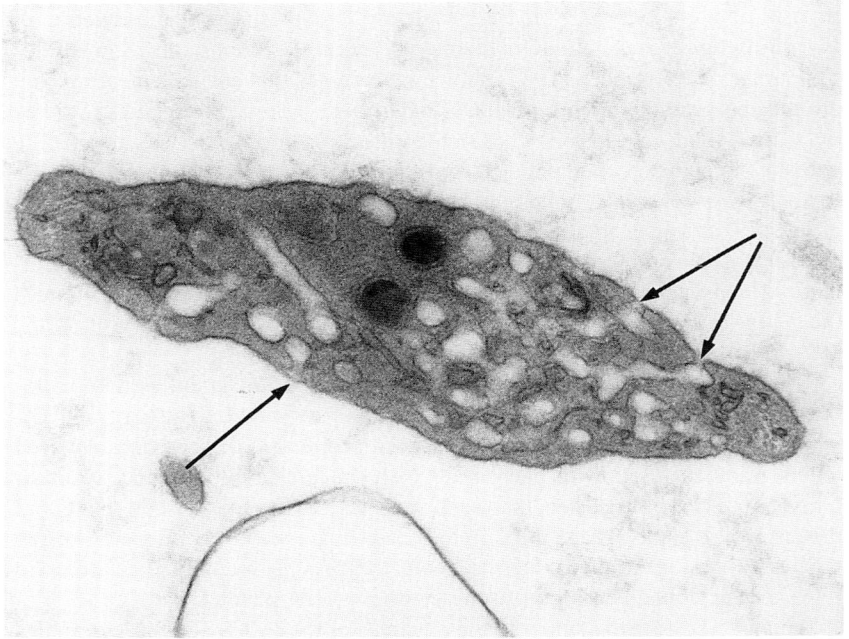

Fig. 5. A platelet in which it is possible to see several channels of the SCCS connecting to the exterior with narrow-necked structures at the channel exit points (arrows). Magnification × 40,500.

II. CURRENT RESEARCH

A. The Mechanism of Granule Membrane Fusion

Whereas the mechanism of granule centralization appears to be well understood, the mechanism(s) producing fusion of the granule membrane with other membranes is poorly characterized. Studies using the phorbol ester phorbol myristate acetate (PMA) provided the first clue that an identifiable process for membrane fusion might exist separate from the regulation of granule centralization. Phorbol myristate acetate produces fusion of granules with the SCCS and fusion of granules with each other to form large intra-platelet vesicular structures without any granule centralization (Fig. 6;

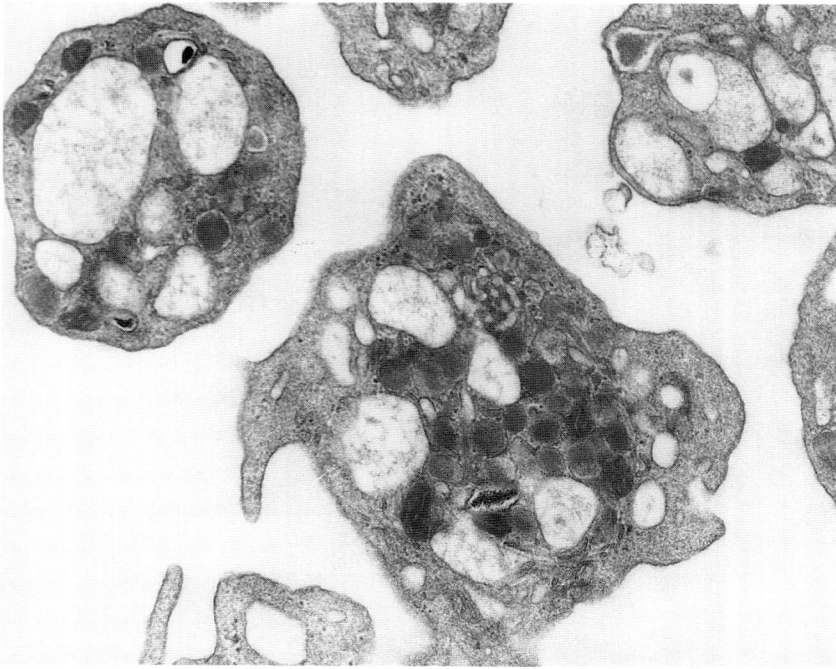

Fig. 6 Platelets exposed to phorbol myristate acetate (15 nM) show the development of large vesicular structures inside the platelet. These particular structures form when the granules dilate and fuse with each other and with channels of the SCCS. Magnification × 30,000.

Estensen and White, 1974). A short-chain unsaturated diglyceride, dicaprin, was later found to have a similar effect (Gerrard *et al.*, 1981). After the initial description of the influence of PMA on platelet function, Castagna *et al.* (1982) showed that PMA specifically bound to and activated protein kinase C (PKC). Subsequent studies using the cell-permeable diglyceride, oleoylacetylglycerol (OAG), also an activator of PKC, provided further clarification and extension of these observations (Friesen and Gerrard, 1985). OAG produced a time- and concentration-dependent fusion of platelet granules (Figs. 7–10), and, at the same time, stimulated the PKC-dependent phosphorylation of pleckstrin and actin-binding protein. It is now known that the endogenous activator of PKC is a diglyceride, primarily stearoylarachidonyl diglyceride, produced when agonists initiate receptor-mediated activation of phospho-

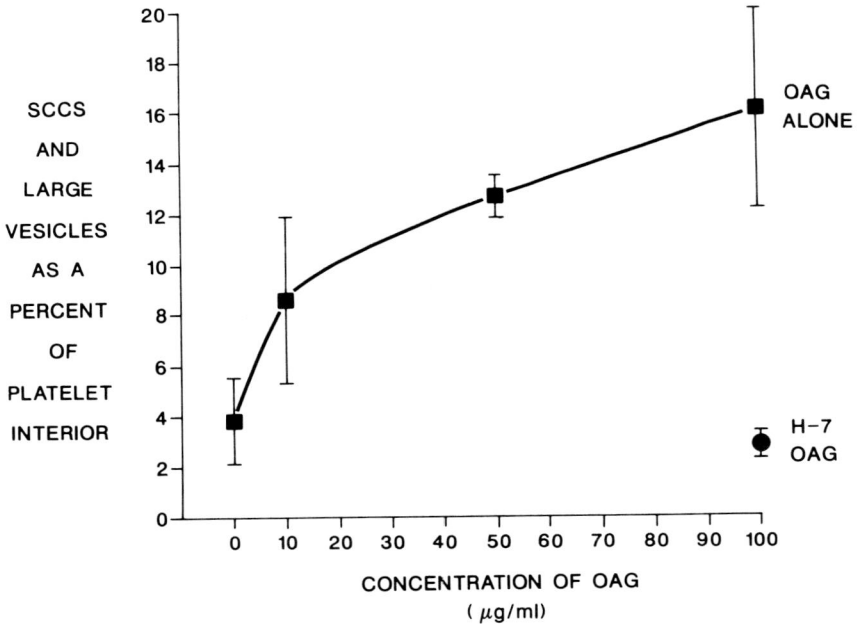

Fig. 7. The influence of the concentration of oleoylacetylglycerol (OAG) on changes in platelet morphology. An increased content of large vesicles and dilated SCCS is seen with an increasing concentration of OAG. When H-7 (110 μM) is added before OAG, it completely blocks the OAG-induced morphological changes.

lipase C (Rittenhouse-Simmons, 1979). Because IP_3, also produced by such phospholipase C activation, simultaneously raises cytoplasmic calcium, for most agonists the dominant morphological change is granule centralization.

Like PMA, OAG does not promote granule centralization (Friesen and Gerrard, 1985). The absence of granule centralization has been attributed to the fact that, although PMA and OAG stimulate a phosphorylation of MLC, the addition of the phosphate group does not occur at the active site phosphorylated by MLC kinase. Therefore, the PMA- and OAG-stimulated MLC phosphorylation do not promote actin–myosin contraction or the formation of the platelet actin–myosin cytoskeleton (Naka *et al.*, 1983; Carroll *et al.*, 1982). Thus, it is the absence of concomitant granule centralization with OAG and PMA that allows the visualization of granule fusion in response to these agents.

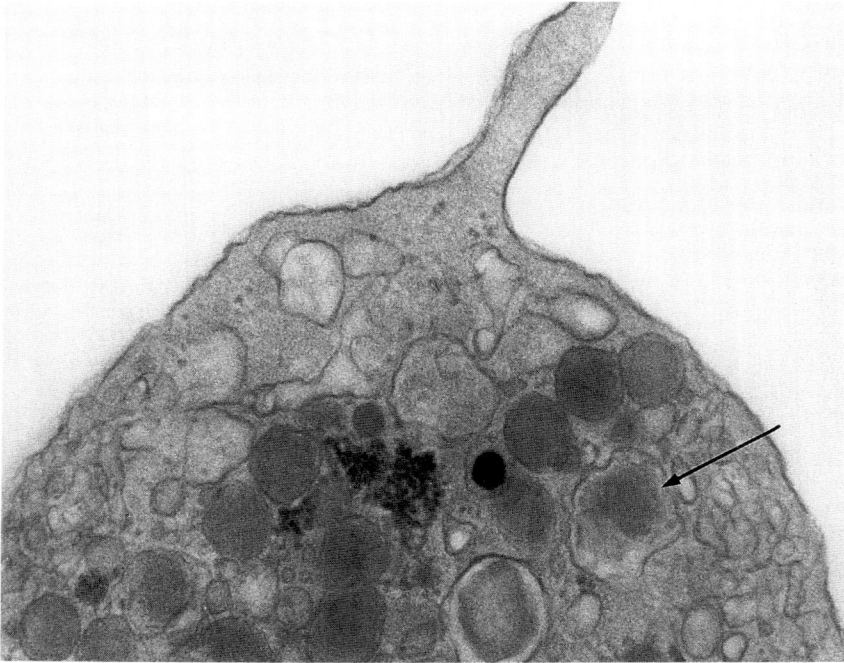

Fig. 8. A platelet 30 sec after exposure to 20 μg/ml OAG. Several of the α granules are now slightly swollen and the contents appear diluted (arrows). It is uncertain whether the perturbation of the membrane needed for fusion leads to an influx of water or whether a pump operates to swell the granules in readiness for fusion. Magnification × 45,000.

Studies using inhibitors of PKC have provided additional support for a role for this enzyme in granule membrane fusion. 1-(5-Isoquinolinylsulfonyl)-2-methylpiperazine (H-7), an inhibitor of PKC, prevented the granule membrane fusion stimulated by PMA and OAG (Fig. 7; Gerrard *et al.*, 1989a). In contrast, the structurally similar *N*-(2-guanidinoethyl)-5-isoquinolinylsulfonamide (HA 1004), with similar properties to H-7 (except that it is not a PKC inhibitor), failed to inhibit granule membrane fusion. Similarly, H-7 inhibited serotonin secretion stimulated by OAG or by OAG combined with the calcium ionophore A23187, whereas HA 1004 was ineffective (Fig. 11). Sphingosine, another inhibitor of PKC, also blocked PMA- and OAG-stimulated granule membrane fusion. Because H-7 and sphingosine may inhibit other protein kinases, including tyrosine kinases, as well as inhibiting PKC, the use of these inhibitors alone does not rule out a role for tyrosine kinases. Neverthe-

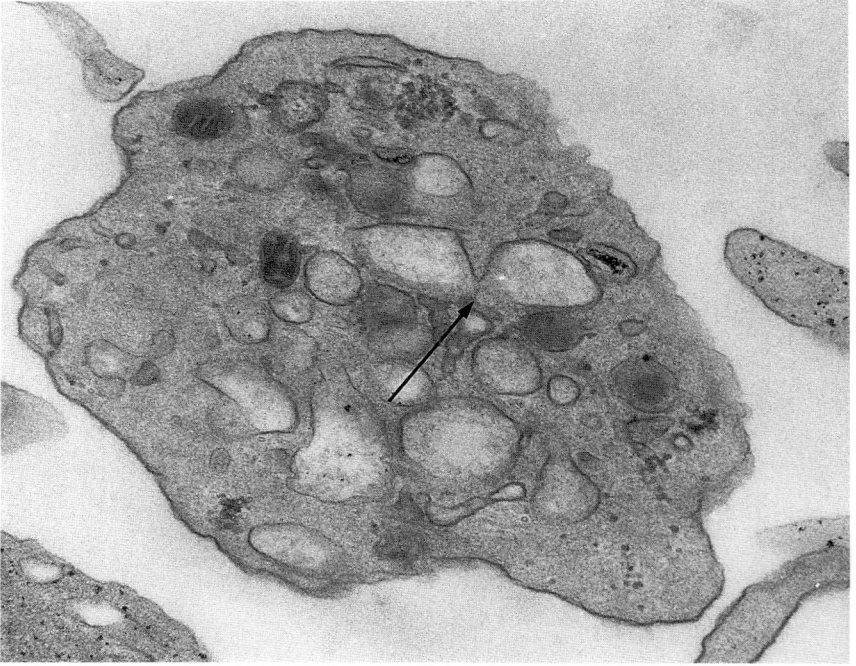

Fig. 9. A platelet exposed to 20 μg/ml OAG for 2 min. The arrow shows fusion occurring between two swollen granules. A very thin remnant of the membrane still separates the two granules. Magnification × 58,000.

less, when viewed together with observations employing two separate and selective PKC activators (PMA and OAG), the results strongly implicate PKC as critical to the fusion of granule and SCCS membranes, which is necessary for platelet granule secretion. Parallel studies using inhibitors of Na^+/H^+ transport (amiloride and dimethyl amiloride) suggest Na^+/H^+ transport is not involved in secretion at the level of granule membrane fusion (Gerrard *et al.*, 1989a).

The mechanism by which PKC activation promotes the membrane fusion necessary for exocytosis is poorly understood. A major substrate for PKC in platelets is pleckstrin, a 40 kDa cytoplasmic protein. Its function is unknown, and its potential relationship to granule membrane fusion uncertain. Furthermore, the time course of pleckstrin phosphorylation is more rapid than the time course for granule membrane fusion (as visualized morphologically

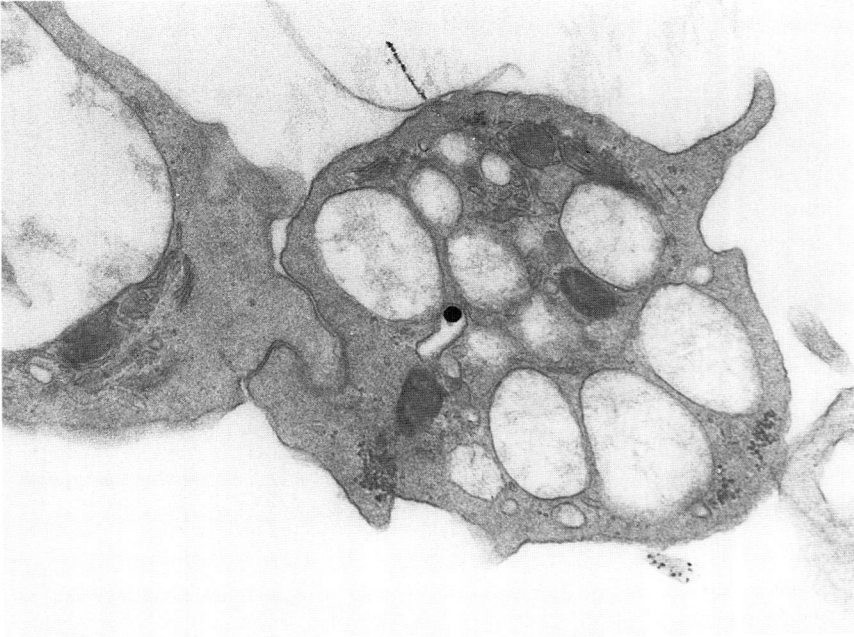

Fig. 10. Platelet exposed to 20 μg/ml OAG for 5 min. At this time, platelets contain multiple large vacuoles, as the fusion of granules has progressed. Magnification × 25,000.

using PMA or OAG), suggesting that additional steps subsequent to pleckstrin phosphorylation may be needed, if indeed pleckstrin is important to membrane fusion. Actin-binding protein is also a substrate for PKC (Carroll *et al.*, 1982). However, a primary role for actin-binding protein in membrane fusion seems unlikely.

B. Protein Kinase C, Histamine, and Granule Membrane Fusion

Recent studies have implicated PKC activation in the stimulation of platelet histamine synthesis (Saxena *et al.*, 1991). Agents that stimulate PKC produce a rise in the cytoplasmic concentration of histamine in platelets (Saxena *et al.*, 1989a,b, 1990). Staurosporine, an inhibitor of PKC, decreases PMA-induced histamine synthesis at similar concentrations to those that inhibit platelet

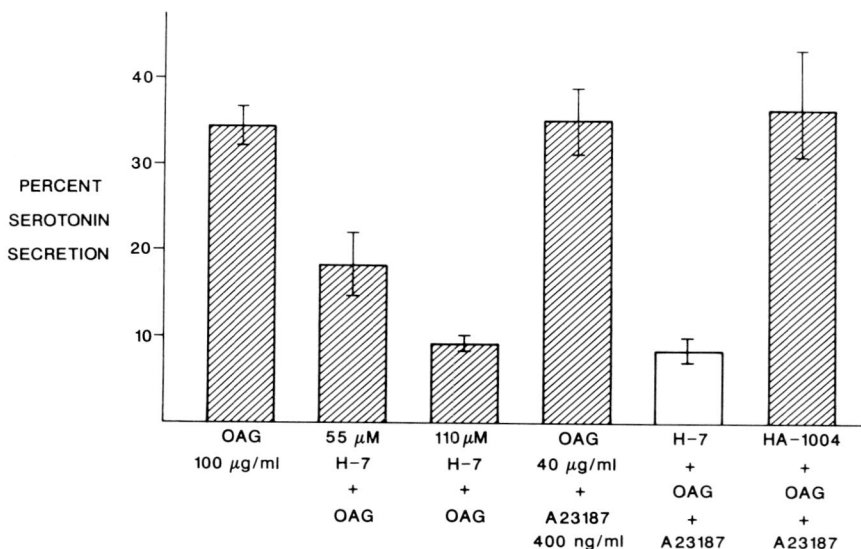

Fig. 11. The PKC inhibitor H-7 produced a concentration-dependent inhibition of serotonin secretion stimulated by OAG (100 μg/ml). H-7 also inhibited secretion induced by the combination of OAG (40 μg/ml) and A23187 (400 ng/ml). The related compound that is not an inhibitor of PKC (HA-1004) was not effective in inhibiting secretion.

aggregation and pleckstrin phosphorylation. Inhibition of the histamine-synthesizing enzyme, histidine decarboxylase, blocks the PMA-stimulated rise in platelet histamine and inhibits PMA-induced aggregation. An antagonist of histamine, binding to its intracellular receptor, *N,N*-diethyl-2-(4-[phenylmethyl]phenoxy)ethanamine HCl (DPPE), blocks PMA-stimulated platelet aggregation (Fig. 12) without affecting histamine synthesis or pleckstrin phosphorylation (Saxena *et al.*, 1989a). DPPE also inhibited granule membrane fusion stimulated by PMA, suggesting that histamine produced in response to PKC activation in platelets may play a role in promoting granule and SCCS membrane fusion (McNicol *et al.*, 1989). Furthermore, the addition of histamine to permeabilized platelets reversed the inhibitory effect of DPPE (Fig. 13). These findings indicate a role for intracellular histamine in promoting membrane fusion. The precise nature of the role of histamine is uncertain, but recent studies suggest histamine may promote the phosphorylation of several platelet proteins by a tyrosine kinase or kinases (Rendu *et al.*, 1991).

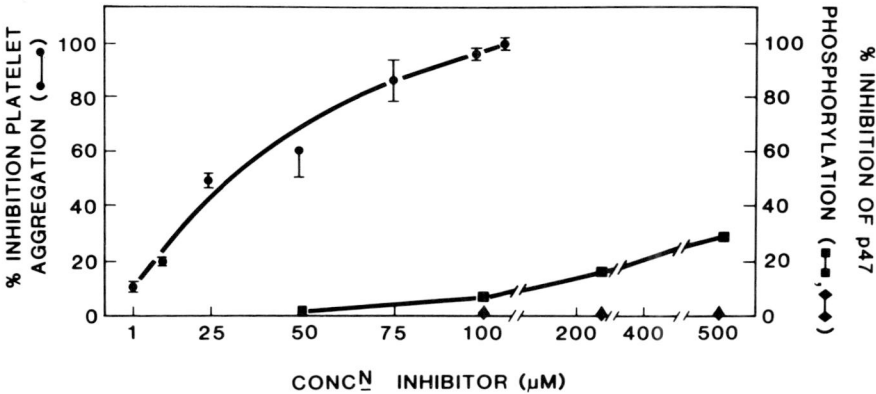

Fig. 12. The influence of *N,N*-diethyl-2-(4-[phenylmethyl]phenoxy)ethanamine HCl (DPPE) on platelet aggregation and pleckstrin (p47) phosphorylation stimulated by phorbol myristate acetate (PMA). Time after addition of PMA to the platelets: ●, 3 min; ■, 30 sec; ♦, 4 min.

C. Further Speculation on the Mechanism by Which Protein Kinase C Promotes Granule Membrane Fusion

Numerous processes may be involved in promoting the fusion of granule and SCCS membranes (Gerrard *et al.*, 1985). Alterations in the lipid or protein components of the two membranes may give rise to a perturbed bilayer, which has a greater tendency to undergo fusion. Alterations in the transport of ions and water across the granule membrane may lead to swelling of the granules and thus promote fusion. Alterations in the proteins on the cytoplasmic face of the SCCS and granule membranes may promote the adherence of granule and SCCS membranes as one step in the fusion process. Which of these processes are actually regulated by PKC is uncertain.

Recent attention has focused on the formation of a fusion pore as a critical step in exocytosis (Breckenridge and Almers, 1987). Understanding this process will require the identification of the proteins involved. Synaptophysin, a 38-kDa protein with four membrane-spanning domains (Sudhof and Jahn, 1991) is a candidate protein for the formation of such a fusion pore in synaptic vesicles. Synaptophysin has several cytoplasmic tyrosine residues and can undergo phosphorylation on tyrosine (Pang *et al.*, 1988). Recently, a protein in platelets has been identified that appears to have domains of homology to

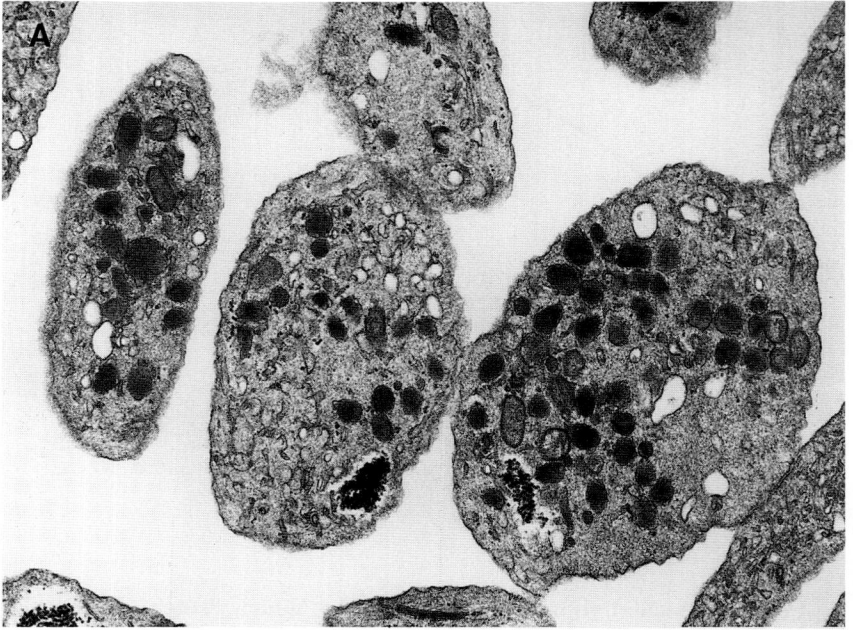

Fig. 13. (A) Platelets incubated with 125 μ*M* DPPE. The DPPE alone had little effect on platelet morphology. Magnification × 20,000. (B) Platelets incubated unstirred with PMA (200 n*M*). Magnification × 15,000. (C) Platelets treated with 125 μ*M* DPPE before exposure to 200 n*M* PMA. The DPPE blocked the granule fusion and vesicle formation found with PMA. Magnification × 23,000. (D) Addition of 4 μ*M* histamine to permeabilized platelets that were pretreated with DPPE and then incubated with PMA partially reverse the inhibitory effect of DPPE, suggesting that histamine may have a role in promoting granule membrane fusion. Magnification × 15,000.

synaptophysin (Gerrard *et al.*, 1991). It is tempting to speculate that this protein, named granulophysin, might be involved in the formation of the exocytotic pore facilitating secretion from platelet-dense granules. Further studies are necessary to probe this possibility.

D. Phospholipase Activation and Platelet Secretion

Perturbation of membrane phospholipids by phospholipases can have a significant effect on membrane architecture and on the fusability of membranes. Histamine produced in response to collagen has been linked to phos-

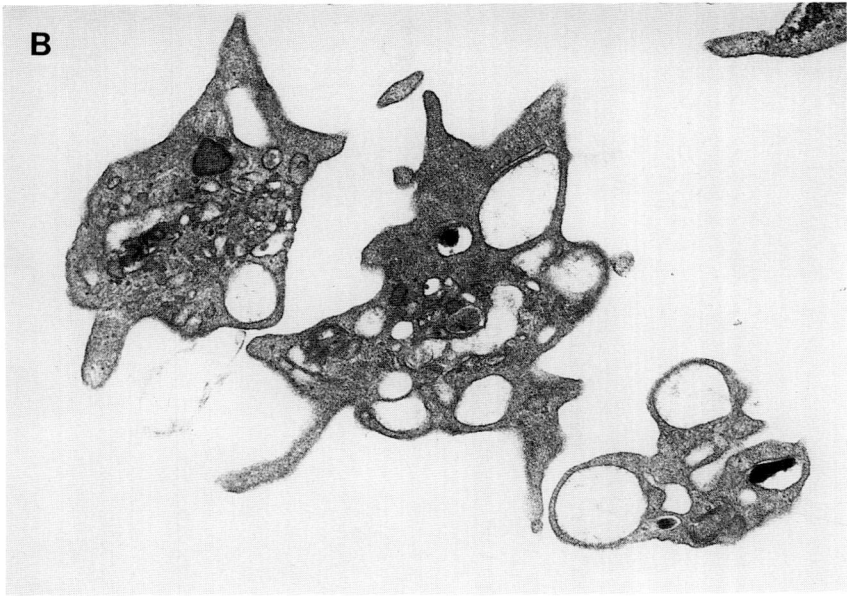

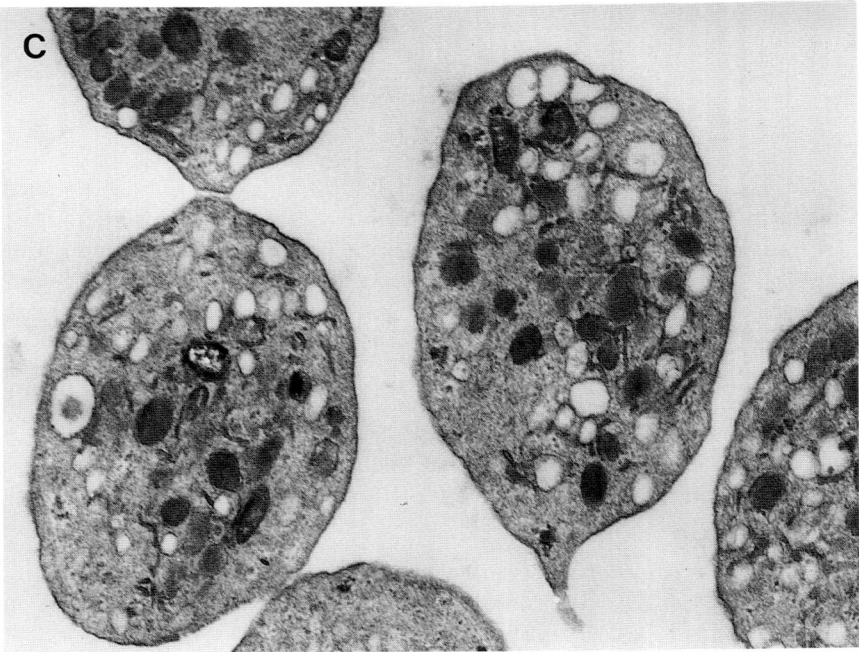

Fig. 13. (*Continued*)

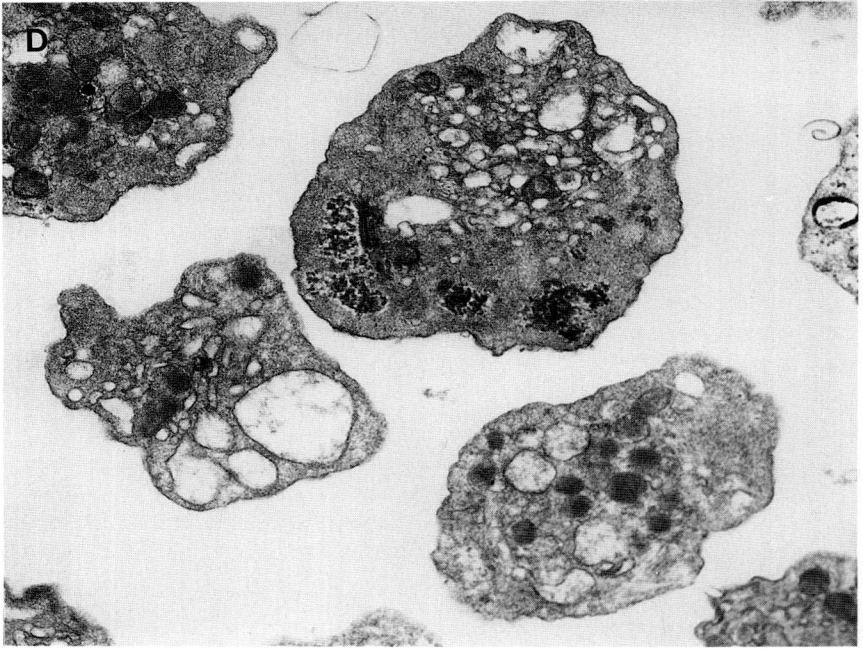

Fig. 13. (*Continued*)

pholipase A_2 activation (Saxena *et al.*, 1990; McNicol *et al.*, 1991). Such activation yields lysophospholipids, which have fusogenic properties. It is, therefore, possible that this process may be important to the promotion of granule fusion. Phospholipase A_2 activation, however, also occurs after stimulation of platelets with A23187 under conditions where there is no increase in platelet histamine (Saxena *et al.*, 1991). Multiple mechanisms may be involved in this case.

E. The Interaction of Calcium and Protein Kinase C in Regulating Secretion

The combination of an agonist that stimulates granule centralization (by increasing cytosolic calcium) and one that promotes granule membrane fusion (by activating PKC) synergistically promotes platelet secretion. This was first

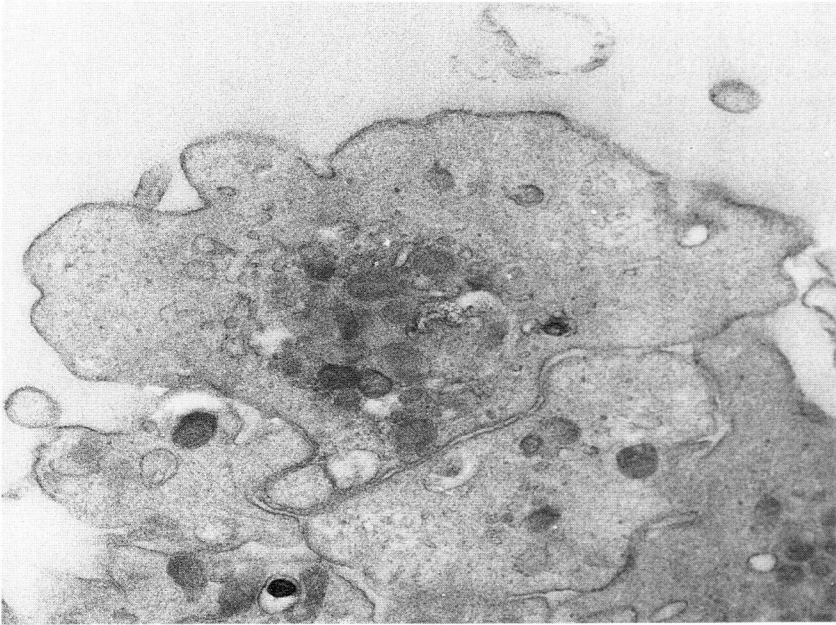

Fig. 14. Platelets exposed to both OAG and A23187 show granule centralization without vesicle formation. This presumably results because the contracting actin and myosin have squeezed the swollen and fused granules so that the contents are extruded and the vesicles are no longer visible. Magnification × 25,000.

shown for platelets in platelet-rich plasma using a combination of prostaglandin G_2 (PGG_2) and PMA (Gerrard *et al.*, 1976). Under the experimental conditions, PGG_2 alone produced 10.1% serotonin secretion; PMA alone produced 11.1% serotonin secretion. Half of the concentration of PGG_2 combined with half of the concentration of PMA, however, gave 42.6% serotonin secretion. Kaibuchi *et al.* (1983) subsequently showed similar synergism when the calcium ionophore A23187 was combined with the PKC activator OAG. These results provide important support for the overall concept that granule centralization and the fusion of granule and SCCS membranes synergize to promote secretion.

One of the puzzling aspects of the morphological changes that accompany secretion is the virtual absence of swollen vesicular structures accompanying secretion seen with ADP, thromboxane A_2, or collagen. Thus, even though fusion of granules with the SCCS can be seen readily with PMA, OAG, and thrombin, it appears to be difficult, morphologically, to visualize the process when it occurs with ADP, thromboxane A_2, or collagen, even when considerable secretion of serotonin occurs. The reason for the difficulty in seeing these vesicular structures appears to be explained by results of experiments that combine OAG and A23187. Under conditions where OAG and A23187 are combined and a synergistic effect to promote secretion occurs, the effect of the OAG to stimulate the formation of large vesicular structures is no longer seen (Fig. 14). It appears that the actin–myosin contraction promoted by the A23187-mediated elevation in cytoplasmic calcium squeezes the vesicular structures so that their contents are extruded to the extracellular milieu through the SCCS channels (Gerrard *et al.*, 1989b). During this process, the vesicular structures are constricted to the point where they are no longer readily visible.

F. Platelet Secretion in Platelet Aggregates

Secretion of platelet granule contents can either be associated with or independent of platelet aggregation. With a strong agonist such as thrombin, the presence of aggregation appears to have little impact on the extent of secretion. With weaker agonists the extent of secretion may vary considerably depending on the conditions used and the extent of aggregation. The precise role of aggregation (or perhaps the occupancy of adhesive protein receptors) in promoting secretion has not been fully explained.

Morphological changes in platelets attached to other platelets in doublets, triplets, or larger aggregates are different from those observed in single cells. In single cells, the granules move to the center of individual platelets in response to agonist stimulation. In contrast, in attached platelets granules are moved to the site of cell-to-cell attachment (Gerrard *et al.*, 1979; Gerrard *et al.*, 1987), where the granules could potentially fuse directly with the plasma membrane during the process of exocytosis.

The change in orientation of the actin–myosin contraction in single versus attached platelets has potentially significant implications for the mechanism of secretion. It is uncertain, for example, whether the secretion of granule contents occurs primarily toward the attached surface, the nonattached surface, or both. Secretion occurring toward the attached surface would tend to drive a high concentration of adhesive proteins (in essence, the "glue") into the site of cell-to-cell attachment. Secretion toward the unattached surface might be important for the recruitment of additional platelets into an aggregate. The

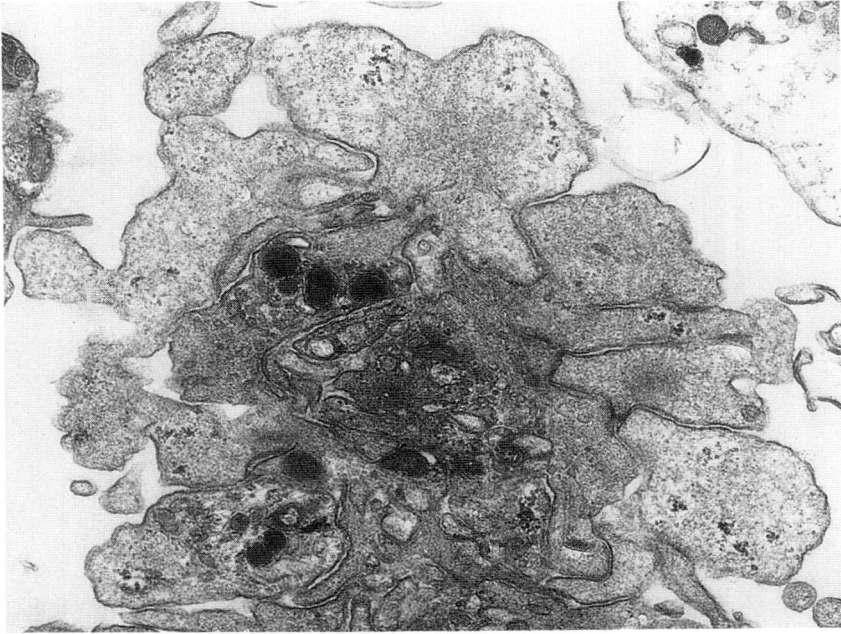

Fig. 15. A small platelet aggregate in platelets stirred with 1.8 μg/ml collagen demonstrates that within an aggregate the peripheral regions are largely depleted of granules, whereas the central region still contains some granules. Magnification × 15,000.

morphology of platelets in small and large clumps of stimulated cells is shown in Figs. 15 and 16.

III. FINAL COMMENTS

A. Are There Multiple Mechanisms Regulating Secretion?

Many biological functions are characterized by duplicate or back-up activation mechanisms. Such duplication may also occur in the mechanism of platelet secretion. For example, the calcium ionophore can stimulate secretion under conditions where there appears to be no change—or even a decrease—

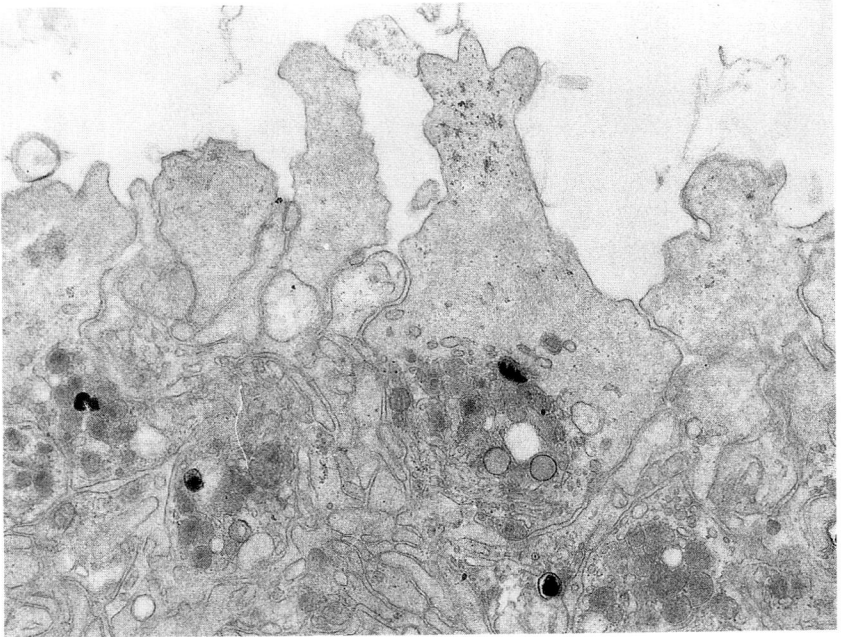

Fig. 16. Platelets stirred with 10 μM IP$_3$ after permeabilization using saponin show a similar movement of granules toward the center of the aggregate while parts of the platelets at the periphery of the aggregate are largely depleted of granules. The results of these observations suggest that the movement of the granules in aggregated platelets is now oriented toward sites of adhesion rather than toward the centers of individual cells. Magnification × 13,000.

in platelet histamine (Saxena *et al.*, 1991). The result suggests the presence of a nonhistamine pathway to stimulate granule membrane fusion. Conditions where histamine appears to play a lesser role may be conditions where a large activation of phospholipases and dramatic membrane perturbation occur independent of histamine synthesis. These and other aspects of the platelet secretory pathway are aspects which await further research.

B. Summary

1. Platelet secretion is facilitated by the combination of an influx of calcium producing actin–myosin contraction and granule redistribution in the platelet.

2. Activation of PKC leads to fusion of granule and SCCS membranes. The mechanism of the effect of PKC is uncertain, but effects of histamine and the activation of phospholipase A_2 are implicated.

ACKNOWLEDGMENT

We thank the MRC for their support through Grant MA7396 to Jon M. Gerrard.

REFERENCES

Bettex-Galland, M., and Luscher, E. F. (1959). Extraction of an actomyosin-like protein from human thrombocytes. *Nature* **184,** 276–277.

Breckenridge, L. J., and Almers, W. (1987). Final steps in exocytosis observed in a cell with giant secretory granules. *Proc. Natl. Acad. Sci. USA* **84,** 1945–1949.

Carroll, R. C., Butler, R. G., Morris, P. A., and Gerrard, J. M. (1982). Separable assembly of platelet pseudopodal and contractile cytoskeletons. *Cell* **30,** 385–393.

Castagna, M., Takai, Y., Kaibuchi, K., Sano, K., Kikkawa, U., and Nishizuka, Y. (1982). Direct activation of calcium-activated phospholipid-dependent protein kinase by tumor-promoting phorbol esters. *J. Biol. Chem.* **257,** 7847–7851.

Dabrowska, R., and Hartshorne, D. J. (1978). A Ca^{2+}- and modulator-dependent myosin light chain kinase from nonmuscle cells. *Biochem. Biophys. Res. Commun.* **85,** 1352–1359.

Estensen, R. D., and White, J. G. (1974). Ultrastructural features of the platelet response to phorbol myristate acetate. *Am. J. Pathol.* **74,** 441–452.

Friesen, L. L., and Gerrard, J. M. (1985). The effects of 1-oleoyl-2-acetylglycerol on platelet protein phosphorylation and platelet ultrastructure. *Am. J. Pathol.* **121,** 79–87.

Gerrard, J. M., and White, J. G. (1976). The structure and function of the platelets with emphasis on their contractile nature. *Pathobiol. Annu.* **6,** 31–58.

Gerrard, J. M., White, J. G., and Rao, G. H. R. (1974). Effects of the ionophore A23187 on blood platelets. II. Influence on ultrastructure. *Am. J. Pathol.* **77,** 151–166.

Gerrard, J. M., Townsend, D., Stoddard, S., Witkop, C. J., and White, J. G., (1976). The influence of prostaglandin G_2 on platelet ultrastructure and platelet secretion. *Am J. Pathol.* **86,** 99–116.

Gerrard, J. M., Schollmeyer, J. V., Phillips, D. R., and White, J. G. (1979). α-Actinin deficiency in Thrombasthenia. Possible identity of α-actinin and glycoprotein III. *Am. J. Pathol.* **94,** 509–528.

Gerrard, J. M., Schollmeyer, J. V., and White, J. G. (1981). The role of contractile proteins in the function of the platelet surface membrane. *Cell Surface Rev.* **7,** 217–252.

Gerrard, J. M., Israels, S. J., and Friesen, L. L. (1985). Protein phosphorylation and platelet secretion. *Nouv. Rev. Franc. Hematol.* **27,** 267–273.

Gerrard, J. M., Carroll, R. C., Israels, S. J. and Beattie, L. L. (1987). Protein phosphorylation. *In* "Platelets in Biology and Pathology III" (D. E. MacIntyre and J. L. Gordon, eds.), pp. 317–351. Elsevier, New York.

Gerrard, J. M., Beattie, L. L., Park, J., Israels, S. J., McNicol, A., Lint, D., and Cragoe, E. J., Jr. (1989a). A role for protein kinase C in the membrane fusion necessary for granule secretion. *Blood* **74**, 2405–2413.

Gerrard, J. M., McNicol, A., Klassen, D., and Israels, S. J. (1989b). Protein phosphorylation and its relation to platelet function. *In* "Blood Cells and Arteries in Hypertension and Atherosclerosis" (P. Meyer and P. Marche, eds.), pp. 93–114. Raven Press, New York.

Gerrard, J. M., Lint, D., Sims, P. H., Wiedmer, T., Fugate, R. D., McMillan, E., Robertson, C., and Israels, S. J. (1991). Identification of a platelet-dense granule membrane protein that is deficient in a patient with the Hermansky-Pudlak syndrome. *Blood* **77**, 101–12.

Ginsberg, M. H., Taylor, L., and Painter, R. G. (1980). The mechanism of thrombin-induced platelet factor 4 secretion. *Blood* **55**, 661–668.

Grette, K. (1962). Studies on the mechanism of thrombin-catalyzed hemostatic reactions in blood platelets. *Acta Physiol. Scand. Suppl.* **195**, 1–93.

Hathaway, D. R., and Adelstein, R. S. (1979). Human platelet myosin light chain kinase requires the calcium-binding protein calmodulin for activity. *Proc. Natl. Acad. Sci. USA* **76**, 1653–1657.

Holmsen, H., and Day, H. J. (1968). Thrombin-induced platelet release reaction and platelet lysosomes. *Nature (London)* **219**, 760–761.

Holmsen, H., and Day, H. J. (1970). The selectivity of the thrombin-induced platelet release reaction: Subcellular localization of released and retained constituents. *J. Lab. Clin. Med.* **75**, 840–855.

Kaibuchi, K., Takai, Y., Sawamura, M., Hosijima, M., Fijikura, T., and Nishizuka, Y. (1983). Synergistic functions of protein phosphorylation and calcium mobilization in platelet activation. *J. Biol. Chem.* **258**, 6701–6704.

Lebowitz, E. A., and Cook, R. (1978). Contractile properties of actomyosin from human blood platelets. *J. Biol. Chem.* **253**, 5443–5447.

McNicol, A., Saxena, S. P., Brandes, L. J., and Gerrard, J. M. (1989). A role for intracellular histamine in the ultrastructural changes induced in platelets by phorbol esters. *Arteriosclerosis* **9**, 684–689.

McNicol, A., Saxena, S. P., Becker, A. B., Brandes, L. J., and Gerrard, J. M. (1991). Further studies on the role of intracellular histamine in platelet function. *Platelets* **2**, 215–221.

Naka, M., Nishikawa, M., Adelstein, R. S., and Hidaka, H. (1983). Phorbol ester-induced activation of human platelets is associated with protein kinase C phosphorylation of myosin light chain. *Nature* **306**, 490–492.

Painter, R. G., and Ginsberg, M. H. (1984). Centripetal myosin redistribution in thrombin-stimulated platelets: Relationship to platelet factor 4 secretion. *Exp. Cell Res.* **155**, 198–212.

Pang, D. T., Wang, J. K. T., Valtorta, F., Benfenati, F., and Greengard, P. (1988). Protein tyrosine phosphorylation in synaptic vesicles. *Proc. Natl. Acad. Sci. USA* **85**, 762–766.

Rendu, F., McNicol, A., Saleun, S., and Gerrard, J. M. (1991). Histamine formed in stimulated human platelets plays a role in tyrosine kinase activation. *Thrombos. Haemostas.* **65**, 729.

Rittenhouse-Simmons, S. (1979). Production of diglyceride from phosphatidylinositol in human platelets. *J. Clin. Invest.* **63**, 580–587.

Saxena, S. P., Brandes, L. J., Becker, A., Simons, K., and Gerrard, J. M. (1989a). Histamine is an intracellular messenger promoting platelet aggregation. *Science* **243**, 1596–1599.

Saxena, S. P., McNicol, A., Brandes, L. J., Becker, A. B., and Gerrard, J. M. (1989b). Histamine formed in stimulated human platelets is cytoplasmic. *Biochem. Biophys. Res. Commun.* **164**, 164–168.

Saxena, S. P., McNicol A., Brandes L. J., Becker A. B., and Gerrard J. M. (1990b). A role for intracellular histamine in collagen-induced platelet aggregation. *Blood* **75**, 407–414.

Saxena, S. P., Robertson, C., Becker, A. B., and Gerrard, J. M. (1991). Synthesis of intracellular histamine in platelets is associated with activation of protein kinase C but not with calcium mobilization. *Biochem. J.* **273**, 405–408.

Stenberg, P. E., Shuman, M. A., Levine, S. P., and Bainton, D. F. (1984). Redistribution of alpha-granules and their contents in thrombin-stimulated platelets. *J. Cell Biol.* **98**, 748–760.

Stenberg, P. E., and Bainton, D. F. (1986). Storage organelles in platelets and megakaryocytes. *In* "Biochemistry of Platelets" (D. R. Phillips and M. A. Shuman, eds.), pp. 257–294. Academic Press, Montreal.

Sudhof, T. C., and Jahn, R. (1991). Proteins of synaptic vesicles involved in exocytosis and membrane recycling. *Neuron* **6**, 665–667.

White, J. G. (1968a). Fine structural alterations induced in platelets by adenosine diphosphate. *Blood* **31**, 604–622.

White, J. G. (1968b). The transfer of thorium particles from plasma to platelets and platelet granules. *Am. J. Pathol.* **53**, 567–575.

White, J. G. (1970). A search for the platelet secretory pathway using electron-dense tracers. *Am. J. Pathol.* **58**, 31–49.

White, J. G. (1972). Uptake of latex particles by blood platelets: Phagocytosis or sequestration? *Am. J. Pathol.* **69**, 439–458.

White, J. G. (1974). Electron microscopic studies of platelet secretion. *Prog. Haemostas. Thrombos.* **2**, 49–98.

White, J. G., and Estensen, R. D. (1972). Degranulation of discoid platelets. *Am. J. Pathol.* **68**, 289–302.

Zucker, M. B., and Borelli, J. (1955). Relationship of some blood clotting factors to serotonin release from washed platelets. *J. Appl. Physiol.* **7**, 432–442.

7

GTP-Binding Proteins and Formation of Secretory Vesicles

ANJA LEYTE, FRANCIS A. BARR, AND WIELAND B. HUTTNER

Institute for Neurobiology
University of Heidelberg
Heidelberg, Germany

SHARON A. TOOZE,

Cell Biology Program
European Molecular Biology Laboratory
Heidelberg, Germany

147

I. INTRODUCTION: SORTING IN AND VESICLE FORMATION FROM
THE TRANS-GOLGI NETWORK

Protein secretion is a common property of most animal cells. Two pathways of protein secretion are known. In the ubiquitous constitutive secretory pathway, proteins are continuously released into the environment without intracellular storage and in the absence of external stimuli. The regulated secretory pathway, on the other hand, functions only in certain specialized cell types such as exocrine cells, endocrine cells, and neurons. It involves the selective sorting of a subset of secretory proteins into highly specialized storage organelles, the secretory granules, which fuse with the plasma membrane to release their content only in response to an external signal (for reviews, see Burgess and Kelly, 1987; Huttner and Tooze, 1989; Miller and Moore, 1990).

Until constitutive and regulated secretory proteins reach the trans-Golgi network (TGN), their intracellular transport route appears to be the same (reviewed in Griffiths and Simons, 1986). In the TGN, the two pathways of secretion diverge, and two distinct populations of secretory vesicles are formed: constitutive secretory vesicles (CSVs) and immature secretory granules (ISGs) (Orci *et al.*, 1987; Tooze *et al.*, 1987; Tooze and Huttner, 1990). This implies the existence of a mechanism by which regulated secretory proteins destined to ISGs are segregated from constitutive secretory proteins destined to CSVs. Various morphological and biochemical lines of evidence suggest that the selective aggregation of regulated secretory proteins in the TGN is a key step in the sorting process (Burgess and Kelly, 1987; Gerdes *et al.*, 1989; Tooze *et al.*, 1989; Chanat and Huttner, 1991). The dense-cored aggregates formed by regulated secretory proteins exclude constitutive secretory proteins.

Besides the sorting of the secretory protein cargo, the formation of CSVs and ISGs from the TGN involves a series of events that occur whenever a vesicle forms from a donor compartment. These include (1) the segregation of cargo proteins from resident proteins of the donor compartment; (2) the assembly of the membrane components characteristic of a given vesicle (certain lipids, membrane proteins which, e.g., determine the intracellular traffic of the vesicle); (3) the formation of a membrane "bud"; (4) the assembly of a coat on the cytoplasmic surface of the bud; and (5) the pinching off of the vesicle from the donor compartment (scission).

This chapter discusses recent data on post-TGN vesicle biogenesis obtained with a strategy based on the biochemical identification of CSVs and ISGs and the use of a cell-free system to study their formation (Tooze and Huttner, 1990). We focus, in particular, on the role of GTP-binding proteins in this process, and speculate on the number and nature of events during ISG and CSV biogenesis in which these proteins participate as regulatory components.

A. Biochemical Characterization of Post-trans-Golgi Network Secretory Vesicles

In the TGN, a subset of secretory proteins is post-translationally modified by sulfation on tyrosines (Baeuerle and Huttner, 1987) and on carbohydrate chains (Kimura *et al.*, 1984). The specificity of these two types of sulfation as TGN modifications has been exploited to selectively label marker molecules for both the constitutive and regulated pathway of secretion as they pass through the TGN, and to detect their exit from the TGN (Tooze and Huttner, 1990). Cells of the neuroendocrine line PC12 contain two major tyrosine-sulfated proteins, chromogranin B (secretogranin I) and secretogranin II (SgII) that have been shown to be efficiently targeted to secretory granules (Lee and Huttner, 1983; Rosa *et al.*, 1985; Rosa *et al.*, 1989; Gerdes *et al.*, 1989). These characteristics make chromogranin B and SgII ideal markers for the regulated secretory pathway. PC12 cells also synthesize one major heparan sulfate proteoglycan (hsPg) (Schubert *et al.*, 1988; Gowda *et al.*, 1989), that can be used as a marker for the constitutive pathway of secretion: it is excluded from secretory granules and its secretion is blocked at 20°C (Tooze and Huttner, 1990). The latter property has been shown to be a hallmark for proteins traveling by the constitutive pathway (Matlin and Simons, 1983). Furthermore, sulfate labeling of secretogranins and the hsPg can be used to selectively monitor their post-TGN transport via the regulated and constitutive pathway, respectively. For this purpose, however, it is essential to separate ISGs and CSVs from the TGN and from each other, for instance, by physical separation of these cellular compartments. An outline for a method to achieve this follows.

To biochemically identify ISGs and CSVs, and to separate them from their donor compartment, the TGN, the following method has been designed (Tooze and Huttner, 1990; Tooze and Huttner, 1992). PC12 cells are pulse-labeled for 5 min with [^{35}S]sulfate, with or without subsequent chase, and then homogenized. Given the time needed for sulfate uptake and activation, plus translocation of activated sulfate (approximately 2 min; Baeuerle and Huttner, 1987), this allows for an effective labeling time of about 3 min. With the use of velocity-controlled sucrose gradient centrifugation, organelles in the postnuclear supernatant (PNS) prepared from the [^{35}S]sulfate-labeled cells are then fractionated to separate the TGN from post-TGN secretory vesicles, and the distribution of the sulfate-labeled marker proteins across the gradient is analyzed by sodium dodecyl sulfate polyacrylamide gel electrophoresis (SDS-PAGE) in conjunction with fluorography.

When no chase has been performed, this procedure yields a single peak of labeled proteins, including SgII and hsPg, in the bottom half of the gradient, coinciding with the peak of the TGN marker enzyme, sialyl transferase (Tooze and Huttner, 1990). [In general, chromogranin B is not analyzed as a marker for the regulated pathway because its migration on SDS-PAGE partly overlaps with that of the constitutively secreted hsPg (Tooze and Huttner, 1990).] This shows that after a 5-min [^{35}S]sulfate pulse, the marker proteins are still present in the TGN. After a 15-min chase, however, most of the labeled SgII and hsPg is found in the top half of the gradient. Apparently, the labeled proteins leave the TGN during this period and are then present in vesicles which, due to their smaller size, sediment more slowly than the TGN. The formation of post-TGN vesicles occurs rapidly, with a half-time of approximately 5 min (Tooze and Huttner, 1990).

B. Separation of Two Classes of Post-trans-Golgi Network Secretory Vesicles

On exit from the TGN, sulfate-labeled SgII and hsPg are packaged directly in two distinct classes of post-TGN vesicles, the ISGs and CSVs (Tooze and Huttner, 1990). These two populations can be separated from one another on the basis of their different buoyant densities in sucrose. For this purpose, vesicle-containing fractions from the velocity gradient just described are subjected to a second, equilibrium sucrose gradient centrifugation. Because ISGs (characterized by dense cores) have a significantly greater density than CSVs, this procedure yields two distinct peaks of labeled proteins: one higher in the gradient, containing the sulfate-labeled hsPg in CSVs, and one lower in the gradient, containing the sulfate-labeled SgII in ISGs (Tooze and Huttner, 1990).

C. Cell-Free Formation of Post-trans-Golgi Network Secretory Vesicles

To allow detailed investigations of the molecular mechanisms underlying post-TGN vesicle formation, a cell-free system has been developed (Tooze and Huttner, 1990). This system is based on the ability to monitor biochemically the transport and sorting of SgII and hsPg from the TGN to ISGs and CSVs, respectively. Protein sorting to the regulated and constitutive secretory pathways is reconstituted *in vitro* when a PNS, derived from PC12 cells pulse-labeled with [^{35}S]sulfate, is prepared in an iso-osmotic, low-ionic strength, sucrose-containing buffer at neutral pH and then supplemented with ATP and an ATP-regenerating system (Tooze and Huttner, 1990). The two types of post-TGN vesicles formed in the cell-free system were found to have similar properties to CSVs and ISGs formed *in vivo*. Unlike most existing cell-free transport assays (for review, see Goda and Pfeffer, 1989), this system measures only the formation of vesicles and not the transfer of a defined marker protein from a donor to an acceptor compartment, which is a multistep process involving both vesicle formation and vesicle fusion.

II. CURRENT RESEARCH: GTP-BINDING PROTEINS AND VESICLE FORMATION

Proteins that bind GTP fulfill a wide range of regulatory functions in all cell types (Bourne *et al.*, 1990). A feature that many of these proteins share is the ability to undergo a cycle of GTP binding and hydrolysis, during which they switch from an inactive (GDP-bound) to an active (GTP-bound) conformational state (Wittinghofer and Pai, 1991). In the case of the small ras-like GTP-binding proteins of the mammalian rab family, which have been implicated in vesicle fusion, it has been postulated that they form part of a proofreading machinery that ensures the correct targeting of vesicles to the appropriate acceptor membrane (for reviews, see Bourne, 1988; Balch, 1990; Goud and McCaffrey, 1991; Pfeffer, 1992). As will be discussed in detail, recent work indicates that GTP-binding proteins are also involved in vesicle formation, and that, in fact, several distinct classes of such proteins participate in this process.

A. Nonhydrolyzable Analogs of GTP Inhibit Cell-Free Vesicle Formation

An important finding obtained with the cell-free system derived from PC12 cells described in the previous section was that the formation of secretory vesicles from the TGN was inhibited by nonhydrolyzable analogs of GTP (Tooze *et al.*, 1990), implying that GTP hydrolysis is required for this process. It was shown that the formation of both ISGs and CSVs is inhibited by GTPγS. Another nonhydrolyzable analog of GTP, GMP-PMP, also inhibited the formation of post-Golgi vesicles, albeit much less effectively than GTPγS on a molar basis. This is in agreement with previous findings on the relative effects of GMP-PNP and GTPγS on cell-free intra-Golgi vesicle traffic (Melançon *et al.*, 1987). The inhibition of post-TGN vesicle formation by nonhydrolyzable GTP analogs was largely prevented by the addition of excess GTP and was specific with respect to the guanine moiety, as a nonhydrolyzable ATP analog, ATPγS did not affect the formation of either class of post-TGN vesicles (Tooze *et al.*, 1990).

With regard to the kinetics of cell-free vesicle formation in the presence of GTPγS, the following observations were made. After 60 min at 37°C in the presence of GTPγS, formation of ISGs was inhibited by approximately 40% and that of CSVs by approximately 80%. After 120 min, the inhibition of formation of both vesicle classes was about 50% (Tooze *et al.*, 1990). For CSVs, this is in agreement with data on the kinetics of inhibition by GTPγS of vesicle formation from other donor membranes (see Section II,C).

B. One Target for GTPγS in Post-trans-Golgi Network Vesicle Formation Is a Heterotrimeric G Protein

Trimeric G proteins, unlike the small ras-like GTPases (Kahn, 1991), are known to be affected by $[AlF_4]^-$ (Higashijima *et al.*, 1991). In the presence of GDP, this compound activates both inhibitory and stimulatory G proteins by mimicking the γ-phosphate group of GTP (Higashijima *et al.*, 1991). Using the cell-free system derived from PC12 cells, it was found that 40 μM $[AlF_4]^-$, but not $AlCl_3$ or KF, inhibited the formation of ISGs and CSVs to the same extent as 10 μM GTPγS (Barr *et al.*, 1991). Given the specificity of $[AlF_4]^-$, this was interpreted as being indicative of a role for trimeric G proteins in the regulation of post-TGN vesicle formation. Moreover, as both GTPγS and $[AlF_4]^-$ activate trimeric G proteins, it was inferred that such proteins exert, directly or indirectly, an inhibitory effect on vesicle formation.

To test this putative role of trimeric G proteins, the effect of purified G protein $\beta\gamma$ subunits on cell-free vesicle formation has been investigated (Barr *et al.*, 1991). The rationale for these experiments was based on the observations that $\beta\gamma$ subunits exert the opposite effect to nonhydrolyzable GTP analogues on the activation state of G protein α subunits (for reviews, see Gilman, 1987; Taylor, 1990; Birnbaumer *et al.*, 1990). This is presumably the result of a shift in the association equilibrium between the α and $\beta\gamma$ subunits toward the trimeric, inactive form. Indeed, it was found that $\beta\gamma$ subunits stimulated the exit of the constitutive and regulated secretory marker proteins from the TGN in a manner that was directly proportional to the amount of subunits added (Barr et al., 1991). Interestingly, the fold stimulation of exit from the TGN was greater for the hsPg, the constitutive secretory marker (4-fold at 400 nM $\beta\gamma$), than for SgII, the regulated secretory marker (2.5-fold at 400 nM $\beta\gamma$). In addition, it was found that the sorting of SgII into ISGs is perturbed by $\beta\gamma$ subunits (400 nM), resulting in the packaging of approximately 40% of SgII into vesicles with a buoyant density characteristic of CSVs. The packaging of hsPg into CSVs was not altered under these conditions. These findings may reflect differences in the mechanisms by which these two classes of vesicles form, or in the way in which their formation is regulated by trimeric G proteins.

The ADP-ribosylating toxins of *Vibrio cholerae* and *Bordetella pertussis* have been extensively used to characterize and distinguish trimeric G proteins. Cholera toxin is known to specifically ribosylate the α subunit of G_s (Cassel and Pfeuffer, 1978), while pertussis toxin acts on the α subunit of G_i and G_o (for review, see Kaziro et al., 1991). Accordingly, these toxins were employed to identify the G protein(s) regulating post-TGN vesicle formation. The approach used was to carry out toxin-dependent ADP ribosylation in a PC12-derived PNS under the conditions of cell-free vesicle formation, and then to analyse the distribution of ADP-ribosylated proteins using velocity sucrose gradient centrifugation (Barr et al., 1991). In the presence of pertussis toxin, an approximately 40 kDa protein was ADP-ribosylated, which on velocity sucrose gradient centrifugation was detected in two peaks; one in the top three fractions of the gradient, which are known to contain various subcellular organelles including post-TGN vesicles (Tooze and Huttner, 1990) and plasma membrane (Régnier-Vigouroux et al., 1991), and the other in the lower part of the gradient known to contain TGN membranes (Tooze and Huttner, 1990). The latter material was found to colocalize with the TGN on subsequent equilibrium sucrose gradient centrifugation. No ADP-ribosylated α subunit comigrating with the TGN was observed when cholera toxin was used (Barr et al., 1991). The conclusion drawn from these data was that a $G\alpha_i$ or $G\alpha_o$, or both, is present on TGN membranes of PC12 cells. In line with previous reports by other investigators (Ercolani et al., 1990; Stow et al., 1991), the pertussis toxin-sensitive G protein on PC12 cell TGN membranes is likely to be G_{i3}.

If pertussis toxin-sensitive G proteins regulate vesicle formation, one might expect that pertussis toxin-catalyzed ADP ribosylation prior to cell-free vesicle formation reduces the inhibition by GTPγS of this process. It has been shown that after ADP ribosylation by pertussis toxin, G protein α subunits are unable to interact with receptors, and thus are able to only undergo the basal, but not the receptor-mediated, GDP/GTP exchange (Birnbaumer *et al.*, 1990). ADP ribosylation by pertussis toxin therefore reduces the activation of α subunits. Preliminary experiments indeed showed that pertussis toxin-catalyzed ADP ribosylation partially prevents the inhibition of post-TGN vesicle formation by GTPγS (Barr *et al.*, 1991).

Work by other investigators also shows that heterotrimeric G proteins regulate secretory protein traffic through the Golgi complex and is consistent with the notion that this regulation occurs, at least in part, at the level of secretory vesicle formation. Thus, the G protein α subunit α_{i3}, which is present in virtually all cell types (Kaziro *et al.*, 1991), has been localized to the Golgi complex (Ercolani *et al.*, 1990; Stow *et al.*, 1991). Overexpression of this α subunit was found to inhibit the constitutive secretion of an hsPg in LLC-PK1 cells (Stow *et al.*, 1991). In addition, a stimulation of hsPg secretion was observed after exposure of the cells to pertussis toxin, the conclusion being that G_{i3} had become inactivated (Stow *et al.*, 1991). Taken together, it appears that pertussis toxin-sensitive G proteins function as part of a signal transducing machinery in the Golgi, mediating an inhibition of vesicle formation.

Trimeric G protein-mediated signal transduction pathways need not be restricted to Golgi membranes, but may operate on other endomembranes as well, and may involve other G proteins in addition to the pertussis toxin-sensitive ones.[1] Possibly, endomembrane signal transduction, in analogy to signal transduction at the plasma membrane (for review, see Gilman, 1987), involves a receptor of the seven transmembrane domain type. At present, we can only speculate on the function of signal transduction processes in vesicle formation; they could, for example, pass information about the lumenal content of an organelle to a cytoplasmic effector system, or they could function in the retention of organelle-specific resident proteins (Machamer, 1991).

C. Other GTP-Binding Proteins Involved in Vesicle Formation

The conclusion that GTP hydrolysis is required for vesicle formation (Tooze *et al.*, 1990) has been confirmed and extended by a more recent report that indicates that GTP hydrolysis plays a role in the formation of transport vesi-

---

[1]Since the submission of this manuscript, we have found (Leyte *et al.*, 1992) that multiple αi/αo and αs G-protein subunits are associated with the TGN. These G-protein subunits exert inhibitory and stimulatory effects on the formation of CSVs and ISGs.

cles from the endoplasmic reticulum (ER). In a cell-free system derived from yeast, this process has been found to be inhibited by nonhydrolyzable GTP analogues (Rexach and Schekman, 1991). The kinetics of inhibition by GTPγS of post-ER vesicle formation in the yeast cell-free system were similar to those observed for the formation of CVSs (Tooze *et al.*, 1990). In the yeast system, too, the degree of inhibition decreased at longer times of incubation. The similarity between the effects of GTPγS in the two cell-free systems supports the idea that GTP-binding proteins carry out a similar role in vesicle formation throughout the secretory pathway.

1. Small Ras-like GTP-Binding Proteins

Further work in the yeast cell-free system revealed that at least part of the requirement for GTP hydrolysis reflected the involvement of the small ras-like GTP-binding protein Sar1 (d'Enfert *et al.*, 1991). Moreover, small ras-like GTP-binding proteins of the mammalian ADP ribosylation factor (ARF) family (Kahn *et al.*, 1991) have been found to be among the components of the nonclathrin coat of intra-Golgi transport vesicles, referred to as coat proteins (Serafini *et al.*, 1991). Although no direct link between nonclathrin coat assembly and vesicle formation has yet to be reported, the interaction of two coat proteins, ARF and β-COP (Duden *et al.*, 1991), with Golgi membranes is affected under conditions known to inhibit vesicle formation, such as the presence of GTPγS (Donaldson *et al.*, 1991a,b; Serafini *et al.*, 1991). This points toward an involvement of ARF in vesicle formation, a suggestion consistent with the observation that a yeast mutant for ARF is defective in invertase secretion (Stearns *et al.*, 1990). Since the yeast GTP-binding protein Sar1 is homologous to mammalian ARF (Nakano and Muramatsu, 1989), it is tempting to speculate that in general, small ras-like GTP-binding proteins of the ARF/Sar1 family are involved in vesicle formation by regulating coat assembly at the level of the donor membrane, and that this regulation is mediated via GTP binding and hydrolysis.

2. GTP-Binding Motor Proteins

Studies on the *shibire* mutant of *Drosophila melanogaster* led to the identification of another class of GTP-binding proteins that may be involved in vesicle formation. The *shibire* mutant was originally identified as a mutant with alterations in the structure of the neuromuscular junction leading to paralysis (Poodry and Edgar, 1979). In this mutant the primary defect is at the level of endocytosis via clathrin-coated pits, which are unable to bud from the plasma membrane (Kosaka and Ikeda, 1983). The *shibire* gene has recently been cloned and sequenced (van der Bliek and Meyerowitz, 1991). The corresponding protein was found to be a homologue of rat dynamin (Chen *et al.*, 1991), a 100 kDa microtubule-binding protein with GTPase activity able to induce microtubule mobility *in vitro* (Shpetner and Vallee, 1989). Together, these findings raise the possibility that GTP-binding motor proteins such as dynamin may be involved in the scission of endocytotic vesicles from the plasma membrane.

Several observations suggest that such a role for these proteins need not be restricted to vesicle formation from the plasma membrane. Dynamin as well as the *shibire* gene product show homology to the yeast Vps1 protein, which is involved in protein sorting from the TGN to the vacuole and is a putative GTPase (Rothman *et al.*, 1990). Clathrin is thought to be involved not only in endocytosis, but also in the formation of vesicles from the TGN, which deliver lysosomal enzymes bound to mannose-6-phosphate receptors to endosomes (Kornfeld, 1987). In addition, patches of clathrin have been found on ISGs (Tooze and Tooze, 1986), and a species of clathrin light chain (LC_b) is enriched in cells and tissues with the regulated secretory pathway (Acton and Brodsky, 1990). If there indeed exists a clathrin-dependent mechanism for vesicle formation from the TGN, it is conceivable that this mechanism would employ a motor protein like dynamin, and therefore, GTP hydrolysis in the final scission event. By analogy, one may speculate on the existence of a similar GTP-binding protein as part of the "scission machinery" that operates in the nonclathrin-dependent mechanism of vesicle formation.

III. FINAL COMMENTS

As mentioned in the Introduction, the formation of vesicles from a donor compartment is a complex cascade of events. Given the distinct classes of GTP-binding proteins that appear to be involved in the overall process of vesicle formation, it is likely, but not proven, that these proteins exert functions in several of the individual steps in vesicle formation. In which step and how do these proteins act? In the following sections, various possibilities are discussed concerning secretory vesicle formation from the TGN.

A. The Role of Heterotrimeric G Proteins in Vesicle Formation

From the data previously described (Barr *et al.*, 1991; Stow *et al.*, 1991; see also footnote 1 and Leyte *et al.*, 1992), it appears that secretory vesicle formation from Golgi membranes is under negative control mediated by G_{i3}.

An inhibitory mechanism for the exit of constitutive and regulated secretory proteins from the TGN might be related to the obvious requirement for a regulatory system that prevents vesicles forming from the TGN unless appropriate cargo is available. Such a system might regulate the rate of vesicle formation to ensure, for example, (1) that secretory proteins destined to undergo post-translational modifications [e.g., sialylation (Roth *et al.*, 1985) and tyrosine sulfation (Baeuerle and Huttner, 1987)] only leave the TGN after having acquired these modifications, which can be crucial for the fate of a protein (e.g., Leyte *et al.*, 1991); or (2) that regulated secretory proteins only leave the TGN after having been segregated via aggregation (Chanat and Huttner, 1991) from constitutive secretory proteins. The lack of completion of these process might somehow be monitored by a transmembrane receptor, for instance by recognition of unmodified sites for posttranslational modification or of nonaggregated regulated secretory proteins. Such a putative transmembrane receptor might then interact with a trimeric G protein, stimulating GDP–GTP exchange on the G protein α subunit and hence promoting its activation. The activated α subunit in turn presumably interacts with an effector on the cytoplasmic side of the TGN membrane (see Fig. 1). Until the activated α subunit hydrolyzes its bound GTP to GDP, this effector will either reduce vesicle formation below the normal rate, or be prevented from maintaining vesicle formation at the normal rate. In this regard, it is interesting to

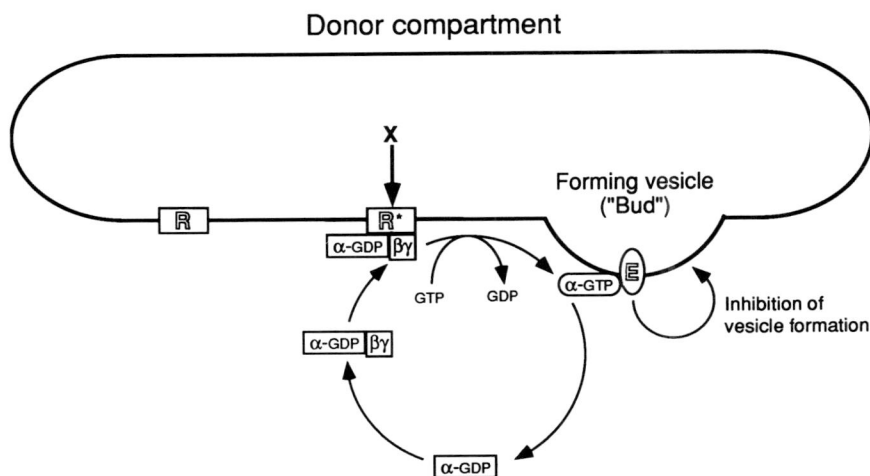

Fig. 1. Schematic representation of a signal transduction machinery in the membrane of the *trans*-Golgi network involved in the regulation of vesicle formation. X, ligand; R, receptor; R*, activated receptor; α, β, and γ, trimeric G protein subunits; E, effector. For detail, see text.

note that inhibition of tyrosine sulfation has been shown to retard the exit of a constitutive secretory protein from the TGN (Friederich *et al.*, 1988).

So far, there are no clues as to the identity of the effector system through which activated α subunits exert an inhibitory effect on post-TGN vesicle formation. However, if one extrapolates from recent observations made with intra-Golgi transport vesicles (Serafini *et al.*, 1991), the effector system may include small ras-like GTP-binding proteins such as ARF. An activated G protein could, for instance, directly or indirectly prevent ARF from recruiting coat proteins to the membrane of a nascent bud, as proposed by Serafini *et al.* (1991). An interaction between trimeric G proteins, ARF, and coat proteins is suggested from the observation that addition of βγ subunits inhibits the association of both ARF and β-COP with Golgi membranes (Donaldson *et al.*, 1991b).

B. Regulation of Coat Assembly

This leads to the regulation of vesicle formation via the control of coat assembly on the cytoplasmic surface of the forming vesicle. As mentioned before, small GTP-binding proteins have been shown to be part of the non-clathrin coat of intra-Golgi transport vesicles (Serafini *et al.*, 1991). This may also be true for post-TGN secretory vesicles. Although the nature of the coats of CSVs and ISGs is not precisely known (Tooze, 1991), the transient presence of these coats implies that their assembly and disassembly before and after scission, respectively, are somehow regulated. Both assembly and disassembly may involve the same small GTP-binding protein that could promote either coating or uncoating, depending on its GTP/GDP-bound state. Free coat subunits of the small GTP-binding protein family may only associate with the membrane when they are in the GTP-bound form. GTP hydrolysis might be linked to a proofreading step related to either coat polymerization or the proper assembly of the various components of the vesicle membrane. Such a "conditional" GTP-hydrolysis could, for instance, ensure the production of vesicles that contain all transmembrane and peripheral molecules required to define its intracellular destination.

C. Promotion of Scission

If scission, the final step in vesicle formation, is promoted by a motorlike protein of the dynamin family (see Section II,C,2), this step, too, might be a GTP-dependent process. Extrapolating from the apparent lack of scission of endocytotic vesicles in the *shibire* mutant of *Drosophila melanogaster*, a similar, though possibly CSV- or ISG-specific motor protein may recognize a completed bud on the TGN and link it to microtubules (Vallee and Shpetner, 1990). Subsequent GTP hydrolysis could then promote movement along these microtubules, and thereby the shearing off of the bud from the donor membrane.

D. Conclusion

It is becoming apparent that GTP-binding proteins (heterotrimeric G proteins, small ras-like proteins, and possibly motor proteins) are key components of the regulatory machinery involved in vesicle formation. Future work is needed to characterize these GTP-binding proteins, and a major goal is to identify the various components acting upstream and downstream of them, in particular, the receptor and effector systems.

ACKNOWLEDGMENTS

Anja Leyte was the recipient of an EMBO long-term fellowship. Wieland B. Huttner was supported by the Deutsche Forschungsgemeinschaft (SFB 317).

REFERENCES

Acton, S. L., and Brodsky, F. M. (1990). Predominance of clathrin light chain LC_b correlates with the presence of a regulated secretory pathway. *J. Cell Biol.* **111,** 1419–1426.

Baeuerle, P. A., and Huttner, W. B. (1987). Tyrosine sulfation is a trans-Golgi-specific protein modification. *J. Cell Biol.* **105,** 2655–2664.

Balch, W. E. (1990). Small GTP-binding proteins in vesicular transport. *Trends Biochem. Sci.* **15,** 473–477.

Barr, F. A., Leyte, A., Mollner, S., Pfeuffer, T., Tooze, S. A., and Huttner, W. B. (1991). Trimeric G proteins of the trans-Golgi network are involved in the formation of constitutive secretory vesicles and immature secretory granules. *FEBS Lett.* **294,** 239–243.

Birnbaumer, L., Abramowitz, J., and Brown, A. M. (1990). Receptor–effector coupling by G proteins. *Biochim. Biophys. Acta* **1031,** 163–224.

Bourne, H. R. (1988). Do GTPases direct membrane traffic in secretion? *Cell* **53,** 669–671.

Bourne, H. R., Sanders, D. A., and McCormick, F. (1990). The GTPase superfamily: A conserved switch for diverse cell functions. *Nature* **348,** 125–132.

Burgess, T. L., and Kelly, R. B. (1987). Constitutive and regulated secretion of proteins. *Annu. Rev. Cell Biol.* **3,** 243–293.

Cassel, D., and Pfeuffer, T. (1978). Mechanism of cholera toxin action: Covalent modification of the guanyl nucleotide-binding protein of the adenylate cyclase system. *Proc. Natl. Acad. Sci. USA* **75,** 2669–2673.

Chanat, E., and Huttner, W. B. (1991). Milieu-induced, selective aggregation of regulated secretory proteins in the trans-Golgi network. *J. Cell Biol.* **115**, 1505–1519.

Chen, M. S., Obar, R. A., Schroeder, C. C., Austin, T. W., Poodry, C. A., Wadsworth, S. C., and Vallee, R. B. (1991). Multiple forms of dynamin are encoded by *shibire*, a *Drosophila* gene involved in endocytosis. *Nature* **352**, 583–586.

Donaldson, J. G., Lippincott-Schwartz, J., and Klausner, R. D. (1991a). Guanine nucleotides modulate the effects of brefeldin A in semipermeable cells: Regulation of the association of a 110-kD peripheral membrane protein with the Golgi apparatus. *J. Cell Biol.* **112**, 579–588.

Donaldson, J. G., Kahn, R. A., Lippincott-Schwartz, J., and Klausner, R. D. (1991b). Binding of ARF and β-COP to Golgi membranes: Possible regulation by a trimeric G protein. *Science* **254**, 1197–1199.

Duden, R., Allan, V., and Kreis, T. (1991). Involvement of β-COP in membrane traffic through the Golgi complex. *Trends Cell Biol.* **1**, 14–19.

d'Enfert, C., Wuestehube, L. J., Lila, T., and Schekman, R. (1991). Sec12p-dependent membrane binding of the small GTP-binding protein Sar1p promotes formation of transport vesicles from the ER. *J. Cell Biol.* **114**, 663–670.

Ercolani, L., Stow, J. L., Boyle, J. F., Holtzman, E. J., Lin, H., Grove, J. R., and Ausiello, D. A. (1990). Membrane localization of the pertussis toxin-sensitive G protein subunits α_{i-2} and α_{i-3} and expression of a methallothionein-α_{i-2} fusion gene in LLC-PK1 cells. *Proc. Natl. Acad. Sci. USA* **87**, 4635–4639.

Friederich, E., Fritz, H-J., and Huttner, W. B. (1988). Inhibition of tyrosine sulfation in the trans-Golgi retards the transport of a constitutively secreted protein to the cell curface. *J. Cell Biol.* **107**, 1655–1667.

Gerdes, H.-H., Rosa, P., Phillips, E., Baeuerle, P. A., Frank, R., Argos, P., and Huttner, W. B. (1989). The primary structure of human secretogranin II, a widespread tyrosine-sulfated secretory granule protein that exhibits low pH- and calcium-induced aggregation. *J. Biol. Chem.* **264**, 12,009–12,015.

Gilman, A. G. (1987). G proteins: Transducers of receptor-generated signals. *Annu. Rev. Biochem.* **56**, 615–649.

Goda, Y., and Pfeffer, S. R. (1989). Cell-free systems to study vesicular transport along the secretory and endocytic pathways. *FASEB J.* **3**, 2488–2495.

Goud, B., and McCaffrey, M. (1991). Small GTP-binding proteins and their role in transport. *Curr. Opinion Cell Bio.* **3**, 626–633.

Gowda, D. C., Goossen, B., Margolis, R. K., and Margolis, R. U. (1989). Chondroitin sulfate and heparan sulfate proteoglycans of PC12 pheochromocytoma cells. *J. Biol. Chem.* **264**, 11436–11443.

Griffiths, G., and Simons, K. (1986). The trans-Golgi network: Sorting at the exit site of the Golgi complex. *Science* **234**, 438–443.

Higashijima, T., Graziano, M. P., Suga, H., Kainosho, M., and Gilman, A. G. (1991). ^{19}F and ^{31}P NMR spectroscopy of G protein α-subunits. *J. Biol. Chem.* **266**, 3396–3401.

Huttner, W. B., and Tooze, S. A. (1989). Biosynthetic protein transport in the secretory pathway. *Curr. Opinion in Cell Biol.* **1**, 648–654.

Kahn, R. A. (1991). Fluoride is not an activator of the smaller (20–25 kDa) GTP-binding proteins. *J. Biol. Chem.* **266**, 15595–15597.

Kahn, R. A., Kern, F. G., Clark, J., Gelmann, E. P., and Rulka, C. U. (1991). Human ADP-ribosylation factors. *J. Biol. Chem.* **266**, 2606–2614.

Kaziro, Y., Itoh, H., Kozasa, T., Nakafuku, M., and Satoh, T. (1991). Structure and function of signal-transducing GTP-binding proteins. *Annu. Rev. Biochem.* **60**, 349–400.

Kimura, J. H., Lohmander, L. S., and Hascall, V. C. (1984). Studies on the biosynthesis of cartilate proteoglycan in a model system of cultured chondrocytes from the swarm rat chondrosarcoma. *J. Cell Biochem.* **26**, 261–278.

Kornfeld, S. (1987). Trafficking of lysosomal enzymes. *FASEB J.* **1**, 462–468.

Kosaka, T., and Ikeda, K. J. (1983). Reversible blockage of membrane retrieval and endocytosis in the Garland cell of the temperature-sensitive mutant of *Drosophila melanogaster*, *shibire*[ts1]. *J. Cell Biol.* **97**, 499–507.

Lee, R. W. H., and Huttner, W. B. (1983). Tyrosine-O-sulfated proteins of PC12 pheochromocytoma cells and their sulfation by a tyrosylprotein sulfotransferase. *J. Biol. Chem.* **258**, 11326–11334.

Leyte, A., van Schijndel, H. B., Niehrs, C., Huttner, W. B., Verbeet, M. P., Mertens, K., and van Mourik, J. A. (1991). Sulfation of Tyr1680 of human blood coagulation factor VIII is essential for the interaction of factor VIII with von Willebrand factor. *J. Biol. Chem.* **266**, 740–746.

Leyte, A., Barr, F. A., Kehlenbach, R. H., and Huttner, W. B. (1992). Multiple trimeric G proteins on the *trans*-Golgi network exert stimulatory and inhibitory effects on secretory vesicle formation. *EMBO J.* **11**, 4795–4804.

Machamer, C. E. (1991). Golgi retention signals: Do membranes hold the key? *Trends Cell Biol.* **1**, 141–144.

Matlin, K. S., and Simons, K. (1983). Reduced temperature prevents transfer of a membrane glycoprotein to the cell surface but does not prevent terminal glycosylation. *Cell* **34**, 233–243.

Melançon, P., Glick, B. S., Malhotra, V., Weidman, P. J., Serafini, T., Gleason, M. L., Orci, L., and Rothman, J. E. (1987). Involvement of GTP-binding "G" proteins in transport through the Golgi stack. *Cell* **51**, 1053–1062.

Miller, S. G., and Moore, H.P. (1990). Regulated secretion. *Curr. Opinion Cell Biol.* **2**, 642–647.

Nakano, A., and Muramatsu, M. (1989). A novel GTP-binding protein, Sar1p, is involved in transport from the endoplasmatic reticulum to the Golgi apparatus. *J. Cell Biol.* **109**, 2677–2691.

Orci, L., Ravazzola, M., Amherdt, M., Perelet, A., Powell, S. K. Quinn, D. L., and Moore, H.-P. H. (1987). The *trans*-most cisternae of the Golgi complex: A compartment for sorting secretory and plasma membrane proteins. *Cell* **51**, 1039–1051.

Pfeffer, S. R. (1992). GTP-binding proteins in intracellular transport. *Trends Cell Biol.* **2**, 41–46.

Poodry, C. A., and Edgar, L. (1979). Reversible alterations in the neuromuscular junctions of *Drosophila melanogaster* bearing a temperature-sensitive mutation, *shibire*. *J. Cell Biol.* **81**, 520–527.

Régnier-Vigouroux, A., Tooze, S. A., and Huttner, W. B. (1991). Newly synthesized synaptophysin is transported to synaptic-like microvesicles via constitutive secretory vesicles and the plasma membrane. *EMBO J.* **10**, 3589–3601.

Rexach, M., and Schekman, R. (1991). Distinct biochemical requirements for the budding, targeting, and fusion of ER-derived transport vesicles. *J. Cell. Biol.* **114**, 219–229.

Rosa, P., Hille, A., Lee, R. W. H., Zanini, A., De Camilli, P., and Huttner, W. B. (1985). Secretogranins I and II: Two tyrosine-sulfated secretory proteins common to a variety of cells secreting peptides by the regulated pathway. *J. Cell Biol.* **101**, 1999–2011.

Rosa, P., Weiss, U., Pepperkok, R., Ansorge, W., Niehrs, C., Stelzer, E. H. K., and Huttner, W. B. (1989). An antibody against secretogranin I (chromogranin B) is packaged into secretory granules. *J. Cell Biol.* **109**, 17–34.

Roth, J., Taatjes, D. J., Lucocq, J. M., Weinstein, J., and Paulson, J. C. (1985). Demonstration of an extensive trans-tubular network continuous with the Golgi apparatus stack that may function in glycosylation. *Cell* **43**, 287–295.

Rothman, J. H., Raymond, C. K., Gilbert, T., O'Hara, P. J., and Stevens, T. H. (1990). A putative GTP-binding protein homologous to interferon-inducible Mx proteins performs an essential function in yeast protein sorting. *Cell* **61**, 1063–1074.

Schubert, D., Schroeder, R., LaCorbiere, M., Saitoh, T., and Cole, G. (1988). Amyloid β protein precursor is possibly a proteoglycan core protein. *Science* **241**, 223–241.

Serafini, T., Orci, L., Amherdt, M., Brunner, M., Kahn, R. A., and Rothman, J. E. (1991). ADP ribosylation factor is a subunit of the coat of Golgi-derived COP-coated vesicles: A novel role for a GTP-binding protein. *Cell* **67**, 239–253.

Shpetner, H. S., and Vallee, R. B. (1989). Identification of dynamin, a novel mechanochemical enzyme that mediates interactions between microtubules. *Cell* **59**, 421–432.

Stearns, T., Willingham, M. C., Botstein, D., and Kahn, R. A. (1990). ADP-ribosylation factor is functionally and physically associated with the Golgi complex. *Proc. Natl. Acad. Sci. USA* **87**, 1238–1242.

Stow, J. L., de Almeida, J. B., Narula, N., Holtzman, E. J., Ercolani, L., and Ausiello, D. A. (1991). A heterotrimeric G protein, $G\alpha_{i-3}$, on Golgi membranes regulates the secretion of a heparan sulfate proteoglycan in LLC-PK1 epithelial cells. *J. Cell Biol.* **114**, 1113–1124.

Taylor, C. W. (1990). The role of G proteins in transmembrane signaling. *Biochem. J.* **272**, 1–13.

Tooze, J., and Tooze, S. A. (1986). Clathrin-coated vesicular transport of secretory proteins during the formation of ACTH-containing secretory granules in AtT20 cells. *J. Cell Biol.* **103**, 839–850.

Tooze, J., Tooze, S. A., and Fuller, S. D. (1987). Sorting of progeny coronavirus from condensed secretory proteins at the exit from the trans-Golgi network of AtT20 cells. *J. Cell Biol.* **105**, 1215–1226.

Tooze, J., Kern, H. F., Fuller, S. D., and Howell, K. E. (1989). Condensation-sorting events in the rough endoplasmic reticulum of exocrine pancreatic cells. *J. Cell Biol.* **109**, 35–50.

Tooze, S. A. (1991). Biogenesis of secretory granules. Implications arising from the immature secretory granule in the regulated pathway of secretion. *FEBS Lett.* **285**, 220–224.

Tooze, S. A., and Huttner, W. B. (1990). Cell-free protein sorting to the regulated and constitutive secretory pathways. *Cell* **60**, 837–847.

Tooze, S. A., and Huttner, W. B. (1992). Cell-free formation of immature secretory granules and constitutive secretory vesicles from the trans-Golgi network. *Methods Enzymol.* **219**, 81–93.

Tooze, S. A., Weiss, U., and Huttner, W. B. (1990). Requirement for GTP hydrolysis in the formation of secretory vesicles. *Nature* **347**, 207–208.

Vallee, R. B., and Shpetner, H. S. (1990). Motor proteins of cytoplasmic microtubules. *Annu. Rev. Biochem.* **59**, 909–932.

van der Bliek, A. M., and Meyerowitz, E. M. (1991). Dynamin-like protein encoded by the *Drosophila shibire* gene associated with vesicular traffic. *Nature* **351**, 411–414.

Wittinghofer, A., and Pai, E. F. (1991). The structure of ras protein: A model for a universal molecular switch. *Trends Biochem. Sci.* **16**, 382–387.

8

Signal Transduction during Phagocytosis

**KEITH E. LEWIS, DARREN D. BROWNING,
AND DANTON H. O'DAY**

Department of Zoology
Erindale College
University of Toronto
Mississauga, Ontario, Canada

I. INTRODUCTION

A. Phagocytosis

Historically the study of phagocytosis is over 100 years old. In the latter part of the 19th century, Elie Metchnikoff investigated the activities of motile cells found within starfish larvae (Metchnikoff, 1893). He discovered that when thorns were inserted into the larvae and left for several hours, the thorns

SIGNAL TRANSDUCTION DURING
BIOMEMBRANE FUSION

became encased within ameboid cells. This discovery marked the beginning of research into cellular immunity and the importance of phagocytosis (Metchnikoff, 1893).

Phagocytosis is one of a group of processes involved in the transport of materials across the plasma membrane. In general terms, the internalization of particles and macromolecules can be classified into two cellular activities: phagocytosis and endocytosis, respectively. The diverse types of endocytosis have been extensively reviewed and are covered in other chapters in this volume.

The sequence of events in phagocytosis have been well characterized. When a target particle is bound to the phagocyte surface, the plasma membrane progresses around the particle while simultaneously excluding extracellular fluid from the nascent phagosome. As the advancing plasma membrane closes around the target cell, receptors on the membrane bind uniformly over the particle. This has been termed the "zipper interaction." After the formation of the phagosome, the vesicle migrates into the cytoplasm of the phagocyte and fuses with primary lysosomes for digestion (Griffin *et al.*, 1975).

Phagocytic protozoans are capable of ingesting bacteria as well as other protozoans and use the resulting degraded products as a source of amino acids, lipids, and carbohydrates. Initiation of the phagocytic process in eukaryotic microbes is undoubtedly owing to the stimulus of food particles at the cell surface—but how this is transduced to a cellular response has not been detailed (Fenchel, 1987).

Early protozoans may have nonselectively engulfed quantities of the surrounding water while ingesting useful food particles, but most of the present species demonstrate selective uptake of food particles. This uptake is mediated by specific membrane receptors (Fenchel, 1987). For example, studies of phagocytosis by asexual amebas of *Dictyostelium discoideum* have shown that when yeast is used as a food source its uptake is mediated specifically by wheat germ aggluttinin (WGA) receptors (Hellio and Ryter, 1980). However, during sexual development, as will be discussed, the cannibalistic phagocytosis of amebas by giant cells is mediated by a glucose-type receptor (Lewis and O'Day, 1986). This selectivity of particle uptake by eukaryotic microbes, mediated by specific receptors, is surely the forerunner of self- and nonself-recognition used by the phagocytic cells of the immune system of animals.

Animal phagocytic cells, or phagocytes, use phagocytosis for the removal of pathogenic bacteria, viruses, and antibody–antigen complexes from the tissues (Silverstein *et al.*, 1989). In many cases the uptake is mediated by a group of molecules called opsonins (Lennartz and Brown, 1991). There are two major classes of opsonins, both of which are produced in response to infection (Unkeless and Wright, 1988). One type is the serum antibodies. These globulin molecules bind to specific proteins or carbohydrates on the bacterial or viral outer surfaces. The nonantigen-binding portion of the antibody molecule, termed the Fc portion, in turn binds to receptors on the phagocyte surface, facilitating the uptake of the bound particle (Moraru et al., 1990). A second class of opsonins bind to foreign cells in a nonspecific manner. These are the complement proteins (Aderem et al., 1985). After coating a bacterial cell surface they subsequently adhere to complement receptors on the macrophage surface, again initiating the process of phagocytosis (for a review, see Unkeless and Wright, 1988).

B. Signal Transduction during Phagocytosis

The receptors that bind these two types of opsonins are both transmembrane proteins and are likely involved in initiating the signal transduction events leading to particle update. The Fc receptor is a single polypeptide chain of approximately 55 kDa, and the C3b receptor is a double polypeptide of approximately 190 kDa (Aderem et al., 1985). Although a great deal is known about the nature of the receptors of macrophages and neutrophils, many of the intracellular signal pathways are still poorly understood (Aderem et al., 1985). One of the areas of current interest in these cells is the role of arachidonic acid (AA) and its stimulation and release during ligand binding to receptors known to be involved in particle uptake (Lennartz and Brown, 1991). It is known that activation of the Fc receptor results in the release of AA and it has been proposed that it may be released from membrane phospholipids via two possible pathways, (1) by the action of phospholipase A_2 (PLA_2), or (2) from a metabolic pathway beginning with phospholipase C and followed by diglyceride lipase (Ida et al., 1988).

When Lennartz and Brown (1991) inhibited the release of arachidonic acid with 4-bromophenacyl bromide, this was accompanied by the elimination of IgG-mediated particle uptake. Because these inhibitors are known to act directly on phospholipase A_2, their results indicate the involvement of the PLA_2 pathway. Evidence that AA may act as a second messenger in IgG-mediated phagocytosis was shown by Sakata et al. (1987). They demonstrated that when arachidonate is released it causes the activation of oxygen-generating systems in macrophages, which in turn leads to the activation of the NADPH-oxidase enzyme systems.

A recent review by Suzuki (1991) summarized two possible pathways that differ during the activation of the Fc receptor, one involving receptor Fc2a, the other Fc2b. From experimental evidence it appears that on activation of the Fc2a receptor, casein kinase II phosphorylates myosin leading to its binding with actin. This activation of the cytoskeletal proteins results in the formation of pseudopodia, according to Suzuki (1991). He also suggests that casein kinase II phosphorylates adenylyl cyclase, activating this enzyme without the requirement of GTP-binding proteins. Activation of the second type of IgG receptor, Fc2b, appears to result in the triggering of PLA_2, which in turn degrades phospholipids, causing the release of arachidonic acid. The AA is subsequently converted to prostaglandins and it is the prostaglandins that result in subsequent G protein-mediated adenylyl cyclase induction. IgG binding to either the Fc2a or Fc2b receptors results in an increase in adenylyl cyclase activity, but the pathways leading to this increase differ markedly (Suzuki, 1991).

It has now been shown that the second class of opsonin, C3b protein, does not involve the release or use of arachidonate (Aderem *et al.*, 1985). Blocking of AA synthesis pathways does not result in a reduction of the uptake of C3 protein coated particles. Clearly the signal molecules involved during postactivation of the C3b/bi receptor differ markedly from IgG-mediated particle uptake. Studies on signal transduction pathways operating during C3 protein mediated phagocytosis are ongoing and may involve intracellular inositol trisphosphate (IP_3) and diacylglycerol (Fallman *et al.*, 1989).

There is presently no complete model to explain signal transduction pathways operating during phagocytosis in any organism. As part of our attempts to understand the regulation of this process, this chapter focuses on the chemical signals involved during the phagocytosis of amebas by the phagocytic giant cell of *Dictyostelium discoideum*.

C. Phagocytosis of Amebas by Zygote Giant Cells

Dictyostelium discoideum has been used extensively as a model system for the study of development and cell–cell interactions (Loomis, 1982; O'Day and Lewis, 1981; Spudich, 1987). The asexual pathway has been the focus of the majority of research; however, in recent years several research groups have centered on sexual development (O'Day and Lydan, 1989).

Early sexual development of *D. discoideum* is marked by the appearance of small amebas with condensed nuclei, which have been termed gamete cells

(O'Day *et al.*, 1987). As development continues, gamete cells fuse together to form zygote giant cells (ZGC) (McConachie and O'Day, 1987; O'Day *et al.*, 1987; Szabo *et al.*, 1982). Fusion appears to be regulated by a number of endogenous factors (Saga and Yanagisawa, 1983; O'Day and Lydan, 1989), including an autoinhibitor of cell fusion (Szabo *et al.*, 1982; Lydan and O'Day, 1988). Once they appear, the ZGCs serve as foci for the aggregation of amebas by secreting the chemoattractant cyclic AMP (cAMP) (Abe *et al.*, 1984; O'Day, 1979). On contact with the giant cell, the chemoattracted amebas are engulfed. Classically, the ingested amebas within the ZGC have been referred to as endocytes, but current terminology dictates the more appropriate term, phagosome, be used. Recent evidence supports the idea that the amebas will later be used as a food source by ZGCs. We have found that two lysosomal enzymes, β-glucosidase and α-mannosidase, both increase markedly during the phagocytic phase of sexual development and their appearance is dependent on cell fusion (Browning *et al.*, 1992). It is likely these enzymes are involved in the subsequent digestion of endocytes. Several hundred amebas may be phagocytosed by a single giant cell (Lewis and O'Day, 1985, 1986).

Morphologically, the uptake of amebas by ZGC appears atypical to classical phagocytosis in that it does not require pseudopodal entrapment. Instead, amebas appear to sink into or be pulled into the cytophagic giant cell. Figure 1 shows the sequence of events during this cannibalistic phagocytosis. First, amebas come into contact with the giant cell (Fig. 1, arrow 1), and begin to round-up as they adhere to it (Fig. 1, arrow 2). Inside the cell, endocytes and their nuclei progressively decrease in size (Fig. 1, arrows 3, 4, and 5).

Phagocytosis occurs optimally at 22–24°C, and within a pH range of 5.5–6.8 (Lewis and O'Day, 1985). The addition of various agents to the ZGCs that are known to disrupt cytoskeletal components have suggested the importance of the cytoskeleton to this process (Lewis and O'Day, 1985). The activation of the cytoskeleton in response to cell surface signals has been reviewed by Stossel (1989). Sexual phagocytosis is cannibalistic in nature because ZGCs prefer amebas of their own species. Although the ingestion of amebas and spores from other species of slime molds and of yeasts by *D. discoideum* ZGCs occurs, their uptake is very limited (Lewis and O'Day, 1986).

Work by Vogel *et al.* (1980) has shown that the phagocytic uptake of bacteria by vegetative amebas is mediated by both specific and nonspecific cell surface receptors. The nonspecific receptor participates in all types of phagocytosis and employs hydrophobic interactions, whereas the second type of receptor is highly particle-specific and has been identified as a glycoprotein (Vogel *et al.*, 1980). The specificity of ZGC phagocytosis might also involve

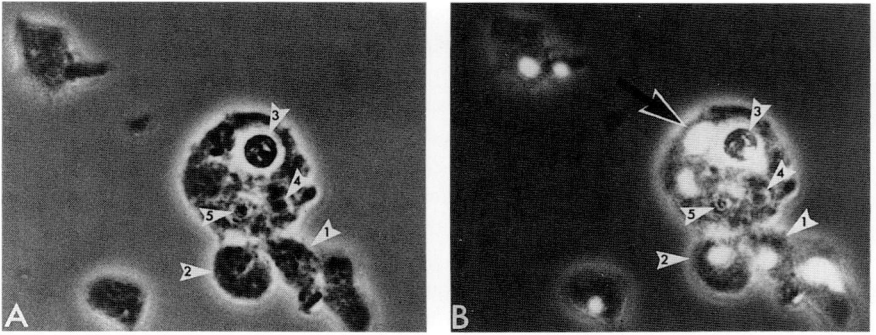

Fig. 1. Phagocytosis of amebas by ZGCs of *D. discoideum*. Cells from phagocytic sexual cultures were stained with Hoechst 33258 and observed under phase contrast (A), or simultaneous phase-contrast fluorescence microscopy to also reveal nuclear changes (B). The black arrow indicates a phagocytic vacuole within the ZGC. The numbered arrows indicate sequential stages of ameboid uptake by the phagocytic giant cell. 1, amebas come into contact with the giant cell surface; 2, amebas begin to round-up as they adhere to the ZGC; 3, an early phagosome (endocyte); and 4,5, sequential stages of phagosome shrinkage showing condensation of both the nucleus and cytoplasm of ingested amebas.

two kinds of cell surface recognition sites: a species-specific amebal receptor and a nonspecific receptor for yeast, and spores and amebas of other species. Since a nonspecific receptor is capable of mediating both yeast and bacterial engulfment (Vogel *et al.*, 1980) during vegetative feeding, it is plausible that similar receptors on the ZGC could be responsible for the low rate of yeast and spore uptake as well as for the ingestion of amebas of other species. It has been suggested that during vegetative growth, self-recognition limits the phagocytic mechanism. It is reasonable to assume, therefore, that this inhibitory mechanism is lost or altered in zygote giant cells.

Several studies have determined that the nonspecific receptor in vegetative cells is a 126-kDa cell surface glycoprotein. In *D. discoideum*, most of the proteins found at the cell surface are glycosylated, a phenomenon that has facilitated the identification of the specific phagocytic receptors by the use of lectins. Lectin inhibition studies have determined that a GlcNAc-containing glycoprotein mediates yeast phagocytosis and a glucose-containing receptor mediates the uptake of *Escherichia coli* by vegetative cells (Vogel *et al.*, 1980).

Lectins and their specific hapten sugars were used to test for their ability to affect cannibalistic phagocytosis and thereby shed light on the receptors involved in this process (Lewis and O'Day, 1985). Only concanavalin A (Con A) was found to effectively inhibit uptake of amebas by ZGCs. Furthermore, when hapten sugars were added exogenously, glucose was extremely effective at inhibiting phagocytosis, whereas other sugars were much less effective or not inhibitory at all. These data are indicative of the involvement of both a glucose-specific lectin activity and a glucosylated receptor in the recognition event leading to cannibalistic engulfment. As noted, this is similar to bacterial phagocytosis during vegetative feeding.

Western blotting followed by analysis throughout development has been carried out as a first step toward identifying specific molecules involved in sexual phagocytosis. The glucose-specific lectin, Con A, recognized a glycoprotein band migrating at about 126 kDa (gp126) which increased in staining intensity following cell fusion (Browning and O'Day, 1991). A WGA-binding band migrating at about 130 kDa (gp130) demonstrated the same developmental kinetics, indicating that it may be the same molecule (i.e., gp126) revealed with Con A. Whether gp126 of sexual cells is related to the 126-kDa nonspecific receptor identified in vegetative amebas remains to be proven, but circumstantial evidence suggests they are different. For example, as seen in Fig. 1, the round morphology that typifies ZGCs is probably associated with a loss of cell substratum adhesion because ZGC do not adhere as efficiently to glass slides or gelatin-coated glass as do nonzygotic cells from the same culture. This fact alone precludes the involvement of the 126-kDa nonspecific receptor which is also responsible for the adhesion of vegetative amebas to glass and gelatin surfaces.

II. CURRENT RESEARCH

A. Developmental Markers of Zygote Giant Cell Differentiation

The initial event of phagocytosis by the cytophagic zygote giant cell is adhesion of the target particle amebas to the cell surface. It is likely this single event triggers the subsequent signal transduction pathways that result in the uptake of the particle. As discussed in other chapters in this volume, cell–cell contact also initiates signal transduction processes leading to other bio-membrane fusion events. As previously detailed, a ConA-binding glycoprotein (gp126) is implicated in this highly selective, species-specific uptake of amebas. Further characterization of the receptor requires isolating ZGCs and localizing gp126 and possibly other specific glycoproteins to the giant cell surface.

By centrifuging mated cultures at low speed through 35% sucrose, highly purified suspensions of ZGCs have been isolated (Browning *et al.*, 1992). This has allowed characterization of cell surface markers that may be important to the phagocytic phase of development, plus demonstration of other zygote-specific molecules. On examination of the isolated ZGC membrane, three WGA-binding proteins are found that increase in concentration during the phagocytic phase of development (Browning *et al.*, 1992). Although a role for WGA-binding proteins in phagocytosis has not been demonstrated, these glycoproteins are currently under investigation.

A common mediator of signal transduction following receptor–ligand binding is a class of trimeric GTP-binding proteins, GS-proteins. A 52-kDA GαS-like protein previously present at a high concentration in cells from early sexual cultures is lost by vegetative amebas due to starvation. Conversely, this 52-kDa GαS-like band remains at high levels in zygote giant cells as development progresses. This G subunit therefore is lost from amebas as they terminate the ingestion of bacteria, but it increases in ZGCs as they embark on their cannibalistic phagocytic phase (Fig. 2). Pharmacological studies suggest G protein function is essential for sexual phagocytosis. At micromolar concentrations, GTPγS, an inhibitor of G protein function, inhibits endocyte formation by ZGCs (Lewis and O'Day, unpublished data). Together these results suggest that Gα52S mediates transduction during amebal uptake by the ZGC during sexual development. We are currently investigating this membrane protein.

B. Caffeine Inhibits Phagocytosis

As reviewed in the introduction, cAMP is intimately associated with phagocytosis in many animal systems. Furthermore, cAMP directs the chemotaxis of amebas to the ZGC surface where they ultimately become ingested (O'Day, 1979). For these reasons, the question of whether cAMP metabolism could be involved in sexual phagocytosis in *D. discoideum* was investigated.

As discussed below, caffeine has been shown to perturb cAMP synthesis in cellular slime molds. The addition of caffeine to cultures of ZGCs and amebas effectively blocks phagocytosis in a dose-dependent manner (Fig. 3). These results are similar to the findings of Gonzalez *et al.* (1990), who found that

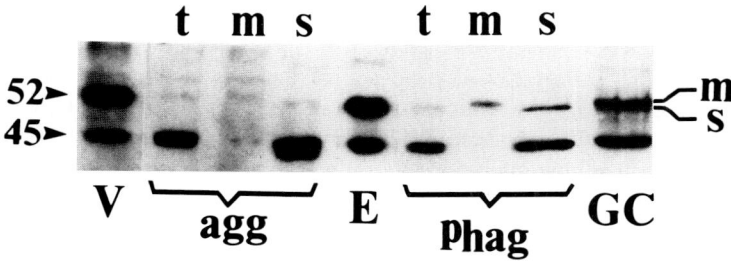

Fig. 2. Gα subunits associated with the phagocytic stages of *Dictyostelium* development. Gα subunits of amebas at different developmental stages were visualized by western blotting with Gα/1 immune serum which is specific to a common domain of the GTP-binding region. Vegetative amebas (V), which phagocytose bacteria as a food source, contain a 52 kDa band that is lost during the aggregation stage of asexual development (agg), in which amebas also lose phagocytic capability. A 52-kDa band is also present in amebas during the early stages of sexual development (E) in which the majority of cells are vegetative, but is then lost later in development during the phase of cannibalistic phagocytosis by zygote giant cells (Phag). When amebas are separated into total cellular (t), membrane (m), and soluble (s) fractions, the 52-kDa band resolves into a larger membrane-bound and a smaller soluble form. The membrane-bound form, which was not detected in cell homogenates from late sexual cultures, is specifically enriched in phagocytic giant cells (GC). The expression of the 52-kDa species contrasts with that of a 45-kDa band, which was p. sent at all stages of development and always localized to the soluble fraction.

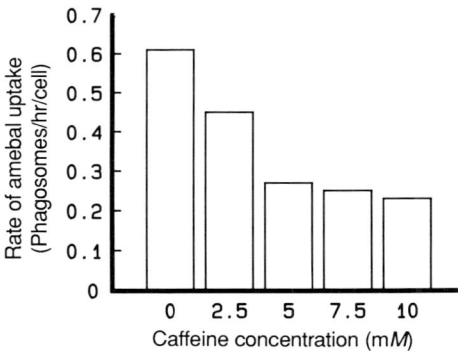

Fig. 3. Effects of caffeine on the phagocytosis of amebas by ZGCs. Caffeine was added to yield final concentrations of 2.5 , 5.0 , 7.5 , and 10 mM. Ten giant cells were scored per slide for total number of phagosomes (endocytes) contained and 10 slides quantified per treatment. All of the data are presented as the rate of ameboid uptake by ZGCs (i.e., number of phagosomes/hr/cell).

caffeine markedly inhibited both pinocytosis and phagocytosis of bacteria by vegetative amebas of *D. discoideum*.

Brenner and Thoms (1984) have shown that the addition of caffeine to aggregating *D. discoideum* amebas causes the inhibition of cAMP synthesis by blocking cAMP-dependent adenylyl cyclase activity. Although the exact mechanism of caffeine inhibition has not been elucidated, one mode of action may involve the disruption of the cytosolic calcium flux, which has been shown to modulate adenylyl cyclase activity in *D. discoideum* (Bominaar and Van Haastert, 1990). By the addition of adenosine, Gonzalez *et al.* (1990) found that they could partially reverse the action of caffeine on pinocytosis. They proposed that the action of adenosine in this situation may be similar to that found in muscle sarcoplasmic reticulum where adenosine has been shown to be a calcium agonist (Endo, 1985). Newell and Ross (1982), however, report that the addition of adenosine to aggregating amebas of *D. discoideum* causes similar effects to the those found with the addition of caffeine (i.e., an increase in aggregation territories).

It has been found that adenosine mimics the effects of caffeine on phagocytosis. When adenosine is added, the uptake of amebas by ZGCs is effectively blocked (Fig. 4). At a concentration of 5 m*M*, phagocytosis is reduced 10-fold. The role of adenosine is currently under investigation, but it has been shown to be an effective inhibitor of phagocytosis in other systems (e.g., mature monocytes in culture; Eppell *et al.*, 1989). Through the addition of

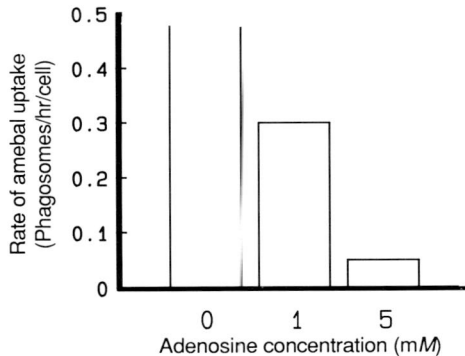

Fig. 4. Effects of adenosine on the phagocytosis of amebas by ZGCs. Adenosine was added to yield final concentrations of 1.0 and 5.0 mM. Ten giant cells were scored per slide for total number phagosomes (endocytes) contained and 10 slides quantified per treatment. The data are presented as the rate of amebic uptake by ZGCs (i.e., number of phagosomes/hr/cell).

adenosine analogs to mature monocytes, Eppell *et al.* (1989) have determined that adenosine binds to the A_2 receptor. On activation of A_2, cAMP is generated as a second messenger resulting in phagocytic inhibition (Eppell *et al.*, 1989; Cronstein *et al.*, 1988). With this in mind, an attempt was made to artificially increase intracellular cAMP levels in ZCGs. To do this, giant cells were treated with dibutyryl-cAMP, a nonhydrolyzable analog of cAMP, in 0.5% dimethyl sulfoxide (DMSO). Figure 5 shows that the addition of dibutyryl-cAMP had a marginal effect in enhancing phagocytosis at all of the concentrations used. This increase indicates that the role of adenosine during phagocytosis by ZGCs is different from that found in monocytes. Investigation is continuing into the effects of adenosine in this system.

C. Calmodulin Is Not Involved

Calcium plays many roles during sexual development of *D. discoideum* and, as indicated earlier, may also function in an, as yet, undefined way during phagocytosis. The importance of calcium was first demonstrated by Chagla *et al.* (1980). They found that 1 mM calcium was the optimum concentration required for sexual development. When calcium was completely

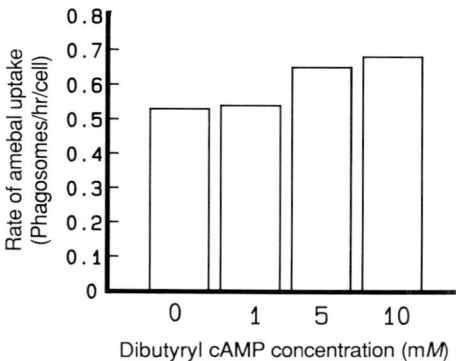

Fig. 5. Effects of dibutyryl-cAMP on the phagocytosis of amebas by ZGCs. Dibutyryl-cAMP was added to yield final concentrations of 1.0 , 5.0 , and 10 mM. Ten giant cells were scored per slide for total number of phagosomes (endocytes) contained and 10 slides quantified per treatment. The data are presented as the rate of ameboid uptake by ZGCs (i.e., number of phagosomes/hr/cell).

removed from the culture by the use of ethylenediamine tetraacetic acid (EDTA) or ethylene glycol-bis(β-aminoethyl ether) *N,N,N', N'*-tetraacetic acid (EGTA), sexual development was completely inhibited. By adding these agents at different times during development, it was discovered that calcium was important for both cell fusion and for pronuclear fusion to complete zygote giant cell formation. From the work of Lydan and O'Day (1988) and Lydan *et al.* (1990), it becomes clear that one of the pivotal molecules involved in these biomembrane fusion events is the calcium-dependent regulatory protein, calmodulin. As seen throughout this volume, calmodulin also functions during many other biomembrane fusion events. In an attempt to determine the role, if any, of calmodulin during sexual phagocytosis, two distinct calmodulin inhibitors were used, trifluoperazine and calmidazolium (R24571), which were added to cultures during phagocytosis. At concentrations known to effect the function of *D. discoideum* calmodulin, neither agent had any effect on ameboid uptake by zygote giant cells. From these results it appears that if calcium is important during phagocytosis, it does not appear to operate through calmodulin (Lewis and O'Day, unpublished data).

III. FINAL COMMENTS

Although research into the sexual cannibalistic phagocytosis of amebas by ZGCs of *D. discoideum* has only just begun, this is an excellent system for the study of signal transduction events leading to particle ingestion by phagocytes.

Phagocytosis in this system is an ordered part of a developmental sequence. Its onset and rate are highly predictable and reproducible. The events are easily visualized at the light-microscopic level, and ZGCs are large enough for studies involving microinjection. Basic methods are available for the rapid isolation of pure ZGCs and amebas. Therefore, it makes an ideal model system for the study of phagocytosis.

Figure 6 summarizes our current basic understanding of the signal transduction mechanisms at work during phagocytosis of *D. discoideum* amebas by zygote giant cells. Amebas are chemoattracted to the surface of the ZGC by the release of cAMP (O'Day, 1979). On contact, the amebas adhere to the ZGC surface. The accumulated data indicate this adhesion is mediated by a glycoprotein with a glucose terminal glycan moiety, possibly gp126. It can be postulated from the data that the interaction at the cell surface between the amebas and ZGCs activates a 52-kDa $G\alpha_s$ protein, possibly leading to a adenylyl cyclase activation.

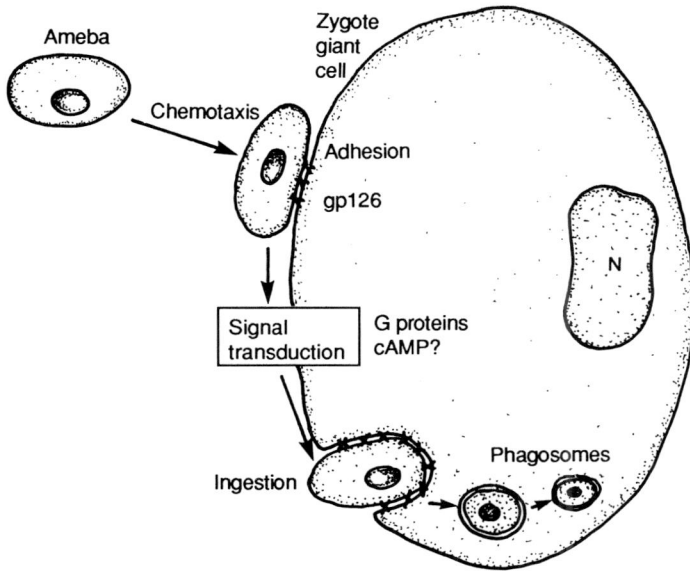

Fig. 6. Amebas are chemoattracted to the ZGC by cAMP. On contact with the ZGC surface, amebas adhere via a glycoprotein with a terminal glucose moiety, possibly gp126. Contact initiates G protein activation, which may stimulate adenylyl cyclase activity leading to an intracellular increase in cAMP. Ultimately, the amebas are ingested and the phagosomes mature.

These research efforts focus on the role of gp126 in sexual phagocytosis, including a detailed study of the glycan moiety of this glycoprotein. A study continues of the role of the 52-kDa $G\alpha_s$ proteins and the consequences of its activation on increasing intracellular levels of cAMP during phagocytosis.

Ultimately, the goal is to comprehend the signal transduction mechanisms at work during the process of phagocytosis. A complete understanding of the *D. discoideum* system should help toward the interpretation of the mechanisms in all phagocytic systems.

REFERENCES

Abe, K., Orii, H., Saga, Y., and Yanagisawa, K. (1984). A novel cyclic AMP metabolism exhibited by giant cells and its possible role in sexual development of *Dictyostelium discoideum. Dev. Biol.* **104,** 477–483.

Aderem, A. A., Wright, S. D., Silverstein, S. C., and Cohn, Z. A. (1985). Ligated complement receptors do not activate the arachidonic acid cascade in resident peritoneal macrophages. *J. Exp. Med.* **161,** 617–622.

Bominaar, A. A. and Van Haastert, P. J. M. (1990). Calcium and the inositol cycle in *Dictyostelium discoideum:* Dynamics and function during chemotaxis. *In* "Calcium as an Intracellular Messenger in Eukaryotic Microbes" (D. H. O'Day, ed), pp. 65–78. American Society for Microbiology, Washington, D.C.

Brenner, M., and Thoms, S. D. (1984). Caffeine blocks activation of cyclic AMP synthesis in *Dictyostelium discoideum. Dev. Biol.* **101,** 136–146.

Browning, D. D., Lewis, K. E., and O'Day, D. H. (1992). Zygote giant cell differentiation in *Dictyostelium discoideum:* Biochemical markers of specific stages of sexual development. *Biochem. Cell Biol.*

Chagla, A. H., Lewis, K. E., and O'Day, D. H. (1980). Ca^{++} and cell fusion during sexual development in liquid cultures of *Dictyostelium discoideum. Exp. Cell Res.* **126,** 501–505.

Cronstein, B. N., Kramer, S. B., Rosenstein, E. D., Korchak, H., Weissmann, G., and Hirschhorn, R. (1988). Occupancy of adenosine receptors raises cyclic AMP alone and in synergy with occupancy of chemoattractant receptors and inhibits membrane depolarization. *Biochem. J.* **252,** 709.

Endo, M. (1985). Regulation of calcium transport across muscle membranes. *In* "Current Topics in Membrane and Transport," (A. E. Shamoo, ed.), Vol. 25, pp. 181–230. Academic Press, New York.

Eppell, B. A., Newell, A. M., and Brown, E. J. (1989). Adenosine receptors are expressed during differentiation of monocytes to macrophages *in vitro. J. Immunol.* **143,** 4141–4145.

Fallman, M., Lew, D. P., Stendahl, O., and Andersson, T. (1989). Receptor-mediated phagocytosis in human neutrophils is associated with increased formation of inositol phosphates and diacylglycerol. *J. Clin. Invest.* **84,** 886–891.

Fenchel, T. (1987). "Ecology of Protozoa: The Biology of Free-Living Phagotrophic Protists." Science Tech Publishers, Madison, Wisconsin.

Gonzalez, C., Gerard, K., and Satre, M. (1990). Caffeine, an inhibitor of endocytosis in *Dictyostelium discoideum* amebas. *J. Cell. Physiol.* **144,** 408–415.

Griffin, F. M., Griffin, J. A., and Silverstein, S. C. (1975). Studies on the mechanisms of phagocytosis. I. Requirements for circumferential attachment of particle-bound ligands to specific receptors on the macrophage plasma membrane *J. Exp. Med.* **142,** 788–809.

Hellio, R., and Ryter, A. (1980). Relationships between anionic sites and lectin receptors in the plasma membrane of *Dictyostelium discoideum* and their role in phagocytosis. *J. Cell Sci.* **41,** 89–104.

Ida, E., Sakata, A., Tominaga, M., Yamasaki, H., and Onoue, K. (1988). Arachidonic release is closely related to Fc receptor-mediated superoxide generation in macrophages. *Microbiol. Immunol.* **32,** 1127.

Lennartz, M. R., and Brown, E. J. (1991). Arachidonic acid is essential for IgG Fc receptor-mediated phagocytosis by human monocytes. *J. Immunol.* **147,** 621–626.

Lewis, K. E., and O'Day, D. H. (1985). The regulation of sexual development in *Dictyostelium discoideum:* Cannibalistic behavior of the giant cell. *Can J. Microbiol.* **31,** 423–435.

Lewis, K. E., and O'Day, D. H. (1986). Phagocytic specificity during sexual development in *Dictyostelium discoideum. Can. J. Microbiol.* **32,** 79–82.

Loomis, W. F. (ed). (1982). "The Development of *Dictyostelium discoideum.*" Academic Press, New York.

Lydan, M. A., and O'Day, D. H. (1988). Different developmental functions for calmodulin in

Dictyostelium: Trifluoperazine and R245571 both inhibit cell and pronuclear fusion but enhance gamete formation. *Exp. Cell Res.* **178**, 51–63.

Lydan, M. A., Browning, D. D., and O'Day, D. H. (1990). Calcium, calmodulin, and the antagonistic action of an endogenous autoinhibitor of cell and pronuclear fusion in *Dictyostelium discoideum. In* "Calcium as an Intracellular Messenger in Eukaryotic Microbes" (D. H. O'Day, ed), pp. 392–409. American Society for Microbiology, Washington, D.C.

McConachie, D. R., and O'Day, D. H. (1987). Pronuclear migration, swelling and fusion during sexual development in *Dictyostelium discoideum. Can. J. Microbiol.* **33**, 1046–1049.

Metchnikoff, E. (1893). "Lectures on the Comparative Pathology of Inflammation. Reprinted by Dover Publications, New York, 1968. [English translation]

Moraru, I. I., Laky, M., Stanuscu, T., Buzila, L., and Popescu, L. M. (1990). Protein kinase C controls Fcγ receptor-mediated endocytosis in human neutrophils. *FEBS Lett.* **274**, 93–95.

Newell, P. C., and Ross, F. M. (1982). Inhibition by adenosine of aggregation center initiation and cyclic AMP binding in *Dictyostelium. J. Gen. Microbiol.* **128**, 2715–2724.

O'Day, D. H. (1979). Aggregation during sexual development in *Dictyostelium discoideum. Can. J. Microbiol.* **25**, 1416–1426.

O'Day, D. H., and Lewis, K. E. (1981). Pheromonal interactions during mating in *Dictyostelium. In* "Sexual Interactions in Eukaryotic Microbes" (D. H. O'Day and P. A. Horgan, eds.), pp. 199–221. Academic Press, New York.

O'Day, D. H., McConachie, D. R., and Rivera, J. (1987). Appearance and developmental kinetics of a unique cell type in *Dictyostelium discoideum:* Is it the gamete phase of sexual development? *J. Exp. Zool.* **242**, 153–159.

O'Day, D. H., and Lydan, M. A. (1989). The regulation of early sexual development in *Dictyostelium discoideum. Biochem. Cell Biol.* **67**, 321–326.

Saga, Y., and Yanagisawa, K. (1983). Macrocyst development in *Dictyostelium discoideum.* III. Cell fusion-inducing factor secreted by the giant cells. *J. Cell Sci.* **62**, 237–248.

Sakata, A., Ida, E., Tominaga, M., and Onoue, K. (1987). Arachidonic acid acts as an intracellular activator of NADPH-oxidase in Fcγ receptor-mediated superoxide generation in macrophages. *J. Immunol.* **138**, 433–439.

Silverstein, S. C., Greenburg, S., Di Virgilio, F., and Steinburg, T. H. (1989). Phagocytosis. *In* "Fundamental Immunology" (W. E. Paul, ed.), pp. 703–720. Raven Press, New York.

Spudich, J. A. (ed.) (1987). Methods in Cell Biology, Vol. 28. Academic Press, New York.

Stossel, T. P. (1989). From signal to pseudopod. *J. Biol. Chem.* **264**, 18,261–18,264.

Suzuki, T. (1991). Signal transduction mechanisms through Fcγ receptors on the mouse macrophage surface. *FASEB J.* **5**, 187–193.

Szabo, S. P., O'Day, D. H., and Chagla, A. H. (1982). Cell fusion, nuclear fusion, and zygote differentiation during sexual development in *Dictyostelium discoideum. Dev. Biol.* **90**, 375–382.

Unkeless, J. C., and Wright, S. D. (1988). Phagocytic cells: Fcγ and complement receptors. *In* "Inflammation: Basic Principles and Clinical Correlates" (J. I. Gallin, I. M. Goldstein, and R. Snyderman, eds.), pp. 343–362. Raven Press, New York.

Vogel, G. L., Scharz, H., and Steinhart, R. (1980). Mechanisms of phagocytosis in *Dictyostelium discoideum:* Phagocytosis is mediated by different recognition sites as disclosed by mutants with altered phagocytic properties. *J. Cell Biol.* **86**, 456–465.

III

Cell Fusion

9

Calcium Signal Transduction Pathway and Myoblast Fusion

JOAV PRIVES

Department of Pharmacological Sciences
State University of New York at Stony Brook
Stony Brook, New York

I. INTRODUCTION

The range and diversity of the topics covered in this volume attest to the prominence of membrane fusion in the behavioral repertoire of eukaryotic cells. Membrane fusion is an essential feature of each cell division and of ubiquitous cellular events such as intracellular membrane protein trafficking, secretion, and endocytosis. Membrane fusion also plays a key role in a variety

181

SIGNAL TRANSDUCTION DURING
BIOMEMBRANE FUSION

of specialized cells, as detailed in several chapters in this book. One of the most striking examples of membrane fusion occurs during myogenesis—the differentiation of skeletal muscle in the embryo. A major event in myogenesis is the fusion of the cell membranes of adjacent mononucleated precursor cells, called myoblasts, to form multinucleated syncytia called myotubes. Subsequently, the myotubes undergo maturation to form the muscle fibers that characterize skeletal muscle.

A. Events of Myoblast Fusion

Fortunately from the standpoint of the experimentalist, the major features of myogenesis—including myoblast fusion—are faithfully reproduced in cell culture, a state of events that during the past two decades has allowed the detailed characterization of muscle fusion and differentiation using myoblast cultures (for reviews, see Bischoff, 1978; Konigsberg, 1978; Wakelam, 1985, 1988). These experimental preparations include primary cultures of embryonic myoblasts as well as several muscle cell lines. In the case of the primary cultures grown under appropriate conditions, newly plated avian (chick, quail) or mammalian (rat) mononucleated precursor myoblasts typically undergo one or more cycles of cell division, and then withdraw from mitosis before spontaneously undergoing fusion and biochemical differentiation, giving rise to multinucleated skeletal muscle fibers. In the case of the myogenic cell lines, both mouse (C2; Yaffe and Saxel, 1977) and rat (L6 and L8; Yaffe, 1968) myoblasts can be induced (by changing to an appropriate culture medium containing less mitogens) to undergo fusion and express a variety of muscle differentiation markers. An additional mouse musclelike line, BC3H1, can be induced, by similar adjustments in the composition of the culture medium, to express several muscle differentiation markers. However, these cells do not express the key muscle-specific regulatory factor MyoD1 and consequently do not fuse; MyoD1-transfected sublines of these cells were shown to undergo fusion (Brennan *et al.*, 1990).

Primary cultures of embryonic chick skeletal muscle have proved to be the preparation of choice for studying biochemical and morphological aspects of myogenesis (Wakelam, 1985, 1988). Cultured chick myoblasts undergo spontaneous and relatively complete differentiation that culminates in well-developed multinucleated muscle fibers, which resemble muscle fibers that develop within the muscles of intact embryos. Moreover, chick myoblasts display rapid and highly synchronous differentiation kinetics in culture, making this system highly suitable for sensitive measurement of developmental changes. This has made it possible to demonstrate with cultured chick myoblasts that the formation of myotubes involves a number of discrete steps. Myogenic precursor cells first withdraw from the cell cycle and acquire a bipolar morphology. Next, these myoblasts undergo alignment and subsequent adhesion (before union of the plasma membranes of adjacent myoblasts) to form multinucleated myotubes (Knudsen and Horwitz, 1977). The adhesive interactions between fusion-competent myoblasts have been shown to be temporally distinct and experimentally separable from the subsequent fusion event that leads to myotube formation (Knudsen and Horwitz, 1977). Recent work has elucidated some of the molecular basis for this prefusional interaction by distinguishing between Ca^{2+}-independent and Ca^{2+}-dependent components of myoblast adhesion (Knudsen, 1990). Two adhesion proteins that have been well characterized in other cell types, NCAM and N-cadherin, both participate in mediating prefusional myoblast adhesion. NCAM is involved in Ca^{2+}-independent interaction between adjacent myoblasts, whereas N-cadherin mediates Ca^{2+}-dependent myoblast adhesion (Knudsen et al., 1990a,b; Pouliot et al., 1990). Myoblast alignment and adhesion are important events in myogensis because they set the stage for membrane fusion. Thus, myoblast fusion is dependent on, but discrete from, the immediately prior recognition, alignment, and adhesion steps.

B. Calcium Dependence of Myoblast Fusion

As is the case with a variety of other examples of biological membrane fusion (for recent reviews, see Burger and Verkleij, 1990; Papahadjopoulos et al., 1990), myoblast fusion displays an absolute requirement for Ca^{2+} (reviewed in Bischoff, 1978; Wakelam, 1985, 1988). This dependence is dramatically shown by the complete block of myoblast fusion that is triggered on depletion of the divalent cation from the culture medium, as was first demonstrated by Shainberg et al. (1969). Ca^{2+} dependence has since been demonstrated in this manner in both avian and mammalian muscle cultures, and in primary cultures as well as cell lines (Wakelam, 1988). As noted by Wakelam (1988), the effect of Ca^{2+} depletion on myoblast fusion is similar between species: fusion of either chick, rat, or calf myoblasts is inhibited by approximately 50% at 500 μM Ca^{2+}. Maximal rates of fusion are attained at 1.4 mM Ca^{2+}. In all cases, myoblast fusion block by Ca^{2+} deprivation is freely reversible; replenishment of the culture medium with Ca^{2+} leads to rapid resumption of fusion. The degree of specificity of the Ca^{2+} dependence of myoblast fusion is such that, whereas Sr^{2+} can replace Ca^{2+} to promote fusion, other cations such as Mg^{2+} and Zn^{2+} are inhibitory to fusion (Schudt *et al.*, 1973).

Primary cultures of embryonic chick myoblasts were used and grown under conditions that promote rapid, highly synchronous differentiation (O'Neill and Stockdale, 1972; Prives and Paterson, 1974). On plating, these cells undergo one- or two-cell division cycles, withdraw from mitosis, and align by 24 hr postplating, then fuse rapidly during the next 15–20 hr, so that maximal levels of fusion (fusion index of between 70 and 80%) are achieved within 2 days in culture. When fusion-competent myoblasts are shifted from normal growth medium containing 1.8 mM Ca^{2+} to very low Ca^{2+} concentrations (medium containing less than 25 μM Ca^{2+}), the Ca^{2+}-deprived myoblasts fail to fuse, and these fusion-arrested myoblasts display a characteristic thin, highly bipolar shape. At Ca^{2+} concentrations of 50 μM–1.4 mM, the rate and extent of myoblast fusion are functions of the external Ca^{2+} concentration.

C. Regulation of Calcium Influx in Fusion-Competent Myoblasts

Substantial experimental findings show that it is the *entry* of Ca^{2+} into prefusional myoblasts that is required for fusion to be triggered. An increase in Ca^{2+} influx immediately preceding myoblast fusion has been documented (David *et al.*, 1981; David and Higginbotham, 1981), and the Ca^{2+} ionophore A23187 has been reported to cause precocious fusion by elevating intracellular Ca^{2+} concentration (David *et al.*, 1981; Przybylski *et al.*, 1989). There is now considerable pharmacological evidence that this influx occurs through transmembrane Ca^{2+} channels. David *et al.* (1981) have shown that the Ca^{2+} channel antagonist D600 (α-isopropyl-α-[(N-methyl-N-homo-veratryl)-α-aminopropyl]-3,4,5-trimethoxyphenylacetonitrile hydrochloride) blocks myoblast fusion. Studies in our laboratory (Rapuano and Prives, 1987; Rapuano *et al.*, 1989) showed that the fusion block by D600 is completely reversed by the Ca^{2+} ionophore A23187, which supports the earlier conclusion (David and Higginbotham, 1981) that D600 exerts its effect by blocking Ca^{2+} influx. Moreover, it was observed that an additional Ca^{2+} channel blocker, nitrendipine, also inhibits fusion, and Bay K 8644, a Ca^{2+} channel activator, accelerates the initial rate of fusion (Rapuano *et al.*, 1989). Together, these findings point to two important features of the Ca^{2+} signal transduction pathway that controls myoblast fusion. To trigger fusion, Ca^{2+}

enters the cell by using transmembrane Ca^{2+} channels to gain access to an intracellular site (or sites) of action. The interaction of Ca^{2+} with its intracellular target(s) then initiates the changes that produce membrane fusion. Defining the role of these channels as a point of regulation of myoblast fusion and identification of the intracellular sites of Ca^{2+} action have been active goals of recent studies.

How is the entry of Ca^{2+} into myoblasts regulated? There is substantial evidence that the D600-sensitive Ca^{2+} channels that are responsible for triggering myoblast fusion are activated by specific ligand–receptor interactions. Several recent findings support the possibility that interaction of prostaglandin E_1 (PGE_1) with its receptor on the surface of fusion-competent myoblasts is the signal that triggers Ca^{2+} entry leading to myoblast fusion. Addition to PGE_1 to chick myoblast cultures induced precocious fusion (Zalin, 1977; Entwistle et al., 1988), and membrane receptors for PGE_1 were found to appear on the myoblast surface shortly (6–8 hr) before the onset of fusion, and to subsequently disappear after rapid fusion is complete (Hausman and Velleman, 1981; Hausman et al., 1986; Elgendy and Hausman, 1990). Moreover, David and Higginbotham (1981) showed that PGE_1 treatment of prefusional myoblasts induced Ca^{2+} entry, and that this uptake as well as fusion were blocked by a PGE_1 antagonist. These findings are consistent with the possibility that the myoblast Ca^{2+} channels are activated by the binding of ligands to surface receptors that are coupled to polyphosphotidylinositide (PIP_2) hydrolysis, as in other documented cases of Ca^{2+} influx induced by receptor activation (Meldolesi et al., 1991).

A functioning polyphosphatidyl-inositol signaling pathway in myoblasts has been demonstrated (Wakelam and Pette, 1982; Wakelam, 1986), and subsequent findings have supported the likelihood that this pathway connects PGE_1 receptor occupancy with myoblast fusion. These findings include (1) that the transient presence of PGE_1 receptors on the myoblast surface precisely corresponds to the stage at which rapid myoblast fusion is initiated (Hausman et al., 1986); (2) that exogenous PGE_1 causes increases in PIP_2 turnover only during this period of time (Elgendy and Hausman, 1990); and (3) that myogenesis is sensitive to both Li^+, which perturbs inositol phosphate metabolism, and the G protein inhibitor pertussis toxin, but only during the interval in which rapid fusion is initiated (Bonincontro et al., 1987; Hausman et al., 1990).

Also consistent with this possibility are recent observations suggesting that activation of the phospholipid- and Ca^{2+}-dependent protein kinase, protein kinase C (PKC), may be part of the link between PGE_1 receptor occupancy and the Ca^{2+} influx that induces myoblast fusion (David et al., 1990). This study reported that exposure of chick myoblasts to pharmacological activators of PKC, including TPA (12-O-tetradecanoylphorbol-13-acetate) and the diacylglycerol analog 1-oleyl-2-acetyl glycerol (OAG), under appropriate experimental conditions resulted in increased Ca^{2+} influx and precocious fusion, and, conversely, a PKC inhibitor, H-7, caused impairment of fusion (David *et al.*, 1990).

The tumor promoter, TPA, exerting its biological actions through activation of PKC (Nishizuka, 1984), has been shown to induce Ca^{2+} influx into myoblasts through D600-sensitive channels (Navarro, 1987; David *et al.*, 1990). A plausible model for the Ca^{2+} influx mechanism is composed of the following sequence of events. Receptor occupany results in G protein-mediated activation of phosphatidylinositide-specific phospholipase C. This enzyme catalyzes phosphoinositol turnover, resulting in production of diacylglycerol, which in turn stimulates PKC, and consequently Ca^{2+} uptake.

D. Role of Intracellular Calcium-Binding Proteins in Myoblast Fusion

Although the intracellular machinery that transduces the entry of Ca^{2+} into the melding of the surface membranes of adjacent myoblasts is still obscure, it is probable that Ca^{2+} will be found to act at multiple sites to activate a number of mechanisms that contribute to fusion. There is experimental evidence to support the expectation that the Ca^{2+}-binding protein, calmodulin, participates in mediating one or more of the Ca^{2+}-dependent events controlling myoblast fusion (Bar-Sagi and Prives, 1983; Kim *et al.*, 1992). Using a pharmacological approach, the effects on Ca^{2+}-entry mediated myoblast fusion of several inhibitors of calmodulin were measured (Bar-Sagi and Prives, 1983). These inhibitors included phenothiazine trifluoperazine (TFP), which binds to the Ca^{2+}-activated form of calmodulin and inhibits its interaction with cellular target proteins (Levin and Weiss, 1977, 1978), chlorpromazine, a phenothiazine that is less active than TFP in this respect, and two additional calmodulin inhibitors, the sulfanilamide derivatives W-5 and W-7, which are structurally unrelated to phenothiazines (Weiss and Levin, 1978; Hidaka *et al.*, 1980). All of these agents were found to inhibit myoblast fusion (Bar-Sagi and Prives, 1983; Kim *et al.*, 1992), and the magnitude of this inhibition was quantitatively related to the relative efficacies of these four compounds as inhibitors of Ca^{2+}–calmodulin (Hidaka *et al.*, 1980, 1981; Weiss and Levin, 1978). Under conditions in which Ca^{2+}-calmodulin was completely blocked, the inhibition of muscle fusion was total, and myoblasts acquired an exaggeratedly elongated morphology similar to that of myoblasts fusion arrested by Ca^{2+} depletion (Bar-Sagi and Prives, 1983). Furthermore, in muscle cultures undergoing normal fusion, immunofluorescence experiments revealed the presence of elevated levels of calmodulin in the tips of fusing myoblasts (Bar-Sagi and Prives, 1983). These findings suggest that calmodulin is an essential component in the Ca^{2+} signal transduction pathway that regulates myoblast fusion.

E. Role of Cytoskeletal-Membrane Interactions in Myoblast Fusion

Shortly before the onset of fusion, the cell membranes of myoblasts undergo a transient remodeling process consisting of changes in biochemical, structural, and topological properties. The surface membranes of fusion-competent myoblasts undergo a marked redistribution of surface proteins with the creation of protein-depleted lipid-enriched membrane domains (Kalderon and Gilula, 1979; Fulton et al., 1981) on the apposing surfaces of adjacent myoblasts. These rearrangements are concomitant with an abrupt increase in membrane fluidity (Prives and Shinitzky, 1977; Herman and Fernandez, 1978), which is thought to be associated with these newly formed lipid domains (Fulton et al., 1981; Herman and Fernandez, 1982). Moreover, these changes that create the sites for membrane–membrane fusion may well be generated through cytoskeletal dynamics.

There is substantial evidence connecting the redistribution of membrane proteins in fusion-competent myoblasts with rearrangements of cytoskeletal proteins immediately under the cell surface. An experimental approach that has proven highly useful for the study of cell surface–cytoskeletal interactions has been the extraction of cultures with the nonionic detergent Triton X-100 under mild conditions, followed by light and electron microscopy (Brown et al., 1976; Osborn and Weber, 1977; Fulton et al., 1981; Prives et al., 1982). The detergent-insoluble cytoskeletal framework revealed by this mild fractionation consisted of the traditional cytoskeletal elements, that is, microfilaments, microtubules, intermediate filaments, as well as a dense surface lamina made up of membrane proteins anchored to underlying cytoskeleton (Ben-Ze'ev et al., 1979). The cytoskeletal framework undergoes extensive rearrangements that are clearly associated with the morphogenesis of myotubes (Fulton et al., 1981). A prominent aspect of this rearrangement is the formation of lacunae, which are essentially openings or holes in the surface lamina of myoblasts (Fulton et al., 1981). These lacunae are absent from both proliferating myoblasts and postfusional myotubes, appearing shortly before the onset of myoblast union, persisting during the period of rapid fusion, and disappearing during subsequent differentiation. Studies using fluorescent and radioactive derivatives of the lectin concanavalin A to label surface glycoproteins on myoblasts indicate that in the intact cell the lacunae underlie surface membrane regions that are devoid of proteins (Fulton *et al.*, 1981). These experiments also showed that these glycoprotein-free surface domains do not result from the loss of surface proteins on detergent extraction; instead, the depletion of proteins from these regions reflects the lateral movement or sequestration of membrane proteins out of these regions and into adjacent areas of the cell surface.

This redistribution of surface proteins results in the formation of the protein-poor lipid domains thought to be the sites of fusion of apposing surface membranes of adjacent adherent myoblasts. The sequestration of proteins away from these domains is thought to be related to the increase in membrane fluidity that occurs immediately before myoblast fusion (Prives and Shinitzky, 1977; Herman and Fernandez, 1978). Shortly after the period of rapid fusion there is a sharp decrease in membrane fluidity (Prives and Shinitzky, 1977) concomitant with the disappearance of surface glycoprotein-deficient domains and the underlying holes in the surface lamina of the cytoskeletal framework (Fulton *et al.*, 1981). The transient increase in membrane fluidity concomitant with the fusion competence of myoblasts was measured in our laboratory as the average fluidity of a whole culture, but subsequent studies using microscopic fluorimetry have extended these findings by demonstrating that the changes appear to be localized to specific regions that directly participate in myoblast fusion (Herman and Fernandez, 1982).

II. CURRENT RESEARCH

Together, these observations support a role for cytoskeletal rearrangements in the molecular mechanisms that bring about the transient appearance of these highly fluid, lipid-enriched, and protein-depleted fusogenic regions on the myoblast surface. Protein-depleted surface domains might be formed by the clearing away of surface proteins from specific regions by lateral movement in the plane of the cell membrane, rather like the movement of surface proteins that constitutes patching and capping in other cell types. This possibility is consistent with a large body of evidence suggesting that changes in surface protein topology can be induced or stabilized by rearrangement of the peripheral cytoskeletal components with which these surface proteins are associated by noncovalent linkages (Carraway and Carraway, 1989; Luna, 1991).

A. Possible Role of Calcium-Regulated Myosin Phosphorylation in Myoblast Fusion

Based on studies of cultured chick myoblast fusion, it can be postulated that calcium-controlled changes in actomyosin contractility produce local changes in myoblast surface properties that contribute to muscle fusion (Rapuano *et al.*, 1989). The earlier light and electron microscopy observations (Fulton *et al.*, 1981) referred to previously indicated that myoblast fusion is associated with coordinated rearrangements of peripheral cytoskeletal elements and surface glycoproteins resulting in the production of lipid-rich, protein-depleted domains. Subsequent studies have indicated that myoblast fusion in these cultures is associated with phosphorylation of the 20-kDa regulatory light chain of nonsarcomeric myosin (MLC; Rapuano *et al.*, 1989). Phosphorylation of this component of the myosin complex at a specific serine residue, catalyzed by a Ca^{2+}–calmodulin dependent enzyme—myosin light chain kinase (MLCK)—is known to cause the activation of myosin interaction with F actin with ensuing actomyosin contractility (Adelstein *et al.*, 1982; Kamm and Stull, 1985; Sellers and Adelstein, 1987). Thus, the developmentally regulated localized influx of Ca^{2+} into fusion competent myoblasts and its sequestration by calmodulin can trigger actomyosin contractility, which in turn can alter the topological distribution of cell surface proteins. To date, the best-characterized example of the dynamic participation of actin and myosin in surface protein redistribution is the capping of surface receptors induced by external ligands. Actin and myosin have been shown to accumulate and interact under caps induced by Con A binding or antibody binding to surface receptors (Ash and Singer, 1976; Ash *et al.*, 1977; Condeelis, 1979; Bourguignon *et al.*, 1981).

The notion that actomyosin-mediated surface membrane remodeling is a crucial event in myoblast fusion is further supported by experiments involving pharmacological activation of the calcium-activated, phospholipid-dependent kinase, PKC, in chick myoblasts (Rapuano *et al.*, 1989). These experiments used the phorbol ester TPA, which acts at cell membranes to cause sustained activation of PKC (Castagna *et al.*, 1982). Previously, TPA has been used to demonstrate that the same MLCs that are phosphorylated by MLCK are also targets for phosphorylation catalyzed by PKC (Nishikawa *et al.*, 1983; Sellers and Adelstein, 1987). The two kinases catalyze phosphorylations at distinct sites on the polypeptide and have opposing effects. Whereas the MLCK-mediated phosphorylation activates the myosin complex, the PKC-catalyzed phosphorylation has been shown to inhibit myosin–actin interactions and to override the stimulatory effect of the MLCK-catalyzed phosphorylation (Naka *et al.*, 1983; Nishikawa *et al.*, 1983; Sellers and Adelstein, 1987). Several studies have shown that TPA impairs myoblast fusion (Cohen *et al.*, 1977; Dlugosz *et al.*, 1983; Sulakhe *et al.*, 1985; but see also David *et al.*, 1990 for an apparently opposite effect of TPA). Studies carried out by Rapuano and Prives (Rapuano and Prives, 1987; Rapuano *et al.*, 1989) showed that TPA blocks fusion of embryonic chick myoblasts in culture, that this block is a direct consequence of the activation of PKC, and that the arrest of TPA-induced fusion is associated with an increase in phosphorylation of the 20-kDa regulatory light chain of nonsarcomeric myosin.

When fusion-competent chick myoblasts were labeled with [^{32}P]orthophosphate and exposed to low concentration of phorbol ester (10 nM TPA) for brief intervals, we observed when measuring ^{32}P labeling of immunoprecipitated nonsarcomeric myosin that a sizeable TPA-induced increase in MLC phosphorylation was detectable within 15 min of the onset of exposure to the phorbol ester. The inhibition of myoblast fusion was apparent within 2 hr (the shortest practicable interval for the measurement of an effect on fusion by our methods) and sizeable by 4 hr (Rapuano *et al.*, 1989). Structural analogues of TPA that did not activate PKC had no effect on myoblast fusion. Conversely, the diacylglycerol dioctonoylglycerol, a potent PKC activator that is structurally unrelated to TPA, nevertheless mimicked the effects of TPA by inducing the rapid increase in MLC phosphorylation and inhibition of myoblast fusion.

B. Conclusion

Together these studies furnish strong experimental support for the idea that cellular regulation of myoblast fusion includes the use of two antagonistic phosphorylation mechanisms that both target the MLC of nonsarcomeric myosin situated immediately under the myoblast cell surface. The activating pathway transduces Ca^{2+} entry into a Ca^{2+}–calmodulin, MLCK-catalyzed phosphorylation of MLC. This phosphorylation triggers actomyosin interaction and consequent remodeling of the cell surface to produce protein-depleted regions that are the sites of lipid membrane fusion. The inhibitory pathway is triggered under experimental conditions by pharmacological activation of PKC and results in a phosphorylation of MLC at a site that blocks actomyosin interaction, and this effect is associated with the prevention of membrane fusion. The physiological relevance of this inhibitory pathway to the regulation of muscle fusion has yet to be established. The massive, topologically generalized PKC activation induced by exposure of myoblasts to phorbol esters might well have different consequences from the localized, more moderate activation that might arise in response to physiological cues. In addition, David *et al.* (1990) have reported that PKC can exert positive regulation of myoblast fusion by stimulating Ca^{2+} entry.

III. FINAL COMMENTS

It is now widely accepted that the initiation of cell membrane fusion involves the localized removal of membrane proteins to provide a lipid-rich accessible surface patch at the site of fusion (Fulton *et al.*, 1981; Kosower *et al.*, 1983). It is highly likely that Ca^{2+} regulates myoblast fusion not solely through one pathway, but rather through multiple parallel mechanisms. This chapter has focused exclusively on one novel mechanism in which Ca^{2+}-dependent phosphorylation of MLC catalyzed by Ca^{2+}–calmodulin MLCK can trigger actomyosin interaction. This interaction induces myoblast membrane remodeling for fusion by clearing cytoskeletal-associated membrane proteins from specified membrane regions. In addition, there is evidence for the participation in myoblast fusion of an additional Ca^{2+}–calmodulin-dependent kinase (Kim *et al.*, 1992), as well as of an entirely different Ca^{2+}-dependent enzymatic mechanism also acting to produce protein-poor membrane domains in myoblasts. The latter mechanism involves Ca^{2+}-activated neutral proteases (calpains) that have been proposed to catalyze the localized proteolysis of membrane-associated proteins, thus removing steric obstacles to membrane lipid fusion (reviewed in Croall and Demartino, 1991). This type of membrane remodeling through calpain-catalyzed proteolysis has been directly implicated in myoblast fusion (Schollmeyer, 1986a,b). These mechanisms serve to illustrate the likelihood that the Ca^{2+} signal transduction pathways that regulate myoblast fusion can activate a diverse array of molecular mechanisms that act in concert to bring about formation of multinucleated myotubes. It is also likely that myoblast fusion shares some of these mechanisms with other examples of Ca^{2+}-activated membrane fusion. Last, there is a strong likelihood that a more detailed understanding of these mechanisms will emerge from the concerted studies that are currently in progress in a number of laboratories.

ACKNOWLEDGMENT

Work from the author's laboratory is supported by National Institute of Health Grant NS25945.

REFERENCES

Adelstein, R. S., Pato, M. D., Sellers, J. R., DeLanerolle, P., and Conti, M. A. (1982). Regulation of actin–myosin interaction by reversible phosphorylation of myosin and myosin kinase. *Cold Spring Harbor Symp. Quant. Biol.* **46**, 921–928.

Ash, J. F., and Singer, S. J. (1976). Concanavalin A-induced transmembrane linkage of concanavalin A surface receptors to intracellular myosin-containing filaments. *Proc. Natl. Acad. Sci. USA* **73**, 4575–4579.

Ash, J. F., Louvard, D., and Singer, S. J. (1977). Antibody-induced linkages of plasma membrane proteins to intracellular actomyosin-containing filaments in cultured fibroblasts. *Proc. Natl. Acad. Sci. USA* **74**, 5584–5588.

Bar-Sagi, D., and Prives, J. (1983). Trifluoperazine, a calmodulin antagonist, inhibits muscle cell fusion. *J. Cell Biol.* **97**, 1375–1380.

Ben-Ze'ev, A., Duerr, A., Solomon, F., and Penman, S. (1979). The outer boundary of the cytoskeleton: A lamina derived from plasma membrane proteins. *Cell* **17**, 858–865.

Bischoff, R. (1978). Myoblast fusion. *Cell Surface Rev.* **5**, 127–179.

Bonincontro, A., Cametti, C., Hausman, R. E., Indovina, P. L., and Santini, M. T. (1987). Changes in myoblast membrane electrical properties during cell–cell adhesion and fusion *in vitro*. *Biochim. Biophys. Acta* **903**, 89–95.

Bourguignon, L. Y. W., Nagpal, M. L., and Hsing, Y. C. (1981). Phosphorylation of myosin light chain during capping of mouse T-lymphoma cells. *J. Cell Biol.* **91**, 881–894.

Brennan, T. J., Edmondson, D. G., and Olson, E. N. (1990). Aberrant regulation of MyoD1 contributes to the partially defective myogenic phenotype of BC3H1 cells. *J. Cell Biol.* **110**, 929–937 (and erratum: *J. Cell Biol.* **110**, 2231).

Brown, S., Levinson, W., and Spudich, J. (1976). Cytoskeletal elements of chick embryo fibroblasts revealed by detergent extraction. *J. Supramol. Struct.* **5**, 119–130.

Burger, K. N. J., and Verkleij, A. J. (1990). Membrane fusion. *Experientia* **46**, 631–644.

Carraway, K. C., and Carraway, C. A. C. (1989). Membrane–cytoskeleton interactions in animal cells. *Biochim. Biophys. Acta* **988**, 147–171.

Castagna, M., Takai, Y., Kaibuchi, K., Sono, K., Kikkawa, U., and Nishizuka, Y. (1982). Direct activation of calcium-activated phospholipid-dependent protein kinase C by tumor-promoting phorbol esters. *J. Biol. Chem.* **257**, 7847–7851.

Cohen, R., Pacifici, M., Rubinstein, N., Biehl, J., and Holtzer, H. (1977). Effect of a tumor promoter on myogenesis. *Nature (London)* **266,** 538–540.

Condeelis, J. (1979). Isolation of concanavalin A caps during various stages of formation and their association with actin and myosin. *J. Cell Biol.* **80,** 751–758.

Croall, D. E., and Demartino, G. N. (1991). Calcium-activated neutral protease (calpain) system; structure, sunction, and regulation. *Physiol. Rev.* **71,** 813–847.

David, J. D., and Higgenbotham, C. A. (1981). Fusion of chick embryo skeletal myoblasts: Interactions of prostaglandin E_1, cAMP, and calcium influx. *Dev. Biol.* **82,** 307–316.

David, J. D., See, W. M., and Higgenbotham, C. A. (1981). Fusion of chick embryo skeletal myoblasts: Role of calcium influx preceding membrane union. *Dev. Biol.* **82,** 297–307.

David, J. D., Faser, C. R., and Perrot, G. P. (1990). Role of protein kinase C in chick embryo skeletal myoblast fusion. *Dev. Biol.* **139,** 89–99.

Dlugosz, A. A., Tapscott, S. J., and Holtzer, H. (1983). Effects of phorbol 12-myristate 13-acetate on the differentiation program of embryonic chick skeletal myoblasts. *Cancer Res.* **43,** 2780–2789.

Elgendy, H., and Hausman, R. E. (1990). Prostaglandin-dependent phosphatidylinositol signaling during embryonic chick myogenesis. *Cell Differ. Dev.* **32,** 109–116.

Entwistle, A., Zalin, R. J., Bevan, S., and Warner, A. E. (1988). The control of chick myoblast fusion by ion channels operated by prostaglandins and acetylcholine. *J. Cell Biol.* **106,** 1693–1702.

Fulton, A., Prives, J., Farmer, S., and Penman, S. (1981). Developmental reorganization of the skeletal framework and its surface lamina in fusing muscle cells. *J. Cell Biol.* **91,** 103–112.

Hausman, R. E., and Velleman, S. G. (1981). Prostaglandin E_1 receptors on chick embryo myoblasts. *Biochem. Biophys. Res. Commun.* **103,** 213–218.

Hausman, R. E., Dobi, E. T., Woodford, E. J., Petridis, S., Ernst, M., and Nichols, E. B. (1986). Prostaglandin-binding activity and myoblast fusion in aggregates of avian myoblasts. *Dev. Biol.* **113,** 40–48.

Herman, B. A., and Fernandez, S. M. (1978). Changes in membrane dynamics associated with myogenic cell fusion. *J. Cell. Physiol.* **94,** 253–264.

Herman, B. A., and Fernandez, S. M. (1982). Dynamics and topographical distribution of surface glycoproteins during myoblast fusion. *Biochemistry* **21,** 3275–3283.

Hidaka, H., Yamaki, T., Naka, M., Tanaka, T., Hayashi, H., and Kobayashi, R. (1980). Calcium-regulator modulator protein interacting agents inhibit smooth muscle calcium-stimulated protein kinase and ATPase. *Mol. Pharmacol.* **17,** 66–72.

Hidaka, H., Saskai, T., Tanaka, T., Endo, S., Ohno, Y., Fujii, Y., and Nagata, T. (1981). *N*-(6-aminohexyl)-5-chloro-1-naphthalene sulfonamide, a calmodulin antagonist, inhibits cell proliferation. *Proc. Natl. Acad. Sci. USA* **78,** 4354–4357.

Kalderon, N., and Gilula, N. B. (1979). Membrane events involved in myoblast fusion. *J. Cell Biol.* **81,** 411–425.

Kamm, K. E., and Stull, J. T. (1985). The function of myosin and myosin light chain kinase phosphorylation in smooth muscle. *Annu. Rev. Pharmacol. Toxicol.* **25,** 593–620.

Kim, H. S., Lee, I. H., Chung, C. H., Kang, M.-S., and Ha, D. B. (1992). Ca^{2+}/calmodulin-dependent phosphorylation of the 100-kDa protein in chick embryonic muscle cells in culture. *Dev. Biol.* **150,** 223–230.

Knudsen, K. A. (1990). Cell adhesion molecules in myogenesis. *Curr. Opin. Cell Biol.* **2,** 902–906.

Knudsen, K. A., and Horwitz, A. F. (1977). Tandem events in myoblast fusion. *Dev. Biol.* **58,** 328–338.

Knudsen, K. A., McElwee, S. A., and Myers, L. (1990a). A role for the neural cell adhesion molecule, NCAM, in myoblast interaction during myogenesis. *Dev. Biol.* **138,** 159–168.

Knudsen, K. A., Myers, L., and McElwee, S. (1990b). A role for the Ca^{2+}-dependent adhesion molecule, N-cadherin, in myoblast interaction during myogenesis. *Exp. Cell Res.* **188,** 175–184.

Konigsberg, I. R. (1978). Skeletal myoblasts in culture. *In* "Methods in Enzymology" (W. B. Jacoby and T. H. Pastan, eds.), Vol. 58, pp. 511–527. Academic Press, New York.

Kosower, N. S., Glaser, T., and Kosower, E. M. (1983). Membrane-mobility, agent-promoted fusion of erythrocytes: Fusibility is correlated with attack by calcium-activated cytoplasmic proteases on membrane proteins. *Proc. Natl. Acad. Sci. USA* **80,** 7542–7546.

Levin, R. M., and Weiss, B. (1977). Binding of trifluoperazine to the calcium-dependent activator of cyclic nucleotid phosphodiesterase. *Mol. Pharmacol.* **13,** 690–697.

Levin, R. M., and Weiss, B. (1978). Specificity of binding of trifluoperazine to the calcium-dependent activator of phosphodiesterase and to a series of other calcium-binding proteins. *Biochim. Biophys. Acta* **540,** 197–204.

Luna, E. J. (1991). Molecular links between the cytoskeleton and membranes. *Curr. Opin. Cell Biol.* **3,** 120–126.

Meldolesi, J., Clementi, E., Fasolato, C., Zacchetti, D., and Pozzan, T. (1991). Ca^{2+} influx following receptor activation. *TIPS* **12,** 289–292.

Naka, M., Nishikawa, M., Adelstein, R. S., and Hidaka, H. (1983). Phorbol ester-induced activation of human platelets is associated with protein kinase C phosphorylation of myosin light chains. *Nature (London)* **306,** 490–492.

Navarro, J. (1987). Modulation of dihydropyridine receptors by activation of protein kinase C in chick muscle cells. *J. Biol. Chem.* **262,** 4649–4652.

Nishikawa, M., Hidaka, H., and Adelstein, R. S. (1983). Phosphorylation of smooth muscle, heavy meromyosin by calcium-activated phospholipid-dependent protein kinase. *J. Biol. Chem.* **258,** 14069–14072.

Nishizuka, Y. (1984). The role of protein kinase C in cell surface signal transduction and tumor production. *Nature (London)* **308,** 693–698.

O'Neill, M. C., and Stockdale, F. E. (1972). A kinetic analysis of myogenesis *in vitro. J. Cell Biol.* **52,** 52–65.

Osborn, M., and Weber, K. (1977). The detergent-resistant cytoskeleton of tissue culture cells includes the nucleus and microfilament bundles. *Exp. Cell Res.* **106,** 339–349.

Papahadjopoulos, D., Nir, S., and Duzgunes, N. (1990). Molecular mechanisms of calcium-induced membrane fusion. *J. Bioenerg. Biomemb.* **22,** 157–179.

Pouliot, Y., Holland, P. C., and Blaschuk, O. W. (1990). Developmental regulation of a cadherin during the differentiation of skeletal myoblasts. *Dev. Biol.* **141,** 292–298.

Prives, J., and Paterson, B. (1974). Differentiation of cell membranes in cultures of embryonic chick breast muscle. *Proc. Natl. Acad. Sci. USA* **71,** 3209–3211.

Prives, J., and Shinitzky, M. (1977). Increased membrane fluidity precedes fusion of muscle cell. *Nature (London)* **268,** 761–763.

Prives, J., Fulton, A. B., Penman, S., Daniels, E. M., and Christian, C. (1982). Interaction of the cytoskeletal framework with acetylcholine receptor on the surface of embryonic muscle cells in culture. *J. Cell Biol.* **92,** 231–236.

Przybylski, R. J., MacBride, R. G., and Kirby, A. C. (1989). Calcium regulation of skeletal myogenesis. I. Cell content critical to myotube formation. *In Vitro Cell. Dev. Biol.* **25,** 830–838.

Rapuano, M., and Prives, J. (1987). TPA blocks myogenic cell fusion by interfering with the calcium signal transduction pathway. *Ann. N.Y. Acad. Sci.* **494**, 156–159.

Rapuano, M., Ross, A., and Prives, J. (1989). Opposing effects of calcium entry and phorbol esters on fusion of chick muscle cells. *Dev. Biol.* **134**, 271–278.

Schollmeyer, J. E. (1986a). Role of Ca^{2+} and Ca^{2+}-activated protease in myoblast fusion. *Exp. Cell Res.* **162**, 411–422.

Schollmeyer, J. E. (1986b). Possible role of calpain I and calpain II in differentiating muscle. *Exp. Cell Res.* **163**, 413–422.

Schudt, C., van der Bosch, J., and Pette, D. (1973). Inhibition of muscle cell fusion *in vitro* by Mg^{2+} and K^{2+} ions. FEBS Lett. **32**, 296–298.

Shainberg, A., Yagil, G., and Yaffe, D. (1969). Control of myogenesis *in vitro* by Ca^{2+} concentration in nutritional medium. *Exp. Cell Res.* **58**, 163–167.

Sellers, J. R., and Adelstein, R. S. (1987). Regulation of contractile activity. *In* "The Enzymes" (P. D. Boyer and E. G. Krebs, eds.), Vol. 18, pp. 381–418. Academic Press, New York.

Sulakhe, P. V., Johnson, D. D., Phan, N. T., and Wilcox, R. (1985). Phorbol ester inhibits myoblast fusion and activates β-adrenergic receptor coupled adenylate cyclase. *FEBS Lett.* **186**, 281–286.

Wakelam, M. J. O. (1985). The fusion of myoblasts. *Biochem. J.* **228**, 1–12.

Wakelam, M. J. O. (1986). Stimulated inositol phospholipid breakdown and myoblast fusion. *Biochem. Soc. Trans.* **14**, 253–256.

Wakelam, M. J. O. (1988). Myoblast fusion: A mechanistic analysis. *Curr. Top. Membr. Transport* **32**, 87–112.

Wakelam, M. J. O., and Pette, D. (1982). The breakdown of phosphatidylinositol in myoblasts stimulated to fuse by the addition of Ca^{2+}. *Biochem. J.* **202**, 723–729.

Weiss, B., and Levin, R. M. (1978). Mechanism for selectively inhibiting the activation of cyclic nucleotide phosphodiesterase and adenylate cyclase by antipsychotic agents. *Adv. Cyclic Nucleotide Res.* **9**, 285–304.

Yaffe, D. (1968). Retention of differentiation potentialities during prolonged cultivation of myogenic cells. *Proc. Natl. Acad. Sci. USA* **61**, 477–483.

Yaffe, D., and Saxel, O. (1977). Serial passaging and differentiation of myogenic cells isolated from dystrophic mouse muscle. *Nature (London)* **270**, 725–727.

Zalin, R. (1977). Prostaglandins and myoblast fusion. *Dev. Biol.* **59**, 241–248.

10

Phospholipid Metabolism during Calcium-Regulated Myoblast Fusion

VICTOR S. SAURO AND KENNETH P. STRICKLAND

Department of Biochemistry
The University of Western Ontario
London, Ontario, Canada

I. INTRODUCTION

The *in vitro* differentiation of myoblasts has been a popular model for studies of intercellular biomembrane fusion (Bischoff, 1978). Despite exten-

197

sive research in this area, little is known about the nature and mechanism of myoblast fusion (Wakelam, 1988; Knudsen, 1991). However, the expression of the myogenic program and myoblast fusion need not be exclusive independent events as was once thought (Sanwal, 1979). Indeed, it is likely that myoblast fusion is under the control of a myogenic regulatory gene and as such is only one aspect of the activated myogenic program (Davis *et al.*, 1987). Thus, unraveling the mysteries of myoblast fusion may not be possible until there is "a complete understanding of the regulation of muscle-specific gene expression (which) may elude biologists for some time" (Murre *et al.*, 1989). However, significant advances in the current understanding of myoblast fusion can still be made, and may actually facilitate studies of muscle-specific gene expression. The introductory comments that follow present a brief background to myoblast fusion, in particular, focusing on possible signal transduction pathways involved in this phenomenon and a projected mechanism for myoblast fusion.

A. Myoblast Fusion

During skeletal myogenesis, proliferating myoblasts withdraw from the cell cycle, align and adhere to each other, and fuse to become multinucleate myotubes representative of a rudimentary muscle fiber (Wakelam, 1985). Accompanying the morphological differentiation (i.e., the fusion of mononucleate myoblasts to multinucleate myotubes) is the coincident biochemical differentiation characterized by the expression of a number of muscle-specific proteins such as creatine phosphokinase, acetylcholine receptors, and myosin heavy chain (Sanwal, 1979).

The fusion of myoblasts has been dissected into the following stages (Bischoff, 1978; Knudsen, 1991): (1) The acquisition of fusion competence, (2) myoblast adhesion (contact of fusion-competent myoblasts and the subsequent close apposition of myoblast plasma membranes), and (3) plasma membrane union. Studies that specifically focus on only one of these stages are most useful since observations are easier to interpret. For precisely this reason, many reports that identify inhibitors of myoblast fusion are uninformative (Wakelam, 1988) because it is difficult to determine which stage of myoblast fusion is altered.

The acquisition of fusion competence is thought to be a manifestation of the activated myogenic program. It is generally believed that myoblasts withdraw from the cell cycle and undergo specific membrane-associated alterations that impart fusion competence (David and Higginbotham, 1981; Sundler, 1984). There is some evidence to suggest that there may only be one site for membrane union during myoblast fusion (Lipton and Konigsberg, 1972). The putative changes in the myoblast surface that convey fusion competence may be a reflection of prerequisite discrete membrane-associated alterations at these presumably unique fusion sites. If fusion competence is associated with only transient local alterations in the myoblast surface, such changes will indeed be difficult to observe experimentally (Wakelam, 1988). There are numerous reports of transient changes in protein, glycoprotein, and glycolipid profiles of fusing myoblasts, but their role in myoblast fusion remains speculative, if not controversial (Wakelam, 1988; Knudsen, 1991).

Myoblast adhesion has an absolute requirement for calcium-dependent and calcium-independent glycoproteins (Knudsen, 1991). The close apposition of the lipid bilayers may not be a result of myoblast adhesion, but probably follows the adhesion stage. Clearly, the close apposition of the lipid bilayers is an important step in membrane fusion that requires the removal, alteration, or circumvention of significant structural and energetic barriers (Bischoff, 1978). In support of this point, studies have shown major alterations in cell surface molecules (Kalderon and Gilula, 1979; Herman and Fernandez, 1982; Kawasaki et al., 1988) and the cytoskeletal network (Fulton et al., 1981), along with increased membrane fluidity (Prives and Shinitzky, 1977; Herman and Fernandez, 1978), which coincide with myoblast fusion.

The hydrated bilayer probably represents the most significant barrier to the membrane union of closely apposed lipid bilayers (Sundler, 1984; Parsegian et al., 1984). Overcoming or circumventing this repulsive "hydration force" is a necessary step in myoblast fusion and could be accomplished through (1) applied force, (2) displacement or removal of water from the membrane surface, or (3) biochemical modification of the membrane surfaces (Parsegian et al., 1984). Since dehydration alone appears to be insufficient for membrane union, the combination of local dehydration and bilayer disruption have been proposed to be essential steps for membrane union to occur (Hui et al., 1981).

B. Signal Transduction during Myoblast Fusion

Myoblast fusion is probably regulated by an extracellular factor or factors (Konigsberg, 1971; Delain et al., 1981; Zalin, 1987), implicating a crucial role for signal transduction during this event (Wakelam, 1988). Indeed, cyclic AMP (cAMP) (Zalin and Montague, 1974), calcium (David et al., 1981), calmodulin (Bar-Sagi and Prives, 1983), inositol phospholipids (Wakelam and Pette, 1982; Wakelam, 1983), and protein kinase C (PKC; Cohen *et al.*, 1977), have all been proposed to play an essential role during myoblast fusion.

There is little doubt that cAMP and the cAMP-dependent protein kinase are intricately involved in skeletal myogenesis (Rogers *et al.*, 1985; Lorimer *et al.*, 1987; Lorimer and Sanwal, 1989), but their role during myoblast fusion remains controversial (Zalin and Montague, 1974; Schutzle *et al.*, 1984). Indeed, there is a growing consensus that an increase in cAMP is not responsible for initiating myoblast fusion (Entwistle *et al.*, 1986; Knudsen, 1991).

There is an absolute requirement for extracellular calcium during myoblast fusion (Shainberg *et al.*, 1971; van der Bosch *et al.*, 1972). Myoblast fusion is also preceded by an influx of calcium (David *et al.*, 1981) that appears to be essential for membrane union (Rapuano *et al.*, 1989). The calcium-binding protein calmodulin and the calmodulin-dependent protein kinase have been implicated in the rearrangement of the cytoskeleton and possibly the regulation of the calcium influx (Bar-sagi and Prives, 1983). Calcium may activate a specific protease that has been associated with alterations in the myoblast cell surface (Schollmeyer, 1986). These observations suggest that calcium has dual extracellular and intracellular roles and may activate a multitude of biochemical events during myoblast fusion.

Chronic exposure to the potent tumor promoter 12-*O*-tetradecanoylphorbol-13-acetate (TPA) inhibits myoblast fusion (Cohen *et al.*, 1977), suggesting a fundamental but inhibitory role for the calcium and phospholipid-dependent PKC during this event (Nishizuka 1984a, 1986). Surprisingly, recent observations indicate that PKC may play a key role in triggering myoblast fusion (Adamo *et al.*, 1989; Martelly *et al.*, 1989; Farzaneh *et al.*, 1989; David *et al.*, 1990). Thus, PKC also appears to be intricately involved in the mechanism of myoblast fusion, but there is some confusion over its precise function. Wakelam and Pette (1984) have proposed that receptor-mediated phosphatidylinositol 4,5-bisphosphate (PIP_2) breakdown involving a PIP_2-specific phospholipase C initiates myoblast fusion. This is an attractive hypothesis because the products of PIP_2 breakdown increase intracellular calcium (from an internal source) and activate PKC (Nishizuka, 1984b). 1,2-Diacylglycerol (DAG), a product of PIP_2 breakdown, is rapidly converted to phosphatidic acid (PA) by action of DAG kinase (Wakelam, 1983). Therefore, PIP_2 breakdown likely generates both DAG and PA, two known fusogens (Wakelam, 1985), at the site of membrane union (i.e., the plasma membrane). Interestingly, Navarro (1987) has shown that activation of PKC in chick embryonic myoblasts stimulates calcium uptake (probably by increasing the number of voltage-dependent calcium gates), suggesting that there may be significant "cross-talk" between different signal transduction pathways.

A novel signal transduction mechanism involving phosphatidylcholine (PC) hydrolysis has recently been the subject of much attention (Exton, 1990; Billah and Anthes, 1990). The PC-specific phospholipase is either of the C or D type depending on the system studied and both are activated by TPA, implicating a role for PKC in this phenomenon. It has been proposed that PC hydrolysis may be a mechanism for maintaining membrane-associated PKC activity independent of PIP_2 breakdown or calcium fluctuations (Exton, 1990). TPA-stimulated PC hydrolysis, probably involving a phospholipase C, has been reported in chick embryonic myoblasts (Grove and Schimmel, 1981, 1982). The significance of these observations with respect to myoblast fusion remains fascinating but speculative at this time.

C. The Mechanism of Myoblast Fusion: A Hypothesis Implicating a Role for Calcium and Phospholipid Metabolism

In the mechanism hypothesized, myoblast fusion is triggered by an extracellular factor binding to a specific myoblast receptor. Receptor–ligand binding is proposed to initiate PIP_2 breakdown, which in turn increases intracellular calcium. This calcium, in combination with DAG generated by PIP_2 degradation, then activates PKC. Activated PKC is projected to stimulate an influx of calcium, which would then activate a calmodulin-dependent protein kinase and a calcium-dependent protease, and subsequently trigger the structural rearrangement of the myoblast cytoskeleton and cell surface associated with myoblast fusion. Furthermore, PIP_2 breakdown probably generates both DAG and PA, either of which could trigger membrane union if sufficient concentrations at fusion sites were attained. Since the activation of PKC appears to be an essential event during myoblast fusion, it is tempting to propose a mechanism for fusion that involves PC hydrolysis, considering that a PC-specific phospholipase C (which is activated by PKC) is known to be present in chick embryonic myoblasts.

D. Use of the L_6 Myoblast Cell Line for Studies of Myoblast Fusion

The disadvantages of studies involving primary muscle cultures include the presence of contaminating cell types (usually fibroblasts) and the low proportion of fused cells (usually about 70–80%), with the remaining nonfused cells of being of questionable origin (Klamut *et al.,* 1986). As a result, there always remains the uncertainty that reported observations are a manifestation of the contaminating cells rather than a property of skeletal myogenesis. Primarily for these reasons, the permanent L_6 cell line, originally isolated from rat skeletal muscle by Yaffe (1968), was chosen. However, major drawbacks in this system include the probable but unknown genetic alteration that presumably occurs in immortal cell formation, and the slower rate of fusion when compared to that of primary myoblasts. Despite these drawbacks, the L_6 system is an excellent choice for studies of myoblast fusion since greater than 90% of the cells fuse to form myotubes, and nonfusing mutants (potential controls) are readily isolated (Sanwal, 1979).

Fusion can be studied specifically by culturing myoblasts in a fusion-nonpermissive medium (containing less than 50 μM calcium) until they are confluent. Myoblast fusion is then initiated by switching to a fusion-permissive medium (i.e., containing 1.65 mM calcium) as previously described (Sauro *et al.,* 1988). Using this approach, we have studied and continue to study calcium and phospholipid (particularly inositol phospholipid and PC) metabolism to assess their respective functions during calcium-regulated myoblast fusion. Findings emerging from these studies are now presented.

II. CURRENT RESEARCH

During skeletal myogenesis, there is a marked shift in lipid metabolism from triacylglycerol and phospholipid-synthesizing myoblasts to predominately phospholipid-synthesizing myotubes (Smith and Finch, 1979; Sandra and Ionasescu, 1980; Sauro and Strickland, 1987). Examination of this shift has resulted in studies on the regulation of lipid metabolism during L_6 skeletal myogenesis over the last several years (Sauro *et al.,* 1985; Sauro and Strickland, 1987), with special focus on the significant increase in phospholipid (mostly PC) content noted during L_6 myoblast differentiation (Sauro *et al.,* 1988). Much of the increased phospholipid content is probably owing to increased membrane synthesis (e.g., sarcoplasmic reticulum), and therefore a manifestation of the differentiated state of the cell. However, as will be described, there is growing evidence that some of the changes in phospholipid metabolism during L_6 skeletal myogenesis are related to myoblast fusion.

A. Phospholipid Metabolism during L_6 Myoblast Fusion

There have been several studies done on the plasma membrane phospholipid composition during myoblast fusion. These studies have shown that the relative composition of the major phospholipids remains, in general, unchanged during this event (Kent *et al.*, 1974; Boland *et al.*, 1977; Perkins and Scott, 1978). No significant changes in the relative phospholipid composition during L_6 skeletal myogenesis were observed (Sauro *et al.*, 1988), despite noting a four-fold increase in phospholipid content (when expressed relative to DNA). If there are any alterations in the plasma membrane phospholipid composition during myoblast fusion, these changes are subtle and difficult to observe. Previous investigators (Kent *et al.*, 1974; Boland *et al.*, 1977; Perkins and Scott, 1978) may have failed to note the increase in phospholipid content observed during L_6 skeletal myogenesis because they expressed phospholipid content relative to protein. Our group (Sauro *et al.*, 1985) and others (Lorimer *et al.*, 1987) have shown that protein levels increase two- to three-fold compared to DNA contents (which remain relatively constant) during L_6 skeletal myogenesis. The marked increase in phospholipid content may have been masked by the significant increase in protein during skeletal myogenesis. Because DNA contents are a more constant parameter than protein contents during skeletal myogenesis, DNA contents were used to normalize the data.

The increase in phospholipid content during L_6 skeletal myogenesis appears to result from a transient increase in phospholipid synthesis (Fig. 1), followed by a decrease in phospholipid catabolism (Sandra and Ionasescu, 1980), which was also observed in our system. Interestingly, the burst in phospholipid synthesis coincided with myoblast fusion, which typically started on the fourth day and was essentially complete by the sixth day. Figure 1 also shows that the bulk (greater than 80%) of $[^{32}P]PO_4^{3-}$ ($^{32}P_i$) incorporated into L_6 myoblast and myotube phospholipids was distributed between PC and pooled phosphatidylinositol (PI) and phosphatidylserine (PS). In order to determine if these changes in phospholipid metabolism were related to fusion, phospholipid metabolism was studied during calcium-regulated myoblast fusion as described in the introduction. Myoblasts cultured in calcium-deficient medium (nonpermissive for fusion) were used as controls for these experiments. Furthermore, myotubes were treated similarly to highlight only fusion-related changes in phospholipid metabolism.

Earlier studies showed that $^{32}P_i$ uptake and incorporation into phospholipids were significantly enhanced when cells cultured in calcium-deficient medium were switched to fusion-permissive medium (Sauro *et al.*, 1988). The en-

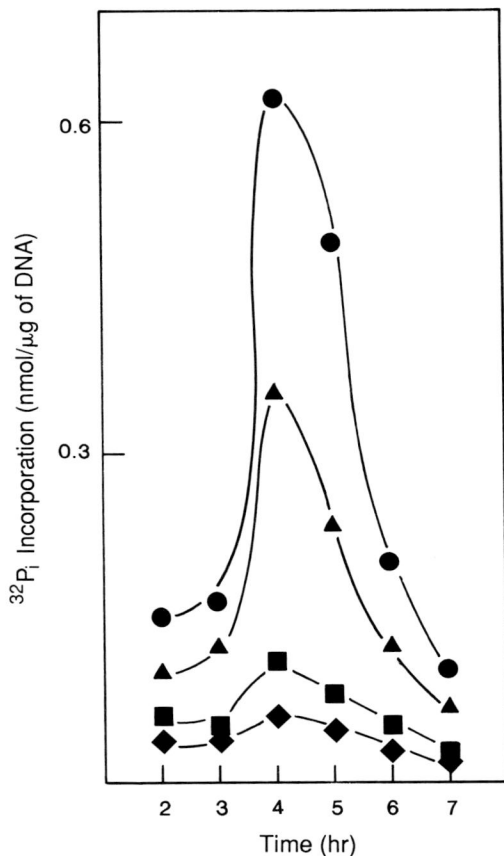

Fig. 1. Phospholipid synthesis during L_6 skeletal myogenesis. L_6 myoblasts were incubated for 6 hr with $^{32}P_i$ (5 μCi/ml of medium) at the same time of day on consecutive days (2–7 days) in culture. Cells were cultured and harvested, and lipids extracted, normalized to DNA, and separated on LK5D thin layer chromatography (TLC) plates as previously described by Sauro *et al.* (1988). Incorporation into PC (●), pooled PI and PS (▲), PE (◆), and sphingomyeliln (■) is shown. Myoblast fusion began on the fourth day and was essentially complete by the sixth day in culture. Results are means of a typical experiment performed in duplicate that varied by less than 15%.

hanced $^{32}P_i$ uptake resulted in a significant increase in radiolabel in all the major phospholipids, but the bulk of the incorporated radiolabel was distributed between PC, pooled PI and PS, and PIP_2. Further studies showed that the predominant phospholipid present in the pooled PI and PS fraction is PI (Sauro *et al.*, 1988), suggesting that the bulk of the incorporated $^{32}P_i$ was distributed between PC and the inositol phospholipids only. Interestingly, $^{32}P_i$ uptake and incorporation into myoblast phospholipids (five-fold over controls) was much greater than that into myotube phospholipids (two-fold over controls).

Since $^{32}P_i$ incorporation into all myoblast and myotube phospholipids was enhanced with the addition of fusion-permissive medium, it seemed probable that switching from calcium-deficient medium to fusion-permissive medium stimulates $^{32}P_i$ uptake. In order to determine if there was an increase in phospholipid synthesis independent of activated $^{32}P_i$ uptake, a similar experiment was performed, but using cells prelabeled with $^{32}P_i$ for the final hour in calcium-deficient medium. Clearly, phospholipid synthesis is stimulated by switching to the fusion-permissive medium (Fig. 2). Furthermore, Fig. 2 shows that PC, pooled PI and PS (but predominantly PI), and PIP_2 synthesis in myoblasts are enhanced to a much greater degree than that seen in myotubes, suggesting that there may be fusion-related changes in these particular phospholipids. In similar experiments using chick embryonic myoblasts, significant changes were noted in the inositol phospholipids only (Wakelam and Pette, 1982; Wakelam, 1983). This apparent contradiction could be due to the different system used in this study (a permanent rat myoblast cell line) compared to that in other studies (primary chick myoblasts). These observations then led to exploration of the metabolism of PC, PI, and PIP_2 (which have all been implicated in signal transduction pathways) during L_6 myoblast fusion.

B. PIP_2 Metabolism during L_6 Myoblast Fusion

Polyphosphoinositide and PI breakdown (via phospholipase C action) occurs rapidly in chick embryonic myoblasts when calcium is added to a calcium-deficient medium (Wakelam and Pette, 1982; Wakelam, 1983). $^{32}P_i$ rapidly incorporates into PIP_2 when L_6 myoblasts are switched from calcium-deficient medium to fusion-permissive medium (Fig. 3). No significant change in $^{32}P_i$ incorporation into PIP_2 is evident when L_6 myotubes are similarly treated, indicating that the enhanced PIP_2 synthesis noted in myo-

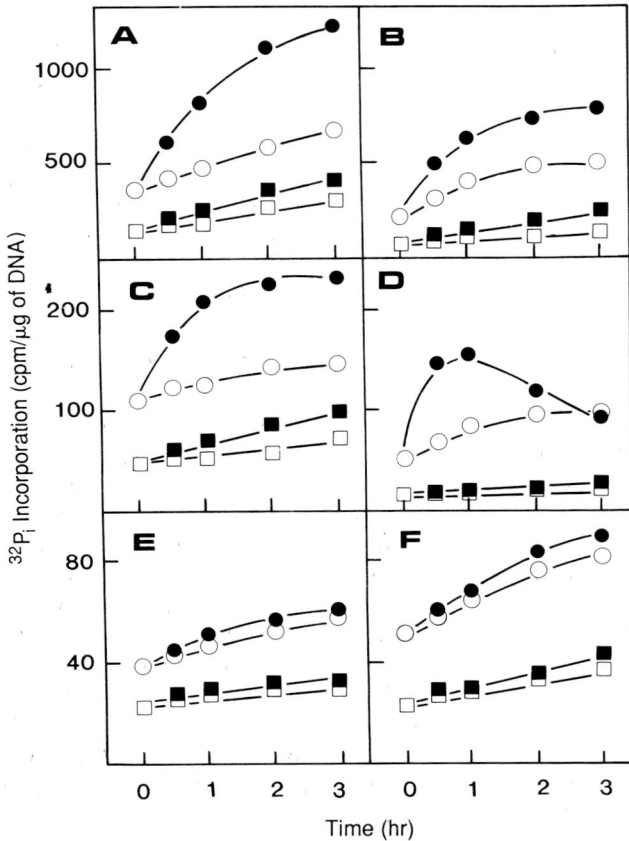

Fig. 2. Fate of $^{32}P_i$-prelabeled lipids of L_6 myoblasts and myotubes incubated in calcium-deficient or fusion-permissive medium. L_6 myoblasts on the third day of culture (●, ○) and myotubes on the sixth day of culture (■, □) were incubated in calcium-deficient medium for 24 hr. $^{32}P_i$ (5 μCi/ml) was added for the final hour of incubation in this medium before cells were switched to either calcium-deficient (○, □) or fusion-permissive (●, ■) medium. Medium was removed, cells were washed and harvested, and phospholipids were extracted and separated following the procedure outlined by Sauro *et al.* (1988). Incorporation into the following lipid fractions is shown in the stated panel: (A) total lipids; (B) PC; (C) pooled PI and PS; (D) PIP_2; (E) PE; (F) sphingomyelin. Results are means of a typical experiment performed in duplicate which varied by less than 10%. The experiment was performed two other times with similar results.

blasts may be a fusion-related event. In addition, this phenomenon appears to be independent of increased protein synthesis because pretreatment with cycloheximide (CHX), which inhibited radioactive leucine incorporation into protein by 93%, did not affect PIP_2 synthesis. Figure 3 also shows that there is little radioactivity associated with phosphatidylinositol 4-monophosphate(PIP), indicating the presence of an active PIP kinase responsible for the near-complete conversion of PIP to PIP_2.

Inositol phospholipids were also prelabeled with $myo[2-^3H]$inositol in order to assess PIP_2 catabolism during calcium-regulated myoblast fusion (Fig. 4). Of the inositol phospholipids studied, only myoblast PIP_2 is significantly degraded (myotube inositol phospholipids are not significantly changed). A somewhat variable decrease was also observed in myoblast PI (results not shown). Furthermore, there was a significant increase in radioactivity associated with PA (to which DAG is rapidly phosphorylated to by action of DAG kinase) in myoblasts, but not myotubes, following the addition of fusion-permissive medium to cells prelabeled with $[U-^{14}C]$glycerol in calcium-deficient medium (Sauro et al., 1988). These observations are consistent with the view that PIP_2 breakdown is mediated by a PIP_2-specific phospholipase C during L_6 myoblast fusion.

PIP_2 turnover appears to be a rapid and early event observed only during calcium-regulated L_6 myoblast fusion (i.e., it is not seen in similarly treated myotubes). With respect to inositol phospholipid metabolism, these observations are consistent with those reported by Wakelam and Pette (1984), and lend support to their hypothesis that PIP_2 breakdown plays a crucial role during myoblast fusion, perhaps even triggering this event. Since little turnover was noted in myotubes, the PIP_2-specific phospholipase C is apparently down-regulated following myoblast fusion, although other more complex interpretations are possible. This conclusion is supported by a recent study that showed that basal hydrolysis of inositol phospholipids is substantially higher in myoblasts than in myotubes (Adamo et al., 1989).

C. Calcium Metabolism during L_6 Myoblast Fusion

Since calcium influx appears to be a prerequisite event for myoblast fusion, an attempt was made to determine if the influx of calcium had any relationship to the observed changes in phospholipid metabolism. Finding showed that intracellular calcium content increased significantly when L_6 myoblasts, cultured in calcium-deficient medium, were switched to fusion-permissive

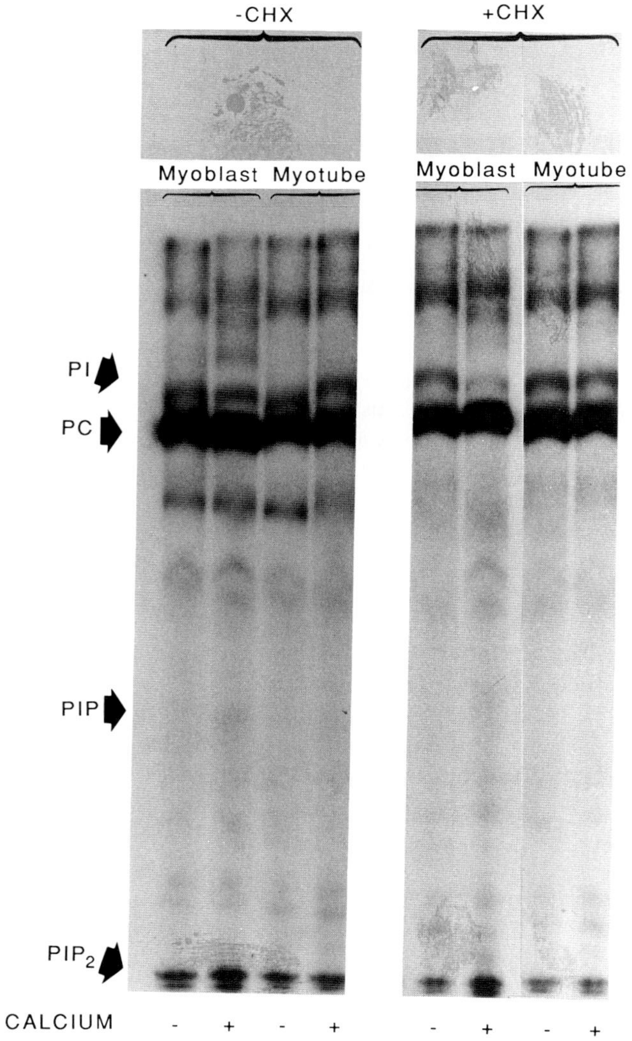

Fig. 3. PIP_2 synthesis in L_6 myoblasts and myotubes incubated in calcium-deficient or fusion-permissive medium. Myoblasts (3 days in culture) or myotubes (6 days in culture) were incubated in calcium-deficient medium for 24 hr before switching either to calcium-deficient (− calcium) or fusion-permissive (+ calcium) medium. Prior to the switching of medium, cells were prelabeled with $^{32}P_i$ (5 μCi/ml of medium) during the last hour of incubation in the calcium-deficient medium, either in the absence (−CHX) or presence (+CHX) of cycloheximide (20 μg/ml of medium). Cells were harvested 15 min after the medium was switched to either calcium-deficient or fusion-permissive medium. Cell culture and harvest, lipid extraction, and polyphosphoinositides separation were performed following the procedures outlined by Sauro *et al.* (1988). The results show an autoradiogram of the TLC. plate after a 24-hr exposure at −80°C using an enhancing screen and Kodak X-Omat X-AR film.

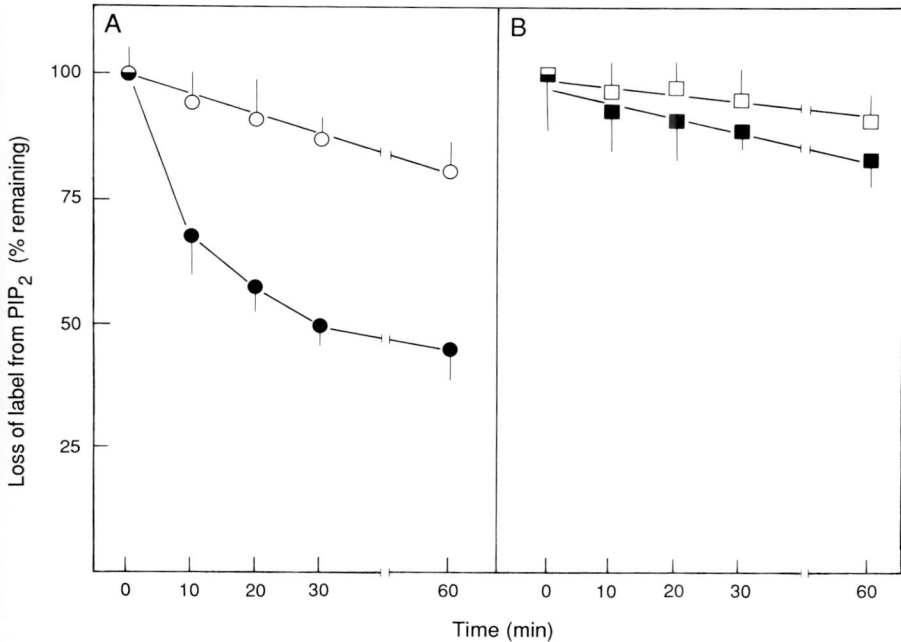

Fig. 4. PIP$_2$ breakdown in L$_6$ myoblasts and myotubes incubated in calcium-deficient or fusion-permissive medium. Myoblasts (●, ○) at day 3 (A) and myotubes (■, □) at day 6 (B) were incubated for 24 hr in calcium-deficient medium containing 1 μCi *myo*[2-^{3}H]inositol/ml of medium. This medium was removed (time zero) and the cells were washed three times with calcium-deficient medium before being replaced by either calcium-deficient (○, □) or fusion-permissive (●, ■) medium. At the indicated times, cells were harvested and lipids were extracted and separated as previously described by Sauro *et al.* (1988). Results are means±SD of five separate determinations.

medium (Sauro *et al.*, 1988). No significant change in intracellular calcium content was observed in similarly treated L$_6$ myotubes. There is a dramatic increase in calcium uptake when L$_6$ myoblasts and myotubes are switched from calcium-deficient medium to fusion-permissive medium (Fig. 5). Calcium uptake reached equilibrium as early as 15 min after the medium was changed (earlier times were studied but yielded highly variable results). Calcium efflux is relatively insignificant during this period, and therefore calcium uptake can act as an approximate measurement of calcium influx. Figure 5 shows that equilibrium-labeling of ^{45}Ca^{2+} in myotubes was approximately twice that of myoblasts, supporting previous observations of increased cal-

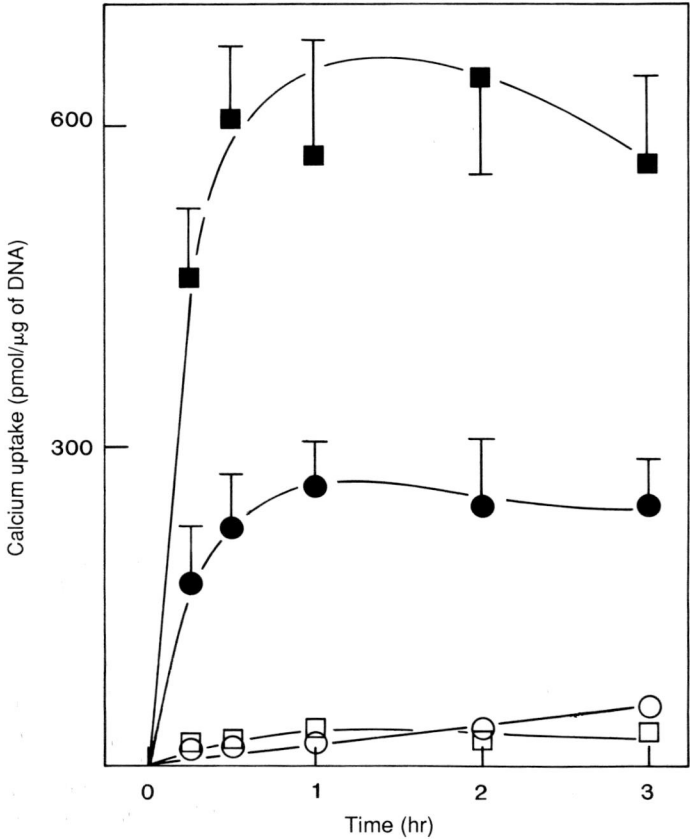

Fig. 5. Calcium uptake during calcium-regulated L_6 myoblast fusion. Myoblasts ($\bullet$, $\bigcirc$) on the third day in culture and myotubes ($\blacksquare$, $\square$) on the sixth day in culture were incubated in calcium-deficient medium for 24 hr. Following this, cells were incubated in either calcium-deficient ($\bigcirc$, $\square$) or fusion-permissive ($\bullet$, $\blacksquare$) medium for the various times stated. $^{45}Ca^{2+}$ (1 μCi/ml of medium) was added immediately to initiate the assay. At the various times indicated, medium was removed, cells were washed and harvested, and calcium uptake was determined following the procedures outlined by Sauro *et al.* (1988). Calcium uptake at 4°C was subtracted from that at 37°C. Results are means±SD of four to seven separate determinations. Where no error bars are shown, errors are less than symbol height.

cium content during skeletal myogenesis (Croop et $al.$, 1982; Sauro et $al.$, 1988). The kinetics of calcium uptake in these experiments are similar to that of PIP_2 breakdown, making it difficult to determine which event occurs first. Although receptor-mediated PIP_2 breakdown is known to elevate intracellular calcium levels, it is also plausible that the influx of calcium could activate the PIP_2-specific phospholipase C, independent of receptor occupancy. Studies with the calcium ionophore A23187 indicate that calcium influx is capable of stimulating pooled PI and PS (but predominantly PI) synthesis to a much greater degree in myoblasts compared to myotubes (Fig. 6). Preliminary studies suggest that similar treatments with A23187 also enhances PIP_2 turnover to a greater degree in myoblasts compared to myotubes. The small stimulatory effect that A23187 has on pooled PI and PS synthesis and PIP_2 turnover in myotubes may be due to the presence of unfused myoblasts (generally, 5–10% of the cells) contaminating the myotube samples. These observations indicate that calcium influx is capable of triggering PIP_2 breakdown probably in myoblasts only, and lend further support to the view that the PIP_2-specific phospholipase C is down-regulated following L_6 myoblast fusion.

D. Phosphatidylcholine Metabolism during L_6 Myoblast Fusion

Because PC contents increase dramatically during L_6 skeletal myogenesis, interest arose in the regulation of PC synthesis during myoblast differentiation. Because PKC is believed to activate PC synthesis (Pelech and Vance, 1989), the burst in phospholipid synthesis shown in Fig. 1 may be the manifestation of a transient activation of PKC during L_6 skeletal myogenesis. This led to an investigation of the role of PKC in the regulation of PC metabolism.

In both myoblasts and myotubes, TPA stimulates a rapid and dramatic increase in $^{32}P_i$ incorporation into PC (and to a lesser degree PA) compared to $^{32}P_i$ incorporation into the remaining phospholipids (Fig. 6). Furthermore, 4α-phorbol-12,13-didecanoate or 4αPDD (a phorbol ester that does not activate PKC) did not significantly alter $^{32}P_i$ incorporation into PC compared to that of untreated control cells. These results support the observations of Hill et $al.$ (1984), and suggest that TPA treatment (probably by PKC activation) stimulates PC synthesis in L_6 cells. However, TPA did not significantly stimulate radioactive choline incorporation into myoblast or myotube PC (results not shown). The apparent contradiction in these observations led to a more thorough examination of the effects of TPA on PC metabolism.

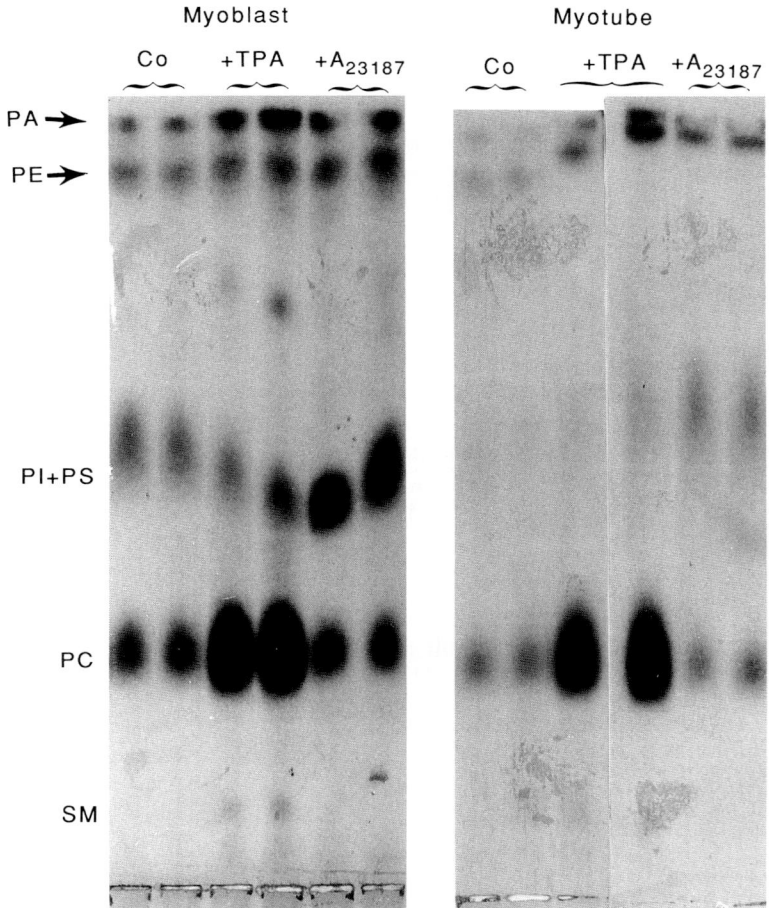

Fig. 6. Effects of TPA and A23187 on the incorporation of $^{32}P_i$ into L_6 myoblast and myotube phospholipids. L_6 myoblasts (3 days in culture) and myotubes (6 days in culture) were prelabeled with $^{32}P_i$ (0.5 μCi/ml of medium) for 60 min (pulse label). The medium was removed and cells were washed twice using phosphate-buffered saline (pH 7.8) before addition of fusion-permissive medium and either 0.1 μM TPA or 5 μM A23187. Cells were cultured and harvested, and lipids were extracted and separated as previously described by Sauro *et al.* (1988). The results show an autoradiogram of the TLC plate after a 12-hr exposure at $-80°C$ using an enhancing screen and Kodak X-Omat X-AR film.

Given that TPA activates PC hydrolysis in a number of different cell types, a study was made on the effect of TPA on both myoblasts and myotubes prelabeled with [methyl-^{3}H]choline. Figure 7 shows that TPA activates PC hydrolysis rapidly and significantly when compared to 4αPDD-treated controls, with the hydrolyzed choline metabolites accumulating in the medium.

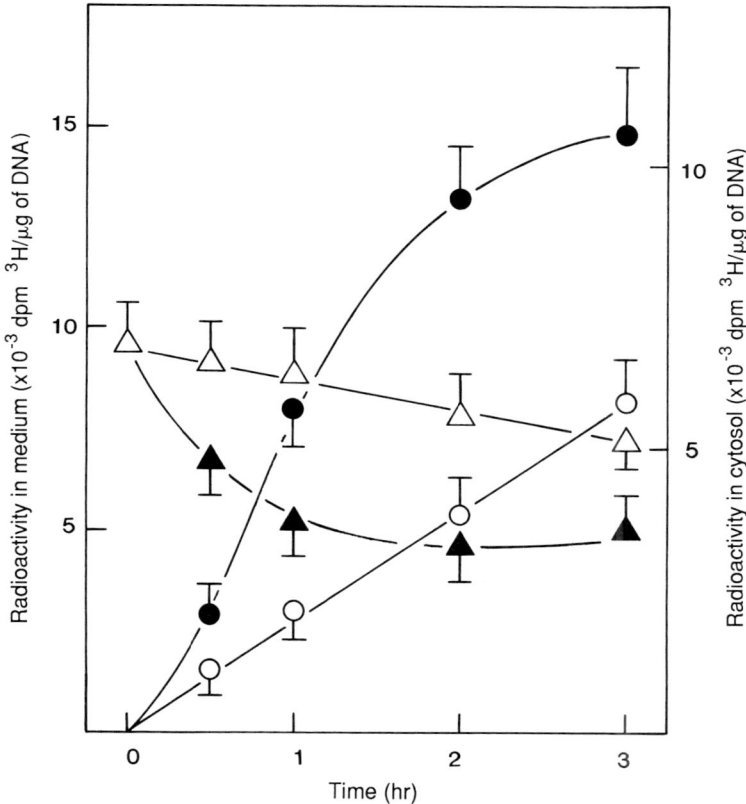

Fig. 7. Effect of TPA on L$_6$ myoblasts prelabeled with [methyl-^{3}H]choline. L$_6$ myoblasts (3 days in culture), were prelabeled with [*methyl*-^{3}H]choline(1 μCi/ml of medium) for 24 hr. Following this pulse-label period, medium was removed and cells were washed twice using phosphate-buffered saline (pH 7.8) before addition of fusion-permissive medium and either 0.1 μM 4αPDD (○, △) or 0.1 μM TPA (●, ▲) for the indicated times. L$_6$ myoblasts were cultured and harvested, lipids were extracted and separated, and radiolabel remaining in the cytosol (▲, △) or secreted into the medium (●, ○) were determined as previously described by Sauro *et al.* (1988).

There was also a significant decrease in the cytosol-associated radioactivity of TPA-treated cells, but the absolute decrease and the kinetics of this decrease preclude the cytosol as a potential source for medium-associated radioactivity. The only pool of radioactivity in this study that could serve as a plausible source for the TPA-stimulated radioactivity in the medium is PC (greater than 80% of the radioactive choline was present in PC at the onset of this experiment). The TPA-stimulated decrease in cytosol-associated radioactivity was probably due to compensatory resynthesis of the degraded PC, although no significant decrease in prelabeled PC was observed. Detecting significant PC breakdown in these studies is difficult, since TPA-stimulated PC hydrolysis over that of controls represents only a small fraction (a maximum of 10–15%) of the total radioactivity initially associated with PC.

The previously described inability of TPA to increase choline incorporation into PC could be explained if PC hydrolysis were tightly coupled to the compensatory resynthesis of PC. Although PC synthesis is activated in TPA-treated cells, choline would fail to accumulate into PC since it was simultaneously being degraded. The increased $^{32}P_i$ incorporation into PC of TPA-treated cells noted in Fig. 6 probably reflects enhanced PC turnover. The addition of purified phospholipase C to chick embryonic myoblasts (Kent, 1979) failed to significantly reduce PC content despite stimulating PC turnover over fivefold, indicating that PC catabolism and compensatory resynthesis are indeed tightly coupled in myoblasts.

Studies using L_6 myoblasts have shown that TPA (at concentrations that stimulate PC hydrolysis) rapidly increases PKC activity associated with the crude particulate fraction, while simultaneously decreasing PKC activity associated with the cytosolic fraction (Narindrasorasak *et al.*, 1987). Preliminary studies have also noted that PC hydrolysis can be activated by 1-oleoyl-2-acetyl-diacylglycerol (a known activator of PKC). Furthermore, TPA-stimulated PC hydrolysis is significantly inhibited by 1-(5-isoquinolinesulfonyl)-2-methyl-piperazine(H7) and sphingosine (known inhibitors of PKC). These preliminary observations strongly implicate PKC in the mechanism of TPA-stimulated PC hydrolysis.

Preliminary work has also shown that the predominant choline metabolite (greater than 80%) isolated from the medium of TPA-treated cells is phosphocholine. Furthermore, there is a significant increase in DAG contents in TPA-stimulated cells compared to 4αPDD-treated cells, similar to that reported by Grove and Schimmel (1981). These observations suggest that TPA stimulates a PC-specific phospholipase C, possibly by PKC-mediated phosphorylation and subsequent activation of this enzyme. PC localized to the outer leaflet of the plasma membrane bilayer appears to be the only logical source for medium-associated phosphocholine. Also, any significant uptake of [methyl-^{3}H]phosphocholine in L_6 cells could not be demonstrated, suggesting that a phosphocholine transporter is not present in this system, and supporting the view that phospho[*methyl*-^{3}H]choline from the cytosol is not the source of medium-associated radioactivity. TPA did not significantly enhance medium-associated radioactivity in cells prelabeled with radioactive inositol, ethanolamine, serine, or 2-deoxyglucose. These observations preclude cytotoxic effects of this treatment as the mechanism responsible for enhanced medium-associated radioactivity, and further support the view that TPA activates a phospholipase C that is specific for PC.

III. FINAL COMMENTS

Myoblast fusion is a complex phenomenon that has intrigued biologists for some time. An understanding of the mechanism of myoblast fusion may not be possible until all the prerequisite biochemical changes that precede membrane union are determined. Furthermore, the mechanism of myoblast fusion may not be universal, a point that could explain the wide range of sometimes-conflicting observations that have been reported in this field. Indeed, primary cultures of chick embryonic myoblasts have an absolute requirement for the presence of embryo extract in the medium to maintain viability; L_6 myoblasts (used in these studies) have no such requirement, suggesting that there are fundamental differences between these systems (Wakelam, 1985; Wakelam, 1988). Therefore, comparisons made between different model systems may not be valid. It is hoped that a better understanding of L_6 myoblast fusion has been attained from these studies, but care must be taken in applying conclusions drawn from these experiments to other model systems.

A. Signal Transduction during L_6 Myoblast Fusion

The mechanism of L_6 myoblast fusion does not appear to involve any obvious changes in the predominant phospholipid composition of myoblast membranes, but this does not preclude the possibility that there may be more subtle changes leading to a more fusogenic membrane. Indeed, these studies suggest that there are marked changes in phospholipid metabolism during L_6 skeletal myogenesis that may be related to myoblast fusion. Wakelam and Pette (1984) have proposed that receptor-mediated PIP_2 breakdown triggers myoblast fusion. Because PIP_2 breakdown activates PKC (apparently an essential event during myoblast fusion) and because PKC activation enhances calcium uptake in myoblasts (Navarro, 1987), it has been suggested that PIP_2 breakdown precedes (and indirectly stimulates) the influx of calcium. However, the kinetics of PIP_2 breakdown and calcium influx during calcium-regulated L_6 myoblast fusion are similar. Thus, these studies failed to determine the sequential order of these events. PIP_2 breakdown appears to be a fusion-related event since it was only observed in myoblasts stimulated to fuse by the addition of fusion-permissive medium, but not in myotubes treated similarly. However, there is no evidence to suggest that PIP_2 breakdown is causally related to myoblast fusion. Indeed, these studies indicate that an influx of calcium is capable of triggering PIP_2 breakdown. Calcium influx is known to be an essential event preceding myoblast fusion (David *et al.*, 1981; Rapuano *et al.*, 1989). It is possible, and perhaps more plausible, that the influx of calcium coincidently triggers PIP_2 breakdown independent of receptor occupancy.

The causal or coincidental relationship between PIP_2 turnover and myoblast fusion remains an open question (Wakelam and Pette, 1984). However, it is believed that PIP_2 breakdown observed during calcium-regulated L_6 myoblast fusion is triggered by the influx of calcium that occurs when calcium-deficient medium is replaced by fusion-permissive medium. Because no significant PIP_2 turnover was noted in L_6 myotubes in these studies, it can be concluded that the PIP_2 specific phospholipase C was down-regulated following myoblast fusion. Adamo *et al.* (1989) have shown that membrane-associated PKC activity is high in myoblasts but decreases significantly following myoblast fusion. The function of PIP_2 breakdown during myoblast fusion (if any) remains speculative, but the observations by Adamo *et al.* (1989) may be a reflection of decreased PIP_2 breakdown during skeletal myogenesis. One function of PIP_2 breakdown in L_6 myoblasts may be to maintain the membrane-associated PKC activity that appears to be essential prior to fusion. When the fusion of myoblasts is complete, the PIP_2-specific phospholipase C is likely down-regulated and subsequently, membrane-associated PKC activity is attenuated.

Wakelam and Pette (1984) have suggested that one function of PIP_2 breakdown may be to generate fusogenic lipids (DAG or PA), which would subsequently trigger the union of closely apposed myoblast plasma membranes. However, a significant increase in DAG content in association with PIP_2 breakdown during calcium-regulated L_6 myoblast fusion was not detected in these studies, suggesting that the quantity of DAG generated is minimal. In stark contrast, TPA-treated myoblasts were found to have significantly greater

DAG contents (nearly double) compared to that of 4αPDD-treated control cells. In relatively high concentrations (20–30 mole %), DAG causes major structural transitions and perturbations of PC bilayers that presumably enhance the fusion of these membranes (Das and Rand, 1984, 1986; Parsegian et al., 1984; De Boeck and Zidovetzki, 1989).

Although only speculation at this time, it is proposed that membrane-associated PKC, maintained by basal PIP_2 breakdown, activates a PC-specific phospholipase C, which in turn stimulates PC hydrolysis (possibly at discrete contact sites on the myoblast cell surface where plasma membranes are in close apposition) and generates measurable quantities of DAG. Phosphatidylcholine hydrolysis could also explain the increased membrane fluidity that is associated with the onset of myoblast fusion (Prives and Shinitzky, 1977). In these studies, a maximum of 10–15% of the total PC is hydrolyzed above basal rates in TPA-treated cells. However, if PC hydrolysis is localized to the outer leaflet of the plasma membrane bilayer (as has been suggested), a much larger fraction of the PC present in this site could be hydrolyzed (perhaps generating a sufficient DAG concentration to trigger membrane union). Furthermore, PC hydrolysis is envisioned to significantly reduce the repulsive hydration barrier by removing polar phosphocholine headgroups from PC of the outer leaflets of juxtaposed myoblast membranes, as well as simultaneously generating sufficient local DAG concentrations to trigger fusion.

Such a view is supported by preliminary findings on PC hydrolysis and the known constitution and location of PC. PC constitutes approximately 50% of the L_6 myoblast plasma membrane phospholipid (Perkins and Scott, 1978) and is the dominant phospholipid in the outer leaflet of the plasma membrane (Sessions and Horwitz, 1983). The preliminary findings, as commented on, strongly suggest that PKC activates a specific phospholipase C that acts on PC to release phosphocholine and DAG.

PKC-activated hydrolysis is intricately involved in the mechanism of L_6 myoblast fusion. However, TPA-stimulated PC hydrolysis has been reported in many different cell types (Exton, 1990; Billah and Anthes, 1990) that do not fuse. Obviously, PC hydrolysis alone is insufficient to impart fusion competence. Rather, PC hydrolysis is envisioned as being a permissive event in the mechanism of myoblast fusion, which requires further biochemical alterations that are unique to fusion-competent myoblasts in order for fusion to occur. Inhibition of myoblast fusion by chronic exposure to TPA (Cohen et al., 1977) may be due to the down-regulation of PKC (Narindrasorasak et al., 1987), and thus need not conflict with the view that PKC plays an essential role during L_6 myoblast fusion.

B. The Mechanism of L_6 Myoblast Fusion: A Revised Hypothesis

Several adaptations must be made to the hypothetical mechanism of myoblast fusion outlined in the introduction. Calcium influx and PIP_2 breakdown are events that clearly precede myoblast fusion. Receptor–ligand binding (alluded to earlier in the introduction) could trigger calcium influx rather than PIP_2 breakdown, as was previously suggested. The influx of calcium would then precede, and subsequently initiate, PIP_2 breakdown. We believe that PIP_2 turnover in myoblasts serves to maintain membrane-associated PKC activity during myoblast fusion, and that PIP_2 breakdown during calcium-regulated myoblast fusion is a consequence of calcium influx, although further studies are needed to confirm this point. The influx of calcium could then stimulate a calcium-dependent protein kinase and a calcium-activated protease, which in turn would generate the structural rearrangements associated with the myoblast cell surface during fusion. Membrane-associated PKC, maintained by PIP_2 breakdown, activates a PC-specific phospholipase C. The resultant PC hydrolysis, in conjunction with the biochemical modifications of the myoblast cell surface, could trigger myoblast fusion. The combination of PIP_2 breakdown and PC hydrolysis is proposed to cause both bilayer disruption (by increasing DAG and PA concentrations) and local dehydration (by stripping off the polar headgroups of PC present in the outer leaflet of closely apposed plasma membranes) at fusion sites, thus meeting the requirements for membrane fusion suggested by Hui *et al.* (1981).

ACKNOWLEDGMENTS

This work was supported by the Medical Research Council of Canada (Grant MT-0617).

REFERENCES

Adamo, S., Caporale, C., Nervi, C., Ceci, R., and Molinaro, M. (1989). Activity and regulation of calcium and phospholipid-dependent protein kinase in differentiating chick myogenic cells. *J. Cell Biol.* **108,** 153–158.

Bar-Sagi, D., and Prives, J. (1983). Trifluoperazine, a calmodulin antagonist, inhibits muscle cell fusion. *J. Cell Biol.* **97,** 1375–1380.

Billah, M. M., and Anthes, J. C. (1990). The regulation and cellular functions of phosphatidylcholine hydrolysis. *Biochem. J.* **269,** 281–291.

Bischoff, R. (1978). Myoblast fusion. *In* "Cell Surface Reviews" (G. Poste and G. L. Nicolson, eds.), Vol. 5, pp. 127–179. Elsevier/North Holland Biomedical Press, New York.

Boland, R., Chyn, T., Roufa, D., Reyes, E., and Martonosi, A. (1977). The lipid composition of muscle cells during development. *Biochim. Biophys. Acta* **489,** 349–359.

Cohen, R., Pacifici, M., Rubinstein, N., Biehl, J., and Holtzer, H. (1977). Effect of a tumor promoter on myogenesis. *Nature* **266,** 538–540.

Croop, J., Dubyak, G., Toyama, Y., Dlugosz, A., Scarpa, A., and Holtzer, H. (1982). Effects of 12-O-tetradecanoylphorbol 13-acetate on myofibril integrity and Ca^{2+} content in developing myotubes. *Dev. Biol.* **89,** 460–474.

Das, S., and Rand, P. R. (1984). Diacylglycerol causes major transitions in phospholipid bilayer membranes. *Biochem. Biophys. Res. Commun.* **124,** 491–496.

Das, S., and Rand, R. P. (1986). Modification by diacylglycerol of the structure and interaction of various phospholipid bilayer membranes. *Biochemistry* **25,** 2882–2889.

David, J. D., and Higginbotham, C. (1981). Fusion of chick embryo skeletal myoblasts: Interactions of prostaglandin E_1, adenosine 3':5' monophosphate, and calcium influx. *Dev. Biol.* **82,** 308–316.

David, J. R., See, W. M., and Higginbotham, C. (1981). Fusion of chick embryo skeletal myoblasts: Role of calcium influx preceding membrane union. *Dev. Biol.* **82,** 297–307.

David, J. D., Faser, C. R., and Perrot, G. P. (1990). Role of protein kinase C in chick embryo skeletal myoblast fusion. *Dev. Biol.* **139,** 89–99.

David, R. L., Weintraub, H., and Lassar, A. (1987). Expression of a single transfected cDNA converts fibroblasts to myoblasts. *Cell* **51,** 987–1000.

De Boeck, H., and Zidovetzki, R. (1989). Effects of diacylglycerols on the structure of phosphatidylcholine bilayers: A ^2H and ^{31}P NMR study. *Biochemistry* **28,** 7439–7446.

Delain, D., Wahrmann, J. P., and Gros, F. (1981). Influence of external diffusible factors on myogenesis of the cells of the line L_6. *Exp. Cell Res.* **131,** 217–224.

Entwistle, A., Curtis, D. H., and Zalin, R. J. (1986). Myoblast fusion is regulated by a prostanoid of the one series independently of a rise in cyclic AMP. *J. Cell Biol.* **103,** 857–866.

Exton, J. H. (1990). Signaling through phosphatidylcholine breakdown. *J. Biol. Chem.* **265,** 1–4.

Farzaneh, F., Entwistle, A., and Zalin, R. J. (1989). Protein kinase C mediates the hormonally regulated plasma membrane fusion of avian embryonic skeletal muscle. *Exp. Cell Res.* **181,** 298–304.

Fulton, A. B., Prives, J., Farmer, S. R., and Penman, S. (1981). Developmental reorganization of the skeletal framework and its surface lamina in fusing muscle cells. *J. Cell Biol.* **91,** 103–112.

Grove, R. I., and Schimmel, S. D. (1981). Generation of 1,2-diacylglycerol in plasma membranes of phorbolester-treated myoblasts. *Biochem. Biophys. Res. Commun.* **102,** 158–164.

Grove, R. I., and Schimmel, S. D. (1982). Effects of 12-O-tetradecanoylphorbol 13-acetate on glycerolipid metabolism in cultured myoblasts. *Biochim. Biophys. Acta* **711,** 272–280.

Herman, B. A., and Fernandez, M. (1978). Changes in membrane dynamics associated with myogenic cell fusion. *J. Cell Physiol.* **94,** 253–264.

Herman, B. A., and Fernandez, M. (1982). Dynamics and topographical distribution of surface glycoproteins during myoblast fusion: A resonance energy transfer study. *Biochemistry* **21,** 3275–3283.

Hill, S. A., McMurray, W. C., and Sanwal, B. D. (1984). Regulation of phosphatidylcholine synthesis and the activity of CTP: Cholinephosphotransferase in myoblasts by 12-*O*-tetradecanoylphorbol-13-acetate. *Can. J. Biochem. Cell Biol.* **62,** 369–374.

Hui, S. W., Stewart, T. P., Boni, L. T., and Yeagle, P. L. (1981). Membrane fusion through point defects in bilayers. *Science* **212,** 921–923.

Kalderon, N., and Gilula, N. B. (1979). Membrane events involved in myoblast fusion. *J. Cell Biol.* **81,** 411–425.

Kawasaki, Y., Wakayama, N., and Seto-Ohshima, A. (1988). Lateral motion of fluorescent molecules embedded into cell membranes of clonal myogenic cells, L_6, changes on cell maturation. *FEBS Lett.* **231,** 321–326.

Kent, C. (1979). Stimulation of phospholipid metabolism in embryonic muscle cells treated with phospholipase C. *Proc. Natl. Acad. Sci. USA* **76,** 4474–4478.

Kent, C., Schimmel, S. D., and Vagelos, P. R. (1974). Lipid composition of plasma membranes from developing muscle cells in culture. *Biochim. Biophys. Acta* **360,** 312–321.

Klamut, H. J., Lin, C. H., and Strickland, K. P. (1986). Normal and dystrophic hamster myoblast and fibroblast growth in culture. *Muscle Nerve* **9,** 597–605.

Knudsen, K. A. (1991). Fusion of myoblasts. *In* "Membrane Fusion" (J. Wilschut and D. Hoekstra, eds.), pp. 601–626. Marcel Dekker, Inc., New York.

Konigsberg, I. R. (1971). Diffusion-mediated control of myoblast fusion. *Dev. Biol.* **26,** 133–152.

Lipton, B. H., and Konigsberg, I. R. (1972). A fine structural analysis of the fusion of myogenic cells. *J. Cell Biol.* **53,** 348–364.

Lorimer, I. A. J., and Sanwal, B. D. (1989). Regulation of cyclic AMP-dependent protein kinase levels during skeletal myogenesis. *Biochem. J.* **264,** 305–308.

Lorimer, I. A. J., Mason, M. E., and Sanwal, B. D. (1987). Levels of type I cAMP-dependent protein kinase regulatory subunit are regulated by changes in turnover rate during skeletal myogenesis. *J. Biol. Chem.* **262,** 17200–17205.

Martelly, I., Gautron, J., and Moraczewski, J. (1989). Protein kinase C activity and phorbol ester binding to rat myogenic cells during growth and differentiation. *Exp. Cell Res.* **183,** 92–100.

Murre, C., McCaw, P. S., Vaessin, H., Caudy, M., Jan, L. Y., Cabrera, C. V., Buskin, J. N., Hauschka, S. D., Lassar, A. B., Weintraub, H., and Baltimore, D. (1989). Interactions between heterologous helix-loop-helix proteins generate complexes that bind specifically to a common DNA sequence. *Cell* **58,** 537–544.

Narindrasorasak, S., Brickenden, A., Ball, E. H., and Sanwal, B. D. (1987). Regulation of protein kinase C by cyclic adenosine 3′:5′-monophosphate and a tumor promoter in skeletal myoblasts. *J. Biol. Chem.* **262,** 10,497–10,501.

Navarro, J. (1987). Modulation of [^{3}H]dihydropyridine receptors by activation of protein kinase C in chick muscle cells. *J. Biol. Chem.* **262,** 4649–4652.

Nishizuka, Y. (1984a). The role of protein kinase C in cell surface signal transduction and tumor promotion. *Nature* **308,** 693–698.

Nishizuka, Y. (1984b). The turnover of inositol phospholipids and signal transduction. *Science* **225,** 1365–1370.

Nishizuka, Y. (1986). Studies and perspectives of protein kinase C. *Science* **233,** 305–312.

Parsegian, V. A., Rand, R. P., and Gingell, D. (1984). Lessons for the study of membrane fusion from membrane interactions in phospholipid systems. *In* Ciba Foundation Symposium: "Cell Fusion," Vol. 103, pp. 9–27. Pitman Books, London.

Pelech, S. L., and Vance, D. E. (1989). Signal transduction via phosphatidylcholine cycles. *Trends Biochem. Sci.* **157,** 28–30.

Perkins, R. G., and Scott, R. E. (1978). Plasma membrane phospholipid cholesterol and fatty acyl composition of differentiated and undifferentiated L_6 myoblasts. *Lipids* **13**, 334–337.

Prives, J., and Shinitzky, M. (1977). Increased membrane fluidity precedes fusion of muscle cells. *Nature* **268**, 761–763.

Rapuano, M., Ross, A. F., and Prives, J. (1989). Opposing effects of calcium entry and phorbol esters on fusion of chick muscle cells. *Dev. Biol.* **134**, 271–278.

Rogers, J. E., Narindrasorasak, S., Cates, G. A., and Sanwal, B. D. (1985). Regulation of protein kinase and its regulatory subunits during skeletal myogenesis. *J. Biol. Chem.* **260**, 8002–8007.

Sandra, A., and Ionasescu, V. V. (1980). Alterations in lipid turnover in developing muscle. *Biochem. Biophys. Res. Commun.* **93**, 898–905.

Sanwal, B. D. (1979). Myoblast differentiation. *Trends Biochem. Sci.* **4**, 155–157.

Sauro, V. S., and Strickland, K. P. (1987). Changes in oleic acid oxidation and incorporation into lipids of differentiating L_6 myoblasts cultured in normal or fatty acid-supplemented growth medium. *Biochem. J.* **244**, 743–748.

Sauro, V. S., Klamut, H. J., Lin, C. H., and Strickland, K. P. (1985). Lysosomal triacylglycerol lipase activity in L_6 myoblasts and its changes on differentiation. *Biochem. J.* **227**, 583–589.

Sauro, V. S., Brown, G. A., Hamilton, M. R., Strickland, C. K., and Strickland, K. P. (1988). Changes in phospholipid metabolism during calcium-regulated myoblast fusion. *Biochem. Cell Biol.* **66**, 1110–1118.

Schollmeyer, J. E. (1986). Role of Ca^{2+} and Ca^{2+}-activated protease in myoblast. *Exp. Cell Res.* **162**, 411–422.

Schutzle, U. B., Wakelam, M. J. O., and Pette, D. (1984). Prostaglandins and cyclic AMP stimulate creatine kinase synthesis but not fusion in cultured embryonic chick muscle cells. *Biochim. Biophys. Acta* **805**, 204–210.

Sessions, A., and Horwitz, A. F. (1983). Differentiation-related differences in the plasma membrane phospholipid asymmetry of myogenic and fibrogenic cells. *Biochim. Biophys. Acta* **728**, 103–111.

Shainberg, A., Yagil, G., and Yaffe, D. (1971). Alterations of enzymatic activities during muscle differentiation *in vitro*. *Dev. Biol.* **25**, 1–29.

Smith, P. B., and Finch, R. A. (1979). Alterations in lipid metabolism of developing muscle cells in culture. *Biochim. Biophys. Acta* **572**, 139–145.

Sundler, R. (1984). Role of phospholipid head group structure and polarity in the control of membrane fusion. *In* "Biomembranes: Membrane Fluidity" (M. Kates and L. A. Manson, eds.), Vol. 12, pp. 563–583. Plenum Press, New York.

van der Bosch, J., Schudt, C., and Pette, D. (1972). Quantitative investigation on Ca^{2+}-and pH-dependence of muscle cell fusion *in vitro*. *Biochem. Biophys. Res. Commun.* **48**, 326–332.

Wakelam, M. J. O. (1983). Inositol phospholipid metabolism and myoblast fusion. *Biochem. J.* **214**, 77–82.

Wakelam, M. J. O. (1985). The fusion of myoblasts. *Biochem. J.* **228**, 1–12.

Wakelam, M. J. O. (1988). Myoblast fusion: A mechanistic analysis. *In* "Current Topics in Membranes and Transport" (F. Bronner and N. Duzganes, eds.), Vol. 32, pp. 87–112.

Wakelam, M. J. O., and Pette, D. (1982). The breakdown of phosphatidylinositolin myoblasts stimulated to fuse by the addition of Ca^{2+}. *Biochem. J.* **202**, 723–729.

Wakelam, M. J. O., and Pette, D. (1984). Myoblast fusion and inositol phospholipid breakdown:

Causal relationship or coincidence? *In* "Ciba Foundation Symposium: Cell Fusion," Vol. 103, pp. 100–118. Pitman books, London.

Yaffe, D. (1968). Retention of differentiation potentialities during prolonged cultivation of myogenic cells. *Proc. Natl. Acad. Sci. USA* **61,** 477–483.

Zalin, R. J. (1987). The role of hormones and prostanoids in the *in vitro* proliferation and differentiation of human myoblasts. *Exp. Cell Res.* **172,** 265–281.

Zalin, R. J., and Montague, W. (1974). Changes in adenylate cyclase, cyclic AMP, and protein kinase levels in chick myoblasts, and their relationship to differentiation. *Cell* **2,** 103–108.

11

Protein Kinase C, Membrane Protein Phosphorylation, and Calcium Influx in Chick Embryo Skeletal Myoblast Fusion

JOHN D. DAVID AND ANN FITZPATRICK

Division of Biological Sciences
University of Missouri
Columbia, Missouri

I. INTRODUCTION

The functional element of differential skeletal muscle, the multinucleate myotube, is formed by cytoplasmic fusion of mononucleated precursor cells called myoblasts. Primary cultures of chick embryo skeletal muscle myoblasts go through four morphologically and biochemically distinct stages (Bischoff, 1978; Wakelam, 1985). (1) Freshly isolated myoblasts initially proliferate independently and are quite motile compared to other cultured cells. Medium composition and plating density determine the time of withdrawal from the

223

SIGNAL TRANSDUCTION DURING
BIOMEMBRANE FUSION

cell cycle. (2) As cell division ceases, migration continues and the single cells align side-by-side and end-to-end in long chains in a process termed recognition-alignment. (3) Alignment is followed by adhesion, defined as the stage immediately prior to membrane union when aggregates are resistant to dispersal by EDTA. (4) Membrane fusion, yielding the multinucleate syncytium or myotube, typically involves 60–80% of the cells, never reaching 100% even when the culture is initiated with a pure clone of myogenic cells. The capacity to fuse and form multinucleate myotubes is a specific and unique property of myoblasts; nuclei from heterologous cells are not incorporated into myotubes. Myogenic cells recognize and speficially adhere to and fuse with myotubes whereas nonmyogenic cells do not adhere or alter their movement when contacting myotubes.

Despite extensive biochemical and morphological investigation of myoblast fusion, the physical mechanism of fusion is as yet not completely understood (reviewed in Wakelam, 1985). However, like many other developmental processes, myoblast fusion is regulated by environmental signals, both extracellular and intracellular in origin, and many of the signal elements have recently been identified. They appear to form a regulatory cascade that includes prostaglandin E_1 (Zalin, 1977; Hausman et al., 1986); phosphatidylinositol metabolites (Elgendy and Hausman, 1990); intracellular sequestered Ca^{2+} (David et al., 1990); protein kinase C (PKC) (David et al., 1990); extracellular Ca^{2+} (Shainberg et al., 1969); and the Ca^{2+}–calmodulain-dependent protein kinase (Bar-Sagi and Prives, 1983).

This chapter (1) provides further evidence that PKC activation is an obligatory step in the signal cascade, (2) identifies two plasma membrane phosphopeptides whose phosphorylation is dependent on stimulation of PKC, (3) demonstrates that intracellular Ca^{2+} release is an obligatory step in the signal cascade, and probably activates PKC, (4) integrates protein phosphorylation and Ca^{2+} influx, and (5) presents a working model of the signal cascade.

II. CURRENT RESEARCH

A. Direct Modulation of Intracellular Protein Phosphorylation also Modulates Myoblast Fusion

1. Alteration of Protein Kinase C Activity Alters the Kinetics of Fusion

A crucial role for PKC in fusion of chick embryo skeletal myoblasts has been demonstrated (David et al., 1990). Addition of the PKC activator 12-O-tetradecanoylphorbol-13-acetate (TPA) at 0.2 nM, after cell alignment and 72 hr after plating, generates precocious fusion (Fig. 1, Table I). The final extent of fusion in a TPA-treated culture is no greater than in a control culture, however the kinetics of fusion are altered as the process is shifted forward approximately 10 hr. Treatment with either higher concentrations of TPA or treatment initiated prior to cell alignment results in inhibition of fusion (David *et al.*, 1990; Cohen *et al.*, 1977; Rapuano and Prives, 1987). This inhibition is consistent with published studies demonstrating down-regulation of the TPA receptor in a variety of cell types (Nishizuka, 1986), including the L_6 rat myoblast cell line (Klip and Ramal, 1987; Narindrasorasak *et al.*, 1987).

1-(5-Isoquinolinesulfonyl)-2-methylpiperazine (H-7) which is a potent and, at low concentrations, specific inhibitor of PKC (Hidaka *et al.*, 1984) dramatically reduces fusion in myoblast cultures (Fig. 1). In addition, 1-oleoyl-2-acetylglycerol (OAG), a diacylglycerol analog that activates PKC, stimulates precocious fusion as does the diacylglycerol kinase inhibitor R59022 (David *et al.*, 1990). TPA, H-7, OAG, or the diacylglycerol inhibitor do not affect

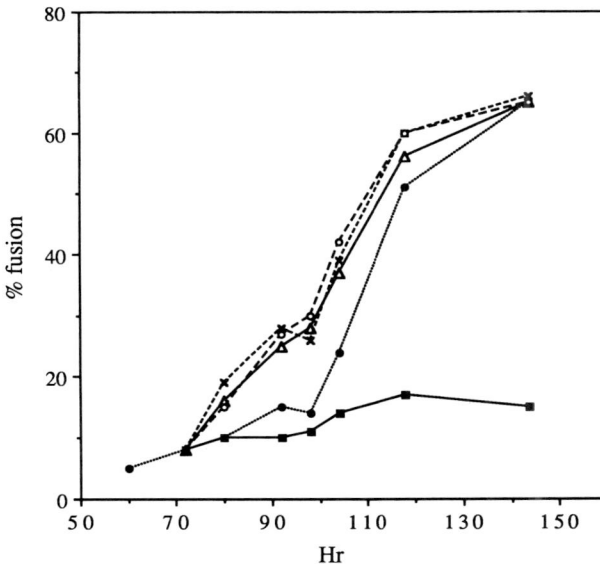

Fig. 1. Myoblast fusion in presence of TPA, okadaic acid, and thapsigargin. Cultures were seeded at 0.4×10^5 cells/35-mm culture dish. TPA (0.1 nM), okadaic acid (1.25 nM), and thapsigargin (8.2 nM) were added separately at 72 hr. H-7 (23 μM) was added at 48 hr. Cultures were fixed, stained, and counted as previously described (David *et al.*, 1981). Average of two experiments. (●) Control; (△) thapsigargin; (x) okadaic acid; (○) TPA; (■) H-7.

TABLE I

TPA, Okadaic Acid, and Thapsigargin Activate Myoblast Fusion

Reagent[a]	Concentration (nM)	Fusion index[b] at 96 hr	Fusion index[b] at 144 hr
—		1.00	1.00
Okadaic acid	0.068	1.12 ± 0.21	
	0.125	1.44 ± 0.066	
	0.68	1.83 ± 0.24	
	1.25	1.88 ± 0.20	1.02 ± 0.17
	6.87	1.83 ± 0.28	
	12.5	1.54 ± 0.07	
Thapsigargin	0.82	1.09 ± 0.07	
	1.53	1.11 ± 0.03	
	8.2	1.67 ± 0.32	0.99 ± 0.04
	15.3	1.29 ± 0.09	
	153.00	0.88 ± 0.09	
TPA[c]	0.01	1.10 ± 0.09	
	0.05	1.47 ± 0.19	
	0.2	1.51 ± 0.22	
	1.0	1.20 ± 0.19	
	2.0	0.95 ± 0.40	

[a]Reagents were added at 72 hr.
[b]Cultures were plated at 0.4×10^5 cells/35-mm culture dish. Fusion index = percent fusion with experimental treatment/percent fusion in control culture. Average of two experiments.
[c]Data from David *et al.* (1990).

cell proliferation or alignment. Final confirmation of PKC as the common target of TPA, H-7, and OAG came with the demonstration that both OAG and TPA are able to significantly attenuate inhibition of fusion by H-7 (David *et al.*, 1990). Farzaneh *et al.* (1989) simultaneously demonstrated that PKC activators TPA and OAG caused a rapid and complete reversal of a fusion block generated by the prostanoid synthesis inhibitor indomethacin, thus confirming a role for PKC in the signal cascade leading to cell fusion.

2. Alteration of Protein Kinase C Activity Affects Membrane Protein Phosphorylation

The intracellular effects of PKC are assumed to be mediated through increased levels of protein phosphorylation, and indeed activation of PKC in a variety of muscle and nonmuscle (Nishizuka, 1984) cell types leads to an alteration in intracellular protein phosphorylation patterns. Toutant and Sobel (1987) identified six cytoplasmic proteins whose phosphorylation was stimulated by TPA in both pituitary and myoblast cell cultures. Phosphorylation of these proteins was strongly reduced in myotubes. Adamo *et al.* (1989) identified four cytoplasmic peptides whose phosphorylation was stimulated by TPA in myotubes but not in myoblasts.

Because myoblast cell–cell fusion is a membrane phenomenon, it was expected that PKC would phosphorylate and thereby activate one or more plasma membrane proteins. Therefore, phosphorylation of plasma membrane was analyzed rather than total cellular protein. As shown in Figs. 2–4 and Table II, two myoblast plasma membrane phosphopeptides are identified (145

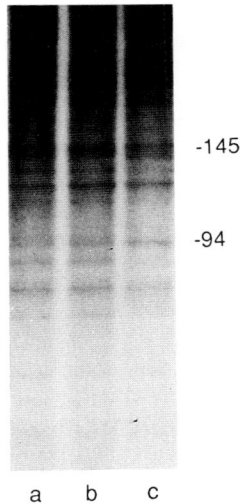

Fig. 2. Developmental regulation of the phosphorylation of 145- and 94-kDa surface polypeptides. Autoradiogram of $^{32}PO_4$-labeled surface peptides from cultures harvested at (a) 60 hr; (b) 84 hr; and (c) 120 hr. Cultures were seeded at 9×10^5 cells/100-mm culture dish. One hour prior to addition of $^{32}PO_4$, the medium was removed and replaced with medium containing 10% the normal level of PO_4. $^{32}PO_4$ was added at 200 μ Ci/ml and incubation continued for 12 hr. Plasma membranes were prepared by a modification of published procedures: membrane proteins were solubilized in SDS, separated in 8–12% SDS-polyacrylamide slab gels, and identified by autoradiography, as described elsewhere (Pauw and David, 1990). Incorporated ^{32}P displayed on the gels was judged to be covalently bound to protein as a result of failure to alter the gel pattern by preextraction of the sample with chloroform/methanol, or predigestion with RNase A or a mixture of DNase I and II.

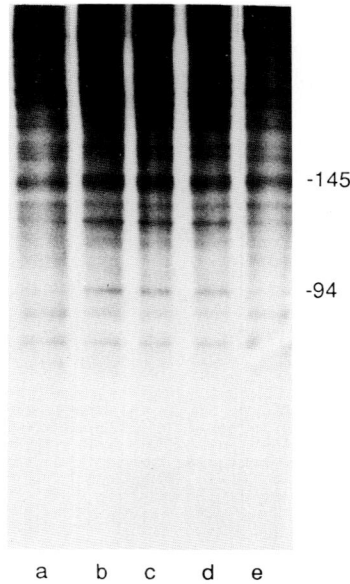

Fig. 3. Stimulation of the phosphorylation of 145- and 94-kDa surface polypeptides by OAG, TPA, and R59022. Autoradiogram of $^{32}PO_4$-labeled surface peptides from (a) control culture; (b) 8.3 μM OAG treated culture; (c) 0.2 nM TPA treated culture; (d) 4 μM R59022 treated culture; and (e) control culture. Drugs were added at 72 hr, $^{32}PO_4$ at 73 hr, and the cultures harvested at 85 hr. All other conditions were as described in Fig. 2. Fusion indices: (a) 1.00; (b) 1.41 $\pm$ 0.10; (c) 1.51 $\pm$ 0.22; (d) 1.68 $\pm$ 0.18; (e) 1.00.

and 94 kDa) whose phosphorylation is both developmentally regulated and coordinately modulated by activators and inhibitors of PKC.

The phosphorylation of both polypeptides increases concomitant with fusion in control cultures (Fig. 2), thus the phosphorylation of these polypeptides is developmentally regulated. Whereas three separate activators of PKC (TPA, OAG, and R59022) stimulated general membrane protein phosphorylation by 20–25%, they stimulated phosphorylation of these two polypeptides at levels 2.5- to 3.5-fold higher than the control level (Fig. 3, Table II). Treatment with the inactive TPA analog 4α-phorbol-12,13-didecanoate did not alter the phosphoprotein profile. Two PKC inhibitors (H-7, and TPA at high concentration) reduced phosphorylation of both the 145- and 94-kDa polypeptides to less than 25% of the control level (Table II).

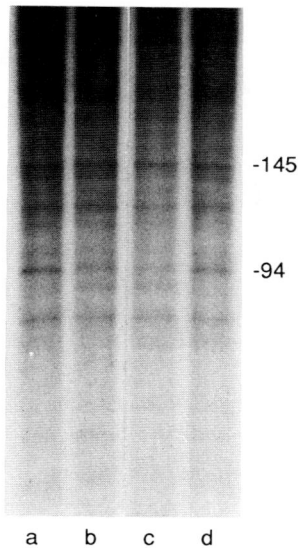

Fig. 4. Inhibition of phosphorylation of 145- and 94-kDa surface peptides by D600 and TFP. Autoradiogram of $^{32}PO_4$-labeled surface peptides from (a) control culture; (b) culture treated with 64 μM D600 at 24 hr; (c) culture treated with 5.55 μM TFP at 60 hr; and (d) control culture. $^{32}PO_4$ was added at 108 hr and all cultures were terminated at 120 hr. Fusion indices: (a) 1.00; (b) 0.39 ± 0.03; (c) 0.29 ± 0.03; (d) 1.00.

3. Blockade of Calcium Influx, or Inhibition of the Calcium- and Calmodulin-Dependent Protein Kinase, Inhibit Myoblast Fusion and Membrane Protein Phosphorylation

It was previously demonstrated that an influx of Ca^{2+} immediately precedes, and is required for, myoblast fusion (David and Higgenbotham, 1981). In part, this demonstration rested on the ability of α-isopropyl-α[(N-methyl-N-homoveratryl) α-aminopropyl]-3,4,5-trimethoxyphenylacetonitrite hydrochloride (D600), which specifically blocks voltage-dependent Ca^{2+} channels (Janis *et al.*, 1987), to block myoblast fusion. It was also demonstrated that trifluoperazine (TFP), which specifically inhibits the Ca^{2+}/CM-dependent protein kinase, also blocks myoblast fusion (unpublished data; see also Bar-Sagi and Prives, 1983). Here it is shown that both D600 and TFP strongly inhibit phosphorylation of the 94 kDa polypeptide, while having minimal

TABLE II

Fusion Modulators Alter Phosphorylation of 145- and 94-kDa Plasma Membrane Proteins

Treatment[a]	Relative phosphorylation[b]	
	145 kDa	94 kDa
—	1.00	1.00
Fusion stimulators		
TPA (0.2 nM)	2.9 ± 0.2 (3)[c]	3.3 ± 0.8 (3)
OAG	2.7 ± 0.6 (4)	2.4 ± 0.8 (4)
R59022	3.0 ± 0.6 (4)	3.2 ± 1.1 (4)
Fusion inhibitors		
TPA (6.7 nM)	0.16 ± 0.06 (8)	0.20 ± 0.08 (8)
H-7	0.20 ± 0.10 (2)	0.29 ± 0.11 (2)
D600	0.85 ± 0.08 (5)	0.21 ± 0.09 (4)
TFP	0.80 ± 0.10 (5)	0.31 ± 0.11 (5)

[a]Reagents were added as indicated in Fig. 2–5. Phosphorylation with stimulators was terminated at early fusion (85 hr) in control and experimental cultures. Phosphorylation with inhibitors was terminated at late fusion (120 hr) in control and experimental cultures.

[b]Relative phosphorylation is calculated as the ratio of autoradiographic band intensity of the experimental treatment to the control. Autoradiograms were quantified as described (Pauw and David, 1990).

[c]Numbers in parentheses indicate the number of separate experiments averaged.

effect on phosphorylation of the 145-kDa polypeptide (Fig. 4, Table II). Thus it can be concluded that PKC is directly responsible for phosphorylation of the 145 kDa peptide only, and that the Ca^{2+}/CM-dependent protein kinase subsequently phosphorylates the 94 kDa polypeptide.

It appears unlikely that the 145-kDa polypeptide is the only substrate for PKC. The limited resolution of the one-dimensional SDS-polyacrylamide gel separation may have hidden other phosphorylated polypeptides. The level of phosphorylation of some polypeptides may have been below detection limits. Finally, only plasma membrane proteins were analyzed, and there is evidence that TPA stimulates phosphorylation of cytoplasmic proteins in myoblasts and and myotubes (Rapuano and Prives, 1987; Toutant and Sobel, 1987).

These results add to, but are certainly not in conflict with, those of Scott and Dousa (1980) who identified two L_6 myoblast plasma membrane polypeptides phosphorylated by a cAMP-dependent protein kinase in a broken-cell preparation; Senechal *et al.* (1982), who identified four L_6 plasma membrane proteins phosphorylated by a Ca^{2+} and cAMP-independent protein kinase *in vitro;* Lognonne and Wahrmann (1988), who identified a 48-kDa L_6 myoblast plasma membrane polypeptide phosphorylated by a Ca^{2+}-dependent ectoprotein kinase; Beguinoit *et al.*, (1989), who reported the tyrosine protein kinase stimulated phosphorylation of a 175-kDa polypeptide in L_6 myoblasts; and Cates *et al.* (1987), who reported that a 46-kDa gelatin-binding glycoprotein from L_6 myoblasts may be phosphorylated by PKC.

Although phosphorylation in primary cultures was examined with the inherent contamination by fibroblasts, neither the 145- or 94-kDa polypeptides are fibroblast in origin. Under the standard culture conditions, fibroblasts represent less than 5% of the total cell population, yet in cultures enriched (95%) in fibroblasts, differential phosphorylation of polypeptides of these molecular weights following treatment with either TPA or H-7 could not be detected.

4. Okadaic Acid, an Inhibitor of Protein Phosphatases That Induces Precocious Fusion

If stimulation of protein phosphorylation by activation of PKC can induce early fusion, then the same effect should be produced by inhibition of the appropriate protein phosphatase(s). Okadaic acid is a specific inhibitor of protein phosphatase 1 (PP1) and protein phosphatase 2a (PP2a), two of the four classes of serine–threonine protein phosphatases (Bialojan and Takai, 1988), which are the chief enzymes that are likely to reverse the actions of PKC. The IC_{50} of okadaic acid for PP1 and PP2 is 0.1 nM and 10 nM, respectively. For PP2b, the IC_{50} is 5 μM, whereas PP2c is essentially not inhibited. Cell membranes are permeable to okadaic acid, and in several cell types the addition of okadaic acid to the growth medium increased the phosphorylation states of many proteins several fold within minutes (for review see Cohen *et al.*, 1990).

Myoblast cultures respond in a dose-dependent fashion to okadaic acid, with half maximal stimulation of fusion at 0.125 nM and maximal stimulation at 1.25 nM (Table I). As with TPA, the kinetics, but not the final extent of fusion, are altered (Fig. 1). As predicted, the time course of fusion in the presence of either okadaic acid or TPA is essentially identical (Fig. 1).

Kim *et al.* (1991) have reported that at 25 nM okadaic acid blocked fusion of chick embryonic myoblasts when the culture was treated prior to cell alignment. Treatment after alignment had no effect. Fusion was scored at the terminal stage of differentiation and thus any precocious fusion would not have been identified. Inhibition of fusion following a chronic administration of okadaic acid is reminiscent of the inhibition generated by the chronic administration of TPA described earlier.

5. A Tyrosine Protein Kinase Inhibitor Does Not Affect Myoblast Fusion

The serine–threonine protein kinase family (PKC, Ca^{2+}/CM-dependent protein kinase, and the cAMP- and cGMP-dependent protein kinases) is apparently responsible for the bulk of observed intracellular protein phosphorylation (Edelman et al., 1987). Tyrosine protein kinase activity is generally found associated with cell surface growth factor receptors (for review, see Ulrich and Schlessinger, 1990) and plays an important role in signal transduction.

The isoflavone genistein inhibits both tyrosine protein kinase autophosphorylation and that kinase's phosphorylation of external substrates with an IC_{50} of 5–15 μM; inhibition of the cAMP-dependent protein kinase is not observed at any concentration of genistein, whereas the IC_{50} for PKC is approximately 400 μM (Akiyama et al., 1987).

The recent availability of this potent and highly specific inhibitor of tyrosine protein kinases allowed investigation of the role of these kinases in the terminal stages of myoblast fusion. Genistein, added at 72 hr in concentrations up to 22 μM, had no effect on either the proliferation or the fusion of chick embryo skeletal myoblasts (Table III). Under the same conditions, in contrast, the PKC inhibitor H-7 was a potent fusion inhibitor (Table III). At 148 μM, genistein still did not inhibit fusion, but it did reduce the final nuclear count to 44% of normal. It can be concluded that tyrosine protein kinases may be involved in the response to myoblast cell proliferation stimuli, but are not involved in the postalignment regulation of myoblast fusion.

B. Modulation of Intracellular Calcium Release Indirectly Modulates Fusion, Probably via Protein Kinase C

1. Thapsigargin Induces Precocious Fusion

It was demonstrated earlier that an increase in Ca^{2+} influx immediately precedees, and is required for, myoblast fusion (David and Higgenbotham, 1981). This conclusion was based in part on inhibition of fusion by the Ca^{2+} channel blocker D600, and stimulation of precocious fusion by the Ca^{2+} ionophore A23187.

TABLE III

A Tyrosine Protein Kinase Inhibitor Does Not Inhibit Myoblast Fusion

Reagent	Concentration (μM)	Time of addition (hr)	Fusion index[a]
—			1.00
Genistein	148.0	72	1.07 ± 0.07
	22.0	72	0.98 ± 0.09
	3.7	72	1.05 ± 0.14
H-7	23.0	72	0.29 ± 0.17
	2.3	72	0.51 ± 0.12
TFP	5.5	60	0.29 ± 0.03

[a]Cultures were seeded at 0.4×10^5 cells/35-mm culture dish. All cultures were terminated at 120 hr. Fusion index = percent fusion with experimental treatment/percent fusion in control culture. Average of two experiments.

More recently (David *et al.*, 1990), it was shown that the calcium antagonist 8-diethylamino-octyl-3,4,5-trimethoxybenzoate (TMB-8) effectively inhibits myoblast fusion but not alignment, and has no effect on cell viability or morphology. TMB-8 blocks the release of Ca^{2+} from intracellular stores, primarily the endoplasmic reticulum (Rahwan *et al.*, 1979), and may generally sequester free Ca^{2+} inside the cell (Charo *et al.*, 1976). Here is further evidence that the release of Ca^{2+} from intracellular stores in the endoplasmic reticulum is a crucial step in the cascade of events that results in myoblast fusion.

Thasrup *et al.* (1990) demonstrated that the sesquiterpene lactone, thapsigargin, releases intracellular calcium stores by inhibition of the endoplasmic reticulum Ca^{2+}-ATPase. The rapid and pronounced increase in free intracellular Ca^{2+} on treatment with thapsigargin is the result of direct discharge of intracellular stored Ca^{2+}. The release is not dependent on hydrolysis of inositolphospholipids. However, of the two intracellular Ca^{2+} compartments distinguishable by their inositol 1,4,5-trisphosphate (IP_3) sensitivity, the IP_3-sensitive Ca^{2+} pool is more sensitive to the Ca^{2+} pump inhibitor thapsigargin than is the IP_3-insensitive Ca^{2+} pool (Takemura *et al.*, 1989). Thapsigargin stimulates dose-dependent precocious fusion of chick skeletal muscle myoblasts (Fig. 1, Table I) when added during the alignment stage of myogenesis. The kinetics of fusion were essentially identical to those in parallel cultures exposed to either TPA or okadaic acid. Thapsigargin had no measurable effect on cell density.

TABLE IV

Thapsigargin Acts Synergistically with TPA and OAG to Stimulate Myoblast Fusion

Treatment A		Treatment B		
Reagent[a]	Concentration	Reagent[a]	Concentration	Fusion index[b]
—		—		1.00
Thapsigargin	1.53 nM	—		1.13 ± 0.07
		TPA	0.01 nM	0.85 ± 0.17
Thapsigargin	1.53 nM	TPA	0.01 nM	1.49 ± 0.07
		OAG	2.5 μM	1.07 ± 0.01
Thapsigargin	1.53 nM	OAG	2.5 μM	1.85 ± 0.35
Thapsigargin	8.2 nM	—		2.00 ± 0.27
—		TPA	0.1 nM	1.93 ± 0.09
Thapsigargin	8.2 nM	TPA	0.1 nM	2.27 ± 0.13
—		OAG	16 μM	1.54 ± 0.19
Thapsigargin	8.2 nM	OAG	16 μM	1.86 ± 0.42

[a]All reagents were added at 72 hr.

[b]Cultures were plated at 0.4×10^5 cells/35-mm culture dish. All cultures were terminated at 96 hr. Fusion index = percent fusion with experimental treatment/percent fusion in control culture. Average of two experiments.

2. Thapsigargin Acts Synergistically with TPA and OAG

Application of thapsigargin in combination with either TPA or OAG, at concentrations of the drugs that alone have no effect on fusion, clearly stimulate the appropriate fusion response (Table IV). In contrast, application of thapsigargin in combination with either TPA or OAG, at concentrations of the drugs that alone produce a maximal fusion response, does not further stimulate fusion (Table IV). It can be concluded that thapsigargin stimulates myoblast fusion by indirect activation of the same intracellular target that is directly activated by TPA and OAG, that is, protein kinase C. This indirect activation of PKC occurs through the release of intracellular sequestered calcium.

3. Thapsigargin Attenuates PY1 and H-7 Generated Inhibition of Fusion

As shown in Table V, thapsigargin completely attenuates inhibition of fusion generated by the prostaglandin antagonist 7-oxa-13-prostynoic acid (PY1), thus indicating that thapsigargin acts at a stage downstream from occupation of the prostaglandin receptor. In contrast, thapsigargin only par-

TABLE V

Thapsigargin Completely Attenuates Inhibition of Myoblast Fusion by PY1, but Only Incompletely Attenuates Inhibition by H-7

Treatment[a]	Fusion index[b]
—	1.00
Thapsigargin	0.97 ± 0.15
PY1	0.24 ± 0.13
Thapsigargin + PY1	1.06 ± 0.06
H-7	0.24 ± 0.10
Thapsigargin + H 7	0.40 ± 0.13

[a]Reagents were added at the following times and concentrations: thapsigargin, 72 hr, 8.2 nM; PY1, 24 hr, 93 μM; and H-7, 48 hr, 23 μM.

[b]Cultures were plated at 0.4 × 10^5 cells 35-mm culture dish. All cultures were terminated at 120 hr. Fusion index = percent fusion with experimental treatment/percent fusion in control culture. Average of two experiments.

tially attenuates the fusion inhibition generated by the PKC inhibitor H-7. Thus thapsigargin and H-7 probably act on the same intracellular target, PKC; H-7 directly inhibiting PKC, and thapsigargin activating PKC indirectly via the release of calcium from the endoplasmic reticulum pool.

C. Integration of Protein Phosphorylation and Extracellular Calcium Influx

1. Protein Kinase C Activators and Inhibitors Modulate Net Calcium Influx

An increase in the net influx of Ca^{2+} immediately precedes myoblast fusion (David and Higgenbotham, 1981). Blockade of voltage-gated calcium channels with D600 both blocks the increase in Ca^{2+} influx and blocks fusion. Inhibition of PKC by H-7 blocks the normal increase in Ca^{2+} influx as well as blocking fusion (David et al., 1990). Stimulation of PKC by TPA or OAG induces a precocious increase in net Ca^{2+} influx that mirrors the precocious fusion induced by the same agents (David et al., 1990). From these results it can be argued that (1) Ca^{2+} influx is either the terminal event or very close to the terminal event preceding fusion and could be the ultimate trigger itself, and (2) PKC activation is both a necessary and sufficient stimulus for the rise in Ca^{2+} influx. Here it is shown that thapsigargin, which also activates PKC, stimulates Ca^{2+} influx (Table VI). Furthermore, the protein phosphatase inhibitor okadaic acid also stimulates Ca^{2+} influx coincident with its stimulation of fusion (Table VI).

2. TPA Induces a Delayed Activation of the D600-Sensitive Calcium Channel

The kinetics of TPA stimulation of Ca^{2+} uptake in myoblasts (Fig. 5) show a lag of approximately 0.5 min following TPA addition. Maximal response is reached between 2 and 3 min after addition of TPA. This lag is consistent with

TABLE VI

Thapsigargin and Okadaic Acid Modulate Ca^{2+} Influx

Treatment[a]	Assay time (hr)	Ca^{2+} influx index[b]
—	48 (prefusion)	1.00
TPA	48	1.56 ± 0.12
Thapsigargin	48	1.63 ± 0.28
Okadaic acid	48	1.55 ± 0.11
—	72 (early fusion)	1.00
TPA	72	1.80 ± 0.39
Thapsigargin	72	1.71 ± 0.43
Okadaic acid	72	1.44 ± 0.34
—	120 (late fusion)	1.00
TPA	120	1.05 ± 0.18
Thapsigargin	120	1.04 ± 0.20
Okadaic acid	120	0.91 ± 0.08

[a]Reagents were added at the following concentrations: thapsigargin, 8.2 nM; okadaic acid, 1.25 nM; TPA, 0.2 nM. Reagents were added at 47 hr for the 48 hr time point, and at 71 hr for the subsequent time points.

[b]At the indicated time, cultures were rapidly rinsed with modified Pucks Saline G (without phenol red and containing 100 nm Ca^{2+}). Ca^{2+} uptake was measured as ^{45}Ca accumulation in 20 sec. Ca^{2+} influx index, ^{45}Ca influx in experimental culture/^{45}Ca influx in control culture. Average of two experiments.

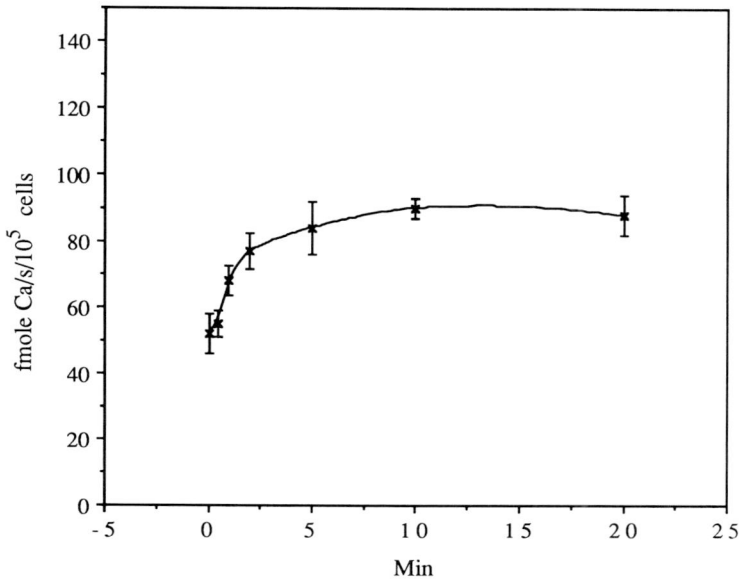

Fig. 5. TPA stimulated uptake of Ca^{2+} into myoblasts. Cultures were seeded at 0.4×10^5 cells/35-mm culture dish. TPA, 0.2 nM, was added at 72 hr (0 time point). At indicated times thereafter, cultures were rapidly rinsed with modified Pucks Saline G (without phenol red and containing 100 μM Ca^{2+}). Ca^{2+} uptake was measured as ^{45}Ca accumulation (50 μ Ci ^{45}Ca; 100 μM total Ca^{2+}) in 20 sec essentially according to David *et al.* (1981). ^{45}Ca uptake was linear for at least 60 sec. Average of two experiments.

one or more intermediate steps, for example, protein phosphorylation, occurring between reception of the TPA signal and opening of the Ca^{2+} channels. Addition of the voltage-sensitive calcium channel blocker, D600, to control cultures inhibits net Ca^{2+} influx 44.5%, and fusion 73.2%. Addition of D600 to TPA-treated cultures completely prevents the normal TPA-generated increase in Ca^{2+} influx (data not shown).

Initial rates of Ca^{2+} influx as a function of external Ca^{2+} concentration and the presence or absence of TPA are plotted in Fig. 6a and b. The addition of TPA to cultures in the early stages of fusion led to an approximate 1.8-fold increase in the rate of Ca^{2+} influx. The kinetics in either the presence or absence of TPA appear to fit Michaelis–Menten criteria, suggesting the presence of a single Ca^{2+} uptake system (channel). Vmax is increased 2.4-fold (7.6 fmol $Ca^{2+}/s/10^5$ cells in the absence of TPA; 18.2 fmol $Ca^{2+}/s/10^5$ cells in its presence), whereas the apparent K_m is increased 1.6-fold (106 μM

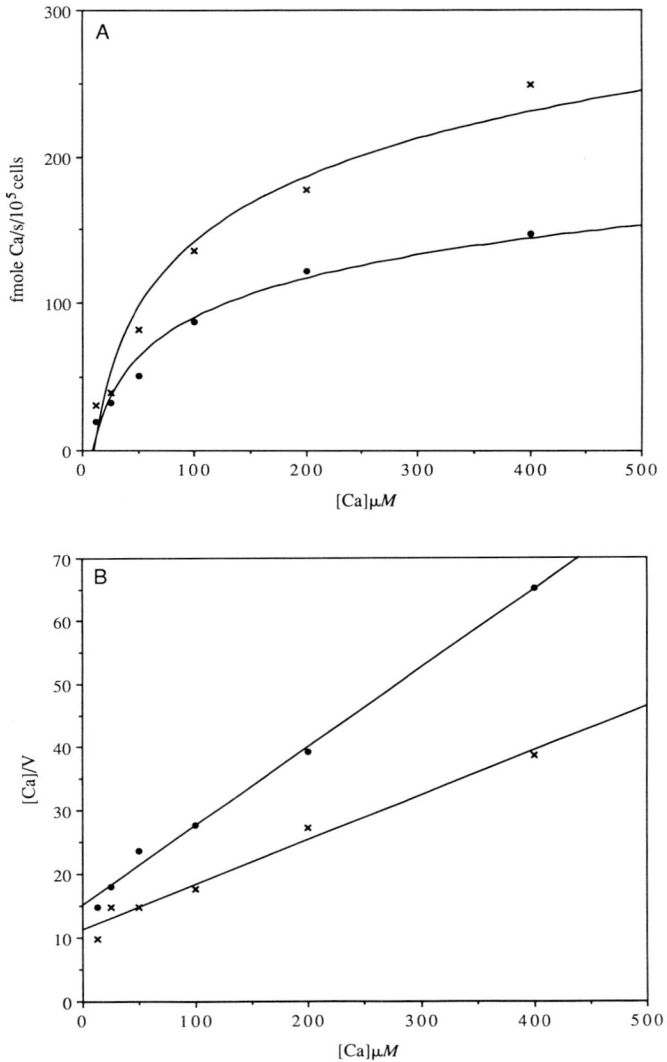

Fig. 6. Kinetic properties of myoblast Ca^{2+} transport ± TPA. Cultures were seeded at 0.4 × 10^5 cells/35-mm culture dish. Some cultures received 0.2 nM TPA at 72 hr as indicated. Thirty minutes after the addition of TPA, cultures were rapidly rinsed with modified Pucks G (without phenol red, and containing 12.5, 25, 50, 100, 200, or 400 μM (Ca^{2+}). (A) The initial rate of Ca^{2+} uptake was measured as ^{45}Ca accumulation in 20 sec; ^{45}Ca uptake was linear for at least 60 sec. Control cultures were treated identically, except for the absence of TPA. Average of two experiments. (B) A Hanes plot (Hanes, 1932) of the data from (A).

Ca^{2+} in control cultures; 169 μM Ca^{2+} in TPA-treated cultures). The data do not allow discrimination between an increased rate of transport through an unaltered number of channels versus the activation of a number of previously inactive channels.

III. FINAL COMMENTS

Although every detail of the casade that converts an extracellular signal to the final intracellular trigger for myoblast fusion is not known, many of the players and their interrelationships have been identified. The involvement of prostaglandins, especially PGE_1, in myoblast fusion has been well-documented. The addition of exogenous PGE_1 causes precocious fusion (Zalin, 1977). PGE_1-specific antagonists block fusion (David et al., 1981). Hausman et al. (1986) have demonstrated (1) synthesis of PGE_1 by myoblasts in culture, (2) the binding of PGE_1 to transient surface receptors on fusion-competent myoblasts, and (3) alterations in myoblast membrane properties subsequent to PGE_1 binding. Signaling between myoblasts must be a normal part of the process that synchronizes muscle development and PGE_1 is the logical candidate for a differentiation staging signal. Because prostaglandins are typically effective over short distances, propagation of the signal throughout a culture requires that responding cells themselves synthesize and release PGE_1 as demonstrated by Entwistle et al. (1988).

Elgendy and Hausman (1990) have demonstrated that binding of PGE_1 to embryonic chick myoblasts induces a rapid increase in inositol phosphate metabolism. Wakelam (1986) had earlier demonstrated phosphatidylinositol turnover in chick skeletal myoblasts, and Adamo et al. (1989) identified differences in phosphatidylinositol metabolism in myoblasts and myotubes. Hausman et al. (1990) have recently reported a requirement for G protein activity at an early stage in myogenesis, following PGE_1 binding but preceding the shift in phosphatidylinositol metabolism. They propose that signal transduction proceeds from an activated PGE_1 receptor, through a G protein and presumed activation of phopholipase C, thus generating intracellular inositol phospholipid metabolites.

Phospholipase C activation would, in addition to generating IP_3, also generate the endogenous PKC activator diacylglycerol. IP_3 has been shown to mobilize intracellular Ca^{2+} stores (Berridge and Irvine, 1984). De Smedt et al. (1991) have demonstrated the presence of an IP_3-sensitive Ca^{2+} pool in myoblasts, which disappears in differentiated muscle cells. Our data indicate that release of Ca^{2+} from an IP_3-sensitive intracellular compartment is an obligatory step in the sequence leading to fusion. The IP_3-generated transitory increase in intracellular Ca^{2+} could potentiate the activation of PKC by diacylglycerol (Wolf et al., 1985). Activation of PKC would then result in specific protein phosphorylation and a stabilization of the fusion signal.

The transduction of many extracellular stimuli is mediated by a rise in intracellular calcium. The release of Ca^{2+} from intracellular stores would produce only a transitory rise in cytosolic Ca^{2+}. Maintenance of elevated cytosolic Ca^{2+} depends on transport from outside the cell due either to activation of voltage-gated (Lee and Tsien, 1983) or receptor-operated (MacDougall et al., 1988) Ca^{2+} channels or inhibition of the plasma membrane Ca^{2+} pump (Reinhart et al., 1984). It has been demonstrated that an increased net influx of Ca^{2+}, through D600-sensitive voltage-gated Ca^{2+} channels is an obligatory step in myoblast fusion (David and Higgenbotham, 1981). This Ca^{2+} influx occurs immediately preceding fusion, and indeed may be the ultimate trigger of the membrane fusion event. Recently, Kano et al. (1989) demonstrated the presence of D600-sensitive Ca^{2+} channels in the plasma membrane of cells in the early stages of myogenesis, and the loss of this specific class of Ca^{2+} channels from the plasma membrane of mature myotubes.

It can be argued on the basis of the data presented in this chapter that PKC is the logical candidate to transduce the extracellular PGE_1 signal into an increase in intracellular Ca^{2+} via activation of plasma membrane voltage-gated Ca^{2+} channels. Many Ca^{2+} channels are regulated by pathways involving phosphorylation of channel submits. Phosphorylation of the 140- to 145-kDa subunit of the D600-sensitive Ca^{2+} channel in cardiac muscle (Sperelakis et al., 1985) and skeletal muscle (for review, see Lazdunski et al., 1988, 1990) by either cAMP-dependent or Ca^{2+}/CM-dependent protein kinases activates Ca^{2+} influx. The T-tubular calcium channel in skeletal muscle is a substrate for PKC (Nastainczyk et al., 1987), as is the D600-sensitive Ca^{2+} channel in Aplysia neurons (DeRiemer et al., 1985). Experiments are in progress to identify the 145- and 94-kDa plasma membrane peptides that are phosphorylated in response to TPA in our cultures.

The results of these experiments point to PKC as an important mediator in myoblast fusion, and reaffirm the central role of Ca^{2+} in the fusion event. Therefore, it can be proposed that PKC is part of a signal cascade that links reception of an extracellular signal, PGE_1, to membrane fusion (Fig. 7). In this model, PGE_1 is the extracellular trigger and synchronizes fusion. The activated PGE_1 receptor, in concert with a G protein, activates phospholipase

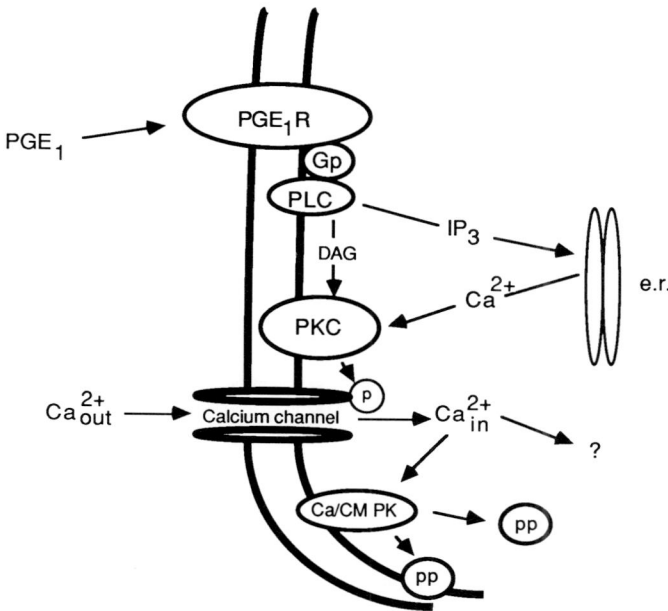

Fig. 7.   Postulated signal transduction cascade. Solid arrows indicate documented interactions; broken arrows postulated events. PGE_1R, PGE_1 receptor; Gp, PGE_1 receptor/phospholipase C-associated G protein; PLC, phospholipase C; PIP_2, phosphatidylinositol 4,5-biphosphate; DAG, diacylglycerol; IP_3, inositol 1,4,5-trisphosphate; ER, endoplasmic reticulum; PKC, protein kinase C; Ca/CM PK, Ca^{2+}–calmodulin-depedent protein kinase; pp, phosphoprotein.

C producing IP_3 and diacylglycerol. IP_3 mobilizes intracellular Ca^{2+} stores, which in concert with diacylglycerol activates PKC. PKC then phosphorylates the Ca^{2+} channel, leading to an increased Ca^{2+} influx, activating the Ca^{2+}/CM-dependent protein kinase and membrane fusion.

ACKNOWLEDGMENTS

This research was supported, in part, by grants from the Graduate School Research Council, University of Missouri (Biomedical Research Support Grant RR07053, NIH) and the Public Health Service (GM 28657) to John David. We thank Dr. Josef Fried for his gift of PY1.

REFERENCES

Adamo, S., Caporale, C., Nervoni, C., Ceci, R., and Molinaro, M. (1989). Activity and regulation of calcium-phospholipid dependent protein kinase in differentiating chick myogenic cells. *J. Cell Biol.* **108**, 153–158.

Akiyama, T., Ishida, J., Ogawara, H., Watanabe, S., Itoh, N., Shibuya, M., and Fukami, Y. (1987). Genestein, a specific inhibitor of tyrosine-specific protein kinases. *J. Biol. Chem.* **262**, 5592–5595.

Bar-Sagi, D., and Prives, J. (1983). Trifluoperazine, a calmodulin antagonist, inhibits muscle cell fusion. *J. Cell Biol.* **97**, 1375–1380.

Beguinoit, F., Kahn, C. R., Moses, A. C., White, M. F., and Smith, R. J. (1989). Differentiation-dependent phosphorylation of a 175,000 molecular weight protein in response to insulin and insulin-like growth factor-I in L_6 skeletal muscle cells. *Endocrinology* **125**, 1599–1605.

Berridge, M. J., and Irvine, R. F. (1984). Inositol trisphosphate, a novel second messenger in cellular signal transduction. *Nature* **312**, 315–321.

Bialojan, C., and Takai, A. (1988). Inhibitory effect of a marine sponge toxin, okadaic acid, on protein phosphatases. *Biochem. J.* **256**, 283–290.

Bischoff, R. (1978). Myoblast fusion. *Cell Surface Rev.* **5**, 127–179.

Cates, G. A., Litchfield, D. W., Narindrasorasak, S., Nandan, D., Ball, E. H., and Sanwal, B. D. (1987). Phosphorylation of a gelatin-binding protein from L_6 myoblasts by protein kinase C. *FEBS Lett.* **218**, 195–199.

Charo, I. F., Feinman, R. D., and Detweiler, T. C. (1976). Inhibition of platelet secretion by an antagonist of intracellular calcium. *Biochem. Biophys. Res. Comm.* **72**, 1462–1467.

Cohen, P., Holmes, C. F. B., and Tsukitani, Y. (1990). Okadaic acid: A new probe for the study of cellular regulation. *Trends Biochem. Sci.* **15**, 98–102.

Cohen, R., Pacifici, M., Rubenstein, N., Biehl, J., and Holtzer, H. (1977). Effect of a tumor promoter on myogenesis. *Nature.* **266**, 538–540.

David, J. D., and Higgenbotham, C. A. (1981). Fusion of chick embryo skeletal myoblasts: Interactions of prostaglandin E1, cAMP, and calcium influx. *Dev. Biol.* **82**, 307–316.

David, J. D., See, W. M., and Higgenbotham, C. A. (1981). Fusion of chick embryo skeletal myoblasts: Role of calcium influx preceding membrane union. *Dev. Biol.* **82**, 297–307.

David, J. D., Faser, C. R., and Perrot, G. P. (1990). Role of protein kinase C in chick embryo skeletal myoblast fusion. *Devel. Biol.* **139**, 89–99.

DeRiemer, S. A., Strong, J. A., Albert, K. A., Greengard, P., and Kazmarek, L. K. (1985). Enhancement of Ca^{2+} current in *Aplysia* neurons by phorbol ester and protein kinase C. *Nature* **313**, 313–316.

DeSmedt, H., Parys, J. B., Himpens, B., Missiaen, L., and Borghgraff, R. (1991). Changes in the mechanism of Ca^{2+} mobilization during the differentiation of BC3H1 muscle cells. *Biochem. J.* **273**, 219–224.

Edelman, A. M., Blumenthal, D. K., and Krebs, E. G. (1987). Protein serine/threonine kinases. *Annu. Rev. Biochem.* **56**, 567–614.

Elgendy, H., and Hausman, R. E. (1990). Prostaglandin-dependent phosphatidylinositol signaling during embryonic chick myogenesis. *Cell Differ. Dev.* **32**, 109–116.

Entwistle, A., Zalin, R. J., Warner, A., and Bevans, S. (1988). A role for acetylcholine receptors in fusion of chick myoblasts. *J. Cell Biol.* **106**, 1703–1712.

Farzaneh, F., Entwistle, A., and Zalin, R. (1989). Protein kinase C mediates the hormonally regulated plasma membrane fusion of avian embryonic skeletal muscle. *Exp. Cell Res.* **181,** 298–304.

Hanes, C. S. (1932). Studies on plant amylases. *Biochem. J.* **26,** 1406–1421.

Hausman, R. E., Dobi, E. T., Woodford, E. J., Petridis, S., Ernst, M., and Nichols, E. B. (1986). Prostaglandin-binding activity and myoblast fusion in aggregates of avian myoblasts. *Dev. Biol.* **113,** 40–48.

Hausman, R. E., Elgendy, H., and Craft, F. (1990). Requirement for G protein activity at a specific time during embryonic chick myogenesis. *Cell Differ. Dev.* **29,** 13–20.

Hidaka, H., Inagaki, M., Kawamoto, S., and Sasaki, Y. (1984). Isoquinalinesulfonamides, novel and potent inhibitors of cyclic nucleotide-dependent protein kinase and protein kinase C. *Biochemistry* **23,** 5036–5041.

Janis, R. A., Silver, P. J., and Triggle, D. J. (1987). Drug action and cellular calcium regulation. *Adv. Drug. Res.* **16,** 309–589.

Kano, M., Wakuta, K., and Satoh, R. (1989). Two components of calcium channel current in embryonic chick skeletal muscle cells developing in culture. *Dev. Brain Res.* **47,** 101–112.

Kim, H. S., Chung, C. H., Kang, M-S., and Ha, D. B. (1991). Okadaic acid blocks membrane fusion of chick embryonic myoblasts in culture. *Bioch. Biophys. Res. Commun.* **176,** 1044–1050.

Klip, A., and Ramal, T. (1987). Protein kinase C is not required for insulin stimulation of hexose uptake in muscle cells in culture. *Biochem. J.* **242,** 131–136.

Lai, Y., Seagar, M. J., Takahashi, M., and Catterall, W. A. (1990). Cyclic AMP-dependent phosphorylation of two size forms of $\alpha 1$ subunits of L-type calcium channels in rat skeletal muscle cells. *J. Biol. Chem.* **265,** 20,893–20,848.

Lazdunski, M., Barhanin, J., Borsotto, M., Cognard, C., Cooper, C., Coppola, T., Fosset, M., Galizzi, J.-P., Hosey, M. M., Mourre, C., Renaud, J-F., Romey, G., Schmid, A., and Vandaele, S. (1988). Molecular properties of structure and regulation of the calcium channel. *Ann. N.Y. Acad. Sci.* **522,** 134–149.

Lee, K. S., and Tsien, R. W. (1983). Mechanism of calcium channel blockade by verapamil, D600, diltiazem, and intrendipine in single dialyzed heart cells. *Nature* **302,** 790–794.

Lognonne, J. L., and Wahrmann, J. P. (1988). A cell surface phosphoprotein of 48 kDa specific for myoblast fusion. *Cell Differ.* **22,** 245–258.

MacDougall, S. L., Grinstein, S., and Gelfaud, E. W. (1988). Detection of ligand-activated conductive Ca^{2+} channels in human B lymphocytes. *Cell* **54,** 229–234.

Narindrasorasak, S., Brickenden, A., Ball, E., and Sanwal, B. D. (1987). Regulation of protein kinase C by cAMP and a tumor promoter in skeletal myoblasts. *J. Biol. Chem.* **262,** 10497–10501.

Nastainczyk, W., Rohrkasten, A., Sieber, M., Rudolph, C., Schachtele, C., Marme, D., and Hoffman, F. (1987). Phosphorylation of the purified receptor for calcium channel blockers by cAMP kinase and protein kinase C. *Eur. J. Biochem.* **169,** 137–142.

Nishizuka, Y. (1984). The role of protein kinase C in cell surface signal transduction and tumor production. *Nature* **308,** 693–698.

Nishizuka, Y. (1986). Studies and perspectives of protein kinase C. *Science* **233,** 305–312.

Pauw, P. G., and David, J. D. (1990). A unique subset of developmentally regulated surface proteins turns over rapidly during fusion of the L_6 rat myoblast cell line. *Exp. Cell Res.* **186,** 74–82.

Rahwan, R. G., Piascik, M. F., and Witiak, D. T. (1979). The role of calcium antagonism in the therapeutic action of drugs. *Can. J. Physiol. Pharmacol.* **57,** 443–460.

Rapuano, M., and Prives, J. (1987). 12-O-tetradecanoylphorbol 13-acetate blocks myogenic cell fusion by interfering with the Ca^{2+} signal transduction pathway. *Ann. N.Y. Acad. Sci.* **494**, 156–159.

Reinhart, P. H., Taylor, W. M., and Bygrave, F. L. (1984). The role of calcium ions in the action mechanism of α-adrenergic agonists in rat liver. *Biochem. J.* **223**, 1–13.

Scott, R. E., and Dousa, T. P. (1980). Differences in the cAMP-dependent phosphorylation of plasma membrane proteins of differentiated and undifferentiated L_6 myogenic cells. *Different.* **16**, 135–140.

Senechal, H., Pichard, A. L., Delain, D., Schapira, G., and Wahrmann, J. P. (1982). Changes in plasma membrane phosphoproteins during differentiation of an established myogenic cell line and a nonfusing α-amanatin resistant mutant. *FEBS Lett.* **139**, 209–213.

Shainberg, A., Yagil, G., and Yaffe, D. (1969). Control of myogenesis *in vitro* by Ca^{2+} concentration in nutritional medium. *Exp. Cell Res.* **58**, 163–167.

Sperelakis, N., Wahler, G. M., and Bkaily, G. (1985). Properties of myocardial Ca^{2+} slow channels and mechanism of action of Ca^{2+} antagonistic drugs. *Curr. Top. Membr. Transport* **25**, 45–76.

Takemura, H., Hughes, A. R., Thastrup, O., and Putney, Jr., T. W. (1989). Activation of calcium entry by the tumor promoter thapsigargin in parotid acinar cells. *J. Biol. Chem.* **264**, 12,266–12,271.

Thastrup, O., Cullen, P. J. Drobak, B. K., Hanley, M. R., and Dawson, A. P. (1990). Thapsigargin, a tumor promoter, discharges intracellular Ca^{2+} stores by specific inhibition of the endoplasmic reticulum Ca^{2+} ATPase. *Proc. Natl. Acad. Sci. USA* **87**, 2466–2470.

Toutant, M., and Sobel, A. (1987). Protein phosphorylation in response to the tumor promoter TPA is dependent on the state of differentiation of muscle cells. *Dev. Biol.* **124**, 370–378.

Ullrich, A., and Schlessinger, J. (1990). Signal transduction by receptors with tyrosine kinase activity. *Cell* **61**, 203–212.

Wakelam, M. J. O. (1985). The fusion of myoblasts. *Biochem. J.* **228**, 1–12.

Wakelam, M. J. O. (1986). Stimulated inositol phospholipid breakdown and myoblast fusion. *Biochem. Soc. Trans.* **4**, 253–256.

Wolf, M., Cuatrecaas, P., and Schyoun, N. (1985) Interaction of protein kinase C with membranes is regulated by Ca^{2+}, phorbol esters, and ATP. *J. Biol. Chem.* **260**, 15718–15722.

Zalin, R. J. (1977). Prostaglandins and myoblast fusion. *Dev. Biol.* **59**, 241–248.

12

Signal Transduction and Cell Fusion in *Dictyostelium:* Calcium, Calmodulin, and an Endogenous Inhibitor

MICHAEL A. LYDAN AND DANTON H. O'DAY

Department of Zoology
Erindale College
University of Toronto
Mississauga, Ontario, Canada

I. INTRODUCTION

The development of sexuality during evolution had profound effects on the ability of a species to increase gene diversity, and hence adaptability. Whereas

245

SIGNAL TRANSDUCTION DURING
BIOMEMBRANE FUSION

the importance of gamete cell production, species–specific gamete recognition, and subsequent fusion is incontrovertible, the intracellular mechanisms that mediate these events are incompletely understood. The sexual cycle of the cellular slime molds in general, and *Dictyostelium discoideum* in particular, is emerging as an excellent tool for studying the biochemical and molecular events during gamete cell differentiation and fusion. This widely used model organism has the same basic cellular architecture as higher eukaryotes, but the ease of culturing, and the rapidity and well-defined kinetics of sexual development in *D. discoideum,* facilitate a detailed examination of the intracellular events that mediate cell fusion, using the tools of both cell and molecular biology.

A. Sexual Development in *Dictyostelium discoideum*

Sexual development in the classical heterothallic strains of *D. discoideum* (NC4 and V12) is typified by a clear temporal segregation of the events of gamete formation, gamete fusion, and subsequent pronuclear fusion (Fig. 1). When equal numbers of heat-shocked NC4 and V12 spores are placed in dark, moist conditions, the spores germinate in about 6 hr. From about 6 to 10 hr of development, gamete cells appear in the culture (O'Day *et al.,* 1987; Lydan and O'Day, 1988a). These cells are small ameboid cells that are highly motile and contain dense nuclei that fluoresce brightly when stained with a fluorescent nuclear stain (O'Day *et al.,* 1987; Lydan and O'Day, 1988a). Gamete levels peak by 10 hr of development, after which their numbers decrease because of their fusion to produce predominantly binucleate cells (Fig. 1; Szabo *et al.,* 1982; O'Day *et al.,* 1987; Lydan and O'Day, 1988a). Binucleate cells are readily distinguished by their increased cytoplasmic volume and the presence of two nuclei (Fig. 1; Szabo *et al.,* 1982; Lydan and O'Day, 1988a). Binucleates are first seen at about 10 hr of development and their numbers increase until about 20 hr (Fig. 1; Lydan and O'Day, 1988a). In the binucleate, there is a period of pronuclear swelling and migration to apposition with a concomitant dramatic increase in cytoplasmic volume (Fig. 1; McConachie and O'Day, 1987). Following their apposition, the pronuclei will fuse, converting the binucleate into a zygote giant cell (Fig. 1; Szabo *et al.,* 1982; McConachie and O'Day, 1987). Zygote giant cells are first seen at about 18 hr of development and peak at 24 hr (Fig. 1; Lydan and O'Day, 1988a). The

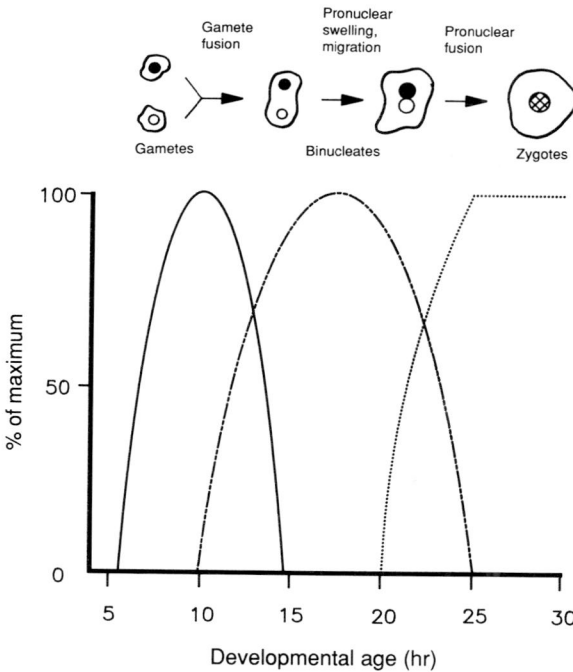

Fig. 1.   Diagrammatic representation of the early events of sexual development in *Dictyostelium discoideum*. During the first 24 hr of development in heterothallic (NC4 × V12) mixed-mating-type cultures, three sexual cell types—gametes, binucleates, and zygote giant cells—are seen with a clear temporal pattern of appearance. Gametes are first seen at about 6 hr of development and increase in numbers until 10 hr of development when their numbers peak. Gametes fuse to form binucleates that first appear at about 10 hr and peak at 20 hr. Pronuclear swelling, migration, and fusion with a concomitant increase in cytoplasmic volume converts the binucleates to zygote giant cells. Zygotes are first seen at 20 hr and peak at about 24–28 hr of development. The data presented represent the percentage of maximum reached by each cell type. [Reproduced with permission from Lydan, M. A., Browning, D. D., and O'Day, D. H. (1990). Calcium, calmodulin, and the antagonistic action of an endogenous inhibitor of cell fusion in *Dictyostelium discoideum*, p. 392–409. *In: Calcium as an Intracellular Messenger in Eukaryotic Microbes*, D. H. O'Day, ed. American Society for Microbiology, Washington, D.C.]

zygote giant cell differentiates into a macrocyst via a period of active phagocytosis, engulfing amebas as a food source, followed by the formation of a cellulosic wall (Lewis and O'Day, 1985, 1986). The later events of sexual development and the regulation of endocytic phagocytosis in the zygote are examined in Chapter 8 in this volume.

B. Extracellular Factors That Mediate Gamete Formation and Fusion in *Dictyostelium discoideum*

1. Exogenous

In addition to the dark, moist environmental conditions previously mentioned, sexual development in *D. discoideum* is dramatically influenced by the ionic composition of the medium. Inclusion of small quantities (ppm) of

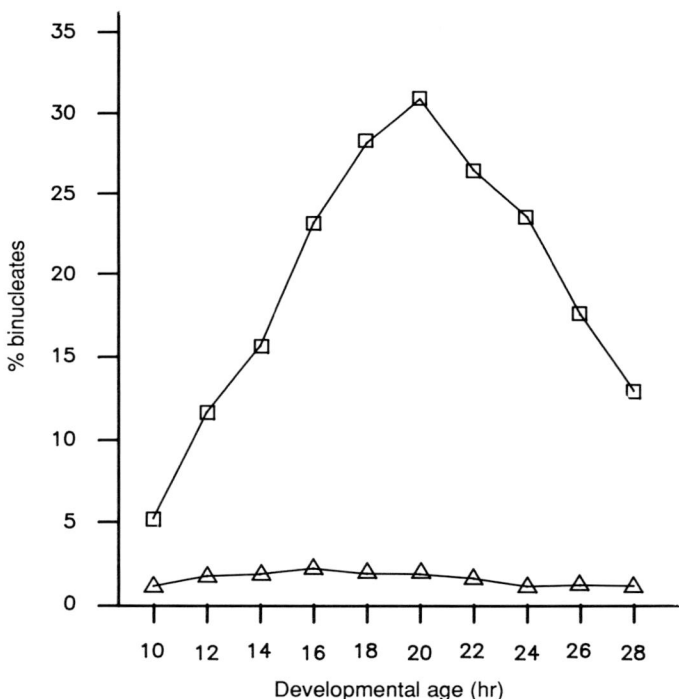

Fig. 2. Effect of exogenous calcium on cell fusion during sexual development in *Dictyostelium discoideum*. Mixed-mating-type (NC4 × V12) cultures were grown in ASP-LPP liquid medium either without (△) or following the addition of 1 m*M* CaCl$_2$ (□). The percentage of binucleates in the culture was determined at various developmental times.

SO_4^{2-}, Cl^-; NO_3^-, and Mg^{2+} in the culture medium enhances both the rate and final extent of cell fusion over equivalent cultures produced without these ions (Lydan and O'Day, 1988a). Of critical importance is the addition of 1 mM Ca^{2+} to the culture medium (Fig. 2; Chagla et al., 1980; Szabo et al., 1982; Lydan and O'Day, 1988a). In cultures without exogenously added Ca^{2+}, gametes will form but not fuse, precluding the formation of macrocysts (Fig. 2; Szabo et al., 1982; O'Day et al., 1987). In equivalent cultures containing 1 mM Ca^{2+}, sexual development proceeds as previously described. In addition, when Ca^{2+} is chelated chemically, development is halted at the stage reached immediately prior to the addition of the chelator (Szabo et al., 1982).

Exogenously added calcium is absolutely required for cell fusion, but not for the acquisition of fusion competence. When calcium is added to cultures previously grown without exogenous calcium, the cells in these cultures fuse rapidly, forming multinucleate syncitia (Saga and Yanagisawa, 1982; McConachie and O'Day, 1986).

2. Endogenous

Two endogenous factors have been shown to mediate sexual development in *D. discoideum*, a sex pheromone that initiates it, and an autoinhibitor that terminates it (MacHac and Bonner, 1975; O'Day and Lewis, 1975; O'Day et al., 1981; Szabo and O'Day, 1984). The sex pheromone is a volatile molecule that is produced by NC4 and that can induce artificial homothallism in V12 (O'Day and Lewis, 1975; Lewis and O'Day, 1977). The autoinhibitor of sexual development is a low molecular weight (<500), hydrophobic molecule that is produced by zygote giant cells and acts to inhibit both gamete fusion to form binucleates and pronuclear fusion to form further zygote giant cells (O'Day et al., 1981; Szabo and O'Day, 1984; Lydan and O'Day, 1989). Although the exact chemical identity of the autoinhibitor remains to be determined, it is neither ammonia nor cyclic AMP (cAMP) (Szabo and O'Day, 1984). The autoinhibitor acts to limit the number of zygote giant cells competing for the finite amebal food source present in the culture that is required during the late events of sexual development. A third endogenous mediator is produced when cells are induced, via markedly different culturing conditions, to fuse rapidly, over a period of about 30 min rather than 10 hr (Saga and Yanagisawa, 1983). This cell fusion inducing factor (CFIF) is distinct from the autoinhibitor because it is a glycoprotein of about 50,000 MW (Saga and Yanagisawa, 1983).

II. CURRENT PROGRESS

The critical role of extracellular calcium as a positive mediator of cell fusion in *D. discoideum,* in conjunction with the increasingly apparent importance of intracellular calcium as a mediator of diverse cellular events, suggests that calcium-dependent signal transduction may play a role during gamete formation and fusion in *D. discoideum.* To this end, examinations concerning the role of calcium at the intracellular level during gamete fusion in *D. discoideum* have employed a progressive approach, starting at the cell surface, to identify the glycoproteins that promote sexual cell recognition and fusion, to identify the role of stimulus-dependent intracellular calcium mobilization, leading to an identification of the intracellular targets for this calcium flux, a calmodulin and calcium/calmodulin-binding proteins that directly mediate gamete fusion.

A. Glycoproteins Associated with Cell Fusion

Cell recognition and adhesion via cell surface glycoproteins is a motif common to a diverse number of cell types, including fusion-competent cells of both gametic and somatic origin (Wakelam, 1985; Wassarman, 1988; Knudsen *et al.,* 1990). Cell fusion in *D. discoideum* can be inhibited by the lectins concanavalin A (Con A) and wheat germ agglutinin (WGA), establishing the importance of D-glucose, D-mannose, and N-acetylglucosamine-containing glycoproteins as positive mediators of gamete fusion (O'Day and Rivera, 1987). SDS-PAGE separation of proteins from sexually developing cells followed by transfer of the proteins to nitrocellulose and probing with ConA and WGA revealed two glycoproteins that were present only during the cell fusion phase of development, one of about 200,000 M_r (gp200), which bound Con A, and a 166,000 M_r WGA-binding protein (gp166). The temporal appearance of these glycoproteins suggested they were the targets of Con A and WGA *in vivo* and thus play a role as essential glycoproteins during cell fusion (Browning and O'Day, 1991; Browning *et al.,* 1992). Support for the role of gp166 as an essential glycoprotein during cell fusion was obtained by an examination of the *in vivo* effects of tunicamycin, an inhibitor of N-linked glycosylation. Tunicamycin inhibited both the detection of gp166 in treated cells and cell fusion itself, providing a causal link between gp166 and cell fusion (Browning and O'Day, 1991). In addition to gp166, a WGA-binding protein of 138,000 M_r has been demonstrated to be involved during cell fusion in *Dictyostelium* (Suzuki and Yanagisawa, 1990). Verification that these two glycoproteins are unique awaits further purification and detailed structural analysis. More importantly, the identification of glycoproteins as positive mediators of cell fusion in *D. discoideum* promises characterization of the cell surface receptors that mediates both gamete recognition and the occupancy of which would trigger an intracellular response leading to cell fusion.

B. Coupling the Cell Surface to an Intracellular Response

In a large number of biological systems, the stimulus of a cell surface receptor is coupled to intracellular enzyme activation via an intermediary GTP-binding or G protein (Gilman, 1987). G proteins are a heteropolymer consisting of several types of α subunits combined with a limited repertoire of β and γ subunits (Gilman, 1987). A Gα subunit of about 52,000 M_r (Gα52) is present maximally during the cell fusion phase of sexual development in *D. discoideum* (Browning *et al.*, 1992, 1993). In addition, a Gα subunit of about 45,000 M_r (Gα45) is present during and following cell fusion (Browning *et al.*, 1992). Based on immunological and cholera- and pertussis-based ADP ribosylation, these Gα subunits are Gαs-like, and thus are potential candidates for essential Gα subunits during cell fusion (Browning *et al.*, 1993). The kinetics of Gα52 appearance makes it an attractive candidate as the cell fusion-specific G protein; however, conclusive evidence will require alteration of the expression of either Gα52 or Gα45, or both, followed by an examination of the consequences of these alterations on sexual development.

C. Intracellular Calcium Mediates Cell Fusion

The presence of cell fusion-specific glycoproteins and G proteins demonstrates that cell fusion in *D. discoideum* could be regulated by receptor-mediated calcium-dependent intracellular signaling. The clear temporal distinction between the phases of gamete formation and fusion, facilitated a pharmacological approach to dissect the role of intracellular calcium during cell fusion in *D. discoideum*. Addition of LaCl$_3$, an agent that blocks calcium uptake in *D. discoideum*, to sexual cultures at the onset of cell fusion inhibited gamete fusion in a dose-dependent manner, identifying the importance of calcium entry during cell fusion (Europe-Finner and Newell, 1985; Lydan and O'Day, 1988b). Calcium influx in *D. discoideum* is presumably required to replenish depleted intracellular calcium stores following an efflux of this ion, rather than playing a direct role as an intracellular message (Europe-Finner and Newell, 1985). Studies employing an agent with an effect directly opposite to that of $LaCl_3$, calcium ionophore A23187, showed that a simple calcium influx was insufficient to promote cell fusion because this agent inhibited cell fusion in *D. discoideum* (Fig. 3; Lydan and O'Day, 1988b). The effect of A23187 here was in contrast to its effect on sea urchin eggs or myoblasts where it either activated the eggs or promoted precocious fusion in myoblasts, respectively (Steinhardt and Epel, 1974; Wakelam, 1985).

    The inability for a direct increase in intracellular calcium via A23187 to

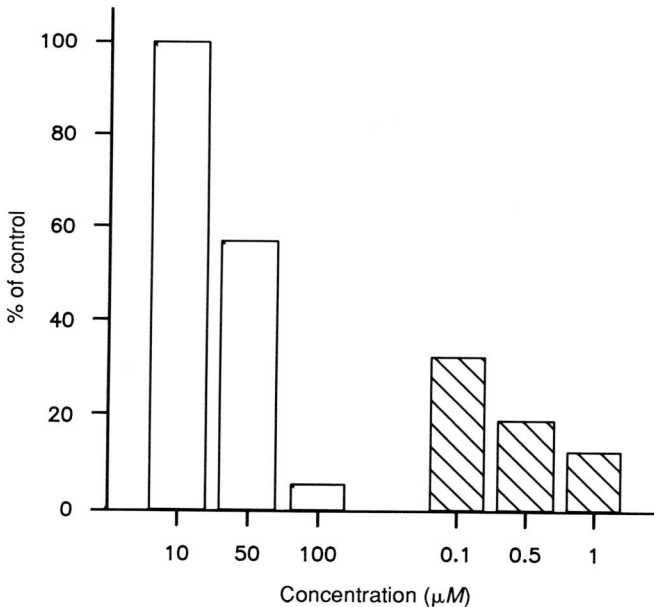

Fig. 3. The effect of $LaCl_3$ and A23187 on gamete fusion in *Dictyostelium discoideum*. Various concentrations of either $LaCl_3$ (▢) or A23187 (▨) were added at 10 hr of development and their effect on binucleate formation was determined at 20 hr of development. The data are presented as a percentage of the binucleates present in equivalent untreated control cultures.

promote cell fusion in *D. discoideum,* suggested that the effect caused by LaCl$_3$ was owing to the inability of the cells to replenish intracellular stores depleted following an intracellular calcium release. In a large number of biological systems, including *D. discoideum,* intracellular calcium transients result from receptor-dependent inositol 1,4,5-trisphosphate (IP$_3$) production followed by IP$_3$-dependent calcium release from intracellular stores (Berridge, 1986; Putney *et al.,* 1989; Europe-Finner and Newell, 1986a,b).

To examine whether the release of calcium, via IP$_3$, plays a role during cell fusion in *D. discoideum,* the effect of two agents, chlortetracycline (CTC) and 3,4,5-trimethoxybenzoic acid 8-(diethylamino)octyl ester (TMB-8), which antagonize the release of calcium from intracellular stores was compared to the effect of exogenous IP$_3$ added to permeabilized cells (Europe-Finner and Newell, 1984; Morin, 1984; Stapleton *et al.,* 1985). Both CTC and TMB-8 inhibited cell fusion in a dose-dependent manner (Fig. 4; Lydan and O'Day, 1988b). In contrast, IP$_3$ caused an immediate, precocious fusion of gametes with a maximal effect seen at about 10 n*M* (Fig. 4; Lydan and O'Day, 1988b). The ultimate extent of fusion after 4 hr in cultures treated with any concentration of IP$_3$ was identical to the control cultures, indicating that IP$_3$ promoted an increased rate of cell fusion in gametes rather than inducing fusion in nonsexual cells (Lydan and O'Day, 1988b). Thus cell fusion in *D. discoideum* is positively mediated by the release of intracellular calcium, probably via an IP$_3$-dependent manner. While work is ongoing to characterize endogenous IP$_3$ levels during cell fusion in *D. discoideum,* pharmacological studies examining the role of calmodulin have supported the hypothesis that intracellular calcium plays a pivotal role during cell fusion in *D. discoideum.*

D. Calmodulin Mediates Sexual Development

Calmodulin is a ubiquitous calcium-dependent regulatory protein that is a central transducer of the message inherent in an intracellular calcium flux in a wide range of cell types and organisms (Cohen and Klee, 1988; Cox, 1988). In addition to its ubiquitous cell and tissue distribution, calmodulin is highly conserved from an evolutionary perspective, making it an attractive molecule to examine during studies of basic biological phenomena such as gamete fusion (Cohen and Klee, 1988). For example, the calmodulin molecules present in such diverse cell types as *D. discoideum* amebas and bovine brain cells are over 90% homologous and have similar biochemical characteristics (Clarke *et al.,* 1980; Bazari and Clarke, 1981; Marshak *et al.,* 1984).

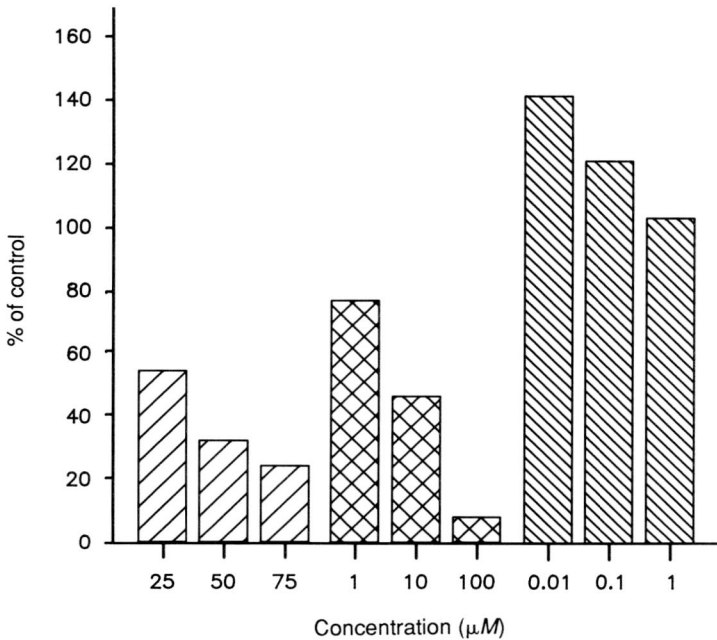

Fig. 4.   The effect of CTC, TMB-8, and IP$_3$ on gamete fusion in *Dictyostelium discoideum*. CTC (▨) and TMB-8 (▨) were added at 10 hr of development and cell-type percentages were determined at 20 hr of development. Various concentrations of IP$_3$ (▨) were added at 10 hr of development and cell-type percentages were determined 1 hr following its addition. The data are presented as a percentage of the binucleates present in equivalent untreated control cultures.

Initial investigations employed a pharmacological approach. Two chemically distinct calmodulin inhibitors were employed to ensure that the developmental effects observed were caused specifically by an inhibition of calmodulin function. Both trifluoperazine and calmidazolium, agents that inhibit the interaction of calcium-activated calmodulin with target proteins, inhibited cell fusion in a dose-dependent manner (Fig. 5; Van Belle, 1981; Weiss *et al.*, 1985; Lydan and O'Day, 1988c). This effect was both reversible and not a result of nonspecific cell lethality (Lydan and O'Day, 1988c). Of particular interest was an unexpected result—a marked increase in gamete levels in those cultures treated with calmodulin inhibitor concentrations, which caused a maximal inhibition of cell fusion (Fig. 5; Lydan and O'Day, 1988c). This

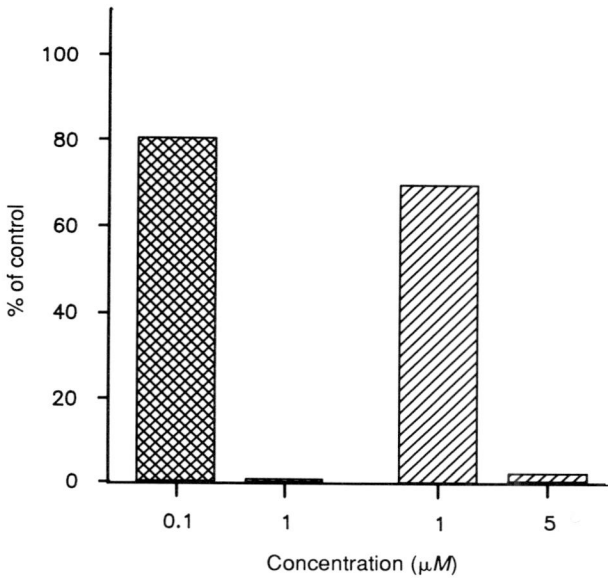

Fig. 5.   The effect of TFP and calmidazolium on binucleate formation during early sexual development in *Dictyostelium discoideum*. TFP (▦) and calmidazolium (▨) were added to sexual cultures at 10 hr of development and cell-type percentages determined at 20 hr of development. The data are presented as a percentage of the binucleates present in equivalent untreated control cultures.

result was distinct from those seen during pharmacological studies employing agents that perturb calcium movement because none of those inhibitors enhanced gamete levels (Section II,A; Lydan and O'Day, 1988b). This suggested that calmodulin has at least two developmental roles during early sexual development in *D. discoideum,* as a negative mediator of gamete formation and a positive mediator of cell fusion.

The demonstration that calmodulin function mediates gamete differentiation and fusion in *D. discoideum* is only a step toward further understanding of the specific calmodulin-dependent mechanisms involved in these events. To examine calmodulin-dependent events in greater detail, recombinant calmodulin was labeled with ^{35}S during its expression in *Escherishia coli,* producing a probe for the detection of calmodulin-binding proteins (CaMBPs; Lydan and O'Day, 1992). Since calmodulin must interact with a protein to perform its function as a regulatory molecule, the presence of unique CaMBPs during either gamete differentiation or fusion may implicate these CaMBPs as effector molecules during these developmental process. Studies employing a direct gel overlay technique have demonstrated that over 25 CaMBPs are present in *D. discoideum,* about 15 of which bind ^{35}S-CaM only when calcium is present (Lydan and O'Day, 1993b).

Of greater interest was the finding that 6 of the 15 calcium-dependent CaMBPs were present only when the cells were developing sexually (Lydan and O'Day, 1993b). Three of these CaMBPs deserve specific mention. A CaMBP of about 91,000 M_r was present only during the gamete differentiation phase of development, and CaMBPs of about 48,000 and 38,000 M_r were present during the onset of and throughout the cell fusion phase of development, respectively (Fig. 6; Lydan and O'Day, 1993b). Whereas the identities and functions of these CaMBPs are unknown, their purification and characterization will provide direct insight into the specific calmodulin-dependent

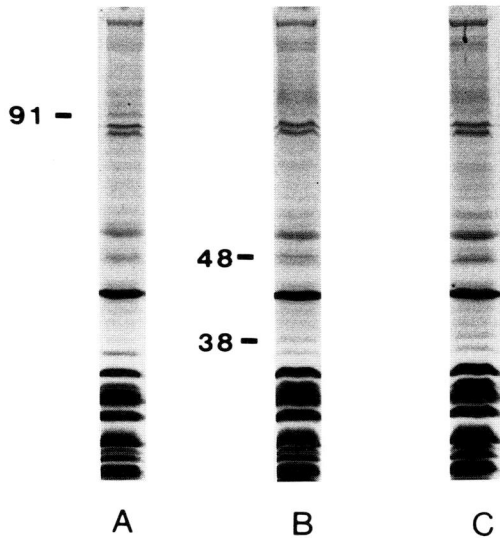

Fig. 6. Calmodulin-binding proteins during cell fusion in *Dictyostelium discoideum.* (A) Whole cell lysates form cells during gamete formation, (B) at the onset of cell fusion, and (C) at the completion of the cell fusion phase of development were separated by SDS-PAGE and probed with ^{35}S-VU-1 CaM as per Lydan and O'Day (1992b). About 25 proteins bound ^{35}S-VU-1 CaM with a 91,000 M_r calmodulin-binding protein (CaMBP91) observed only during gamete formation, a 48,000 M_r CaMBP(48) transiently observed during cell fusion, and a 38,000 M_r CaMBP(38) first seen following gamete formation and throughout cell fusion.

mechanisms that mediate gamete differentiation and fusion during sexual development in *D. discoideum.*

E. The Function of the Autoinhibitor

As previously mentioned, an endogenous mediator of sexual development in *D. discoideum,* the autoinhibitor, acts to inhibit cell fusion. The auto-inhibitor inhibits the developmental events leading to the formation of zygote giant cells while not affecting later events that lead to macrocysts. Purification of the autoinhibitor will allow both a determination of its chemical identity and also provide a chemical agent that has an inherently specific developmental effect.

The autoinhibitor has been partially purified from the culture medium (Lydan and O'Day, 1989). Partially purified autoinhibitor (PPA) inhibits cell fusion, a developmental effect consistent with crude autoinhibitor preparations (Fig. 7; O'Day *et al.,* 1981; Szabo and O'Day, 1984; Lydan and O'Day, 1989). PPA concentrations that almost completely inhibited cell fusion also caused a dramatic increase in the number of gamete cells present in the culture (Lydan and O'Day, 1989). This developmental effect, and inhibition of cell fusion with a concomitant increase in gametes was identical to the effects of trifluoperazine and calmidazolium, characterized inhibitors of calmodulin function (Lydan and O'Day, 1988c, 1989). Thus the autoinhibitor may mediate development by a direct inhibition of calmodulin.

PPA was shown to be a calmodulin inhibitor by its ability to inhibit, *in vitro,* the calmodulin-dependent activation of $3':5'$-cAMP phosphodiesterase (PDE), a calmodulin-dependent enzyme. The activation of PDE by calmodulin was inhibited in a dose-dependent manner by PPA (Fig. 7; Lydan and O'Day, 1989). The effect of PPA was specific to the calmodulin-dependent activation of PDE rather than to an inhibition of its catalytic activity since PPA had no effect on the activity of a calmodulin-independent form of PDE (Fig. 7; Lydan and O'Day, 1989).

III. FINAL COMMENTS

The data accumulated to date have enabled a working model to be proposed for the intracellular events during the initiation and termination of cell fusion in *D. discoideum* (Fig. 8). During cell fusion in *D. discoideum,* a process

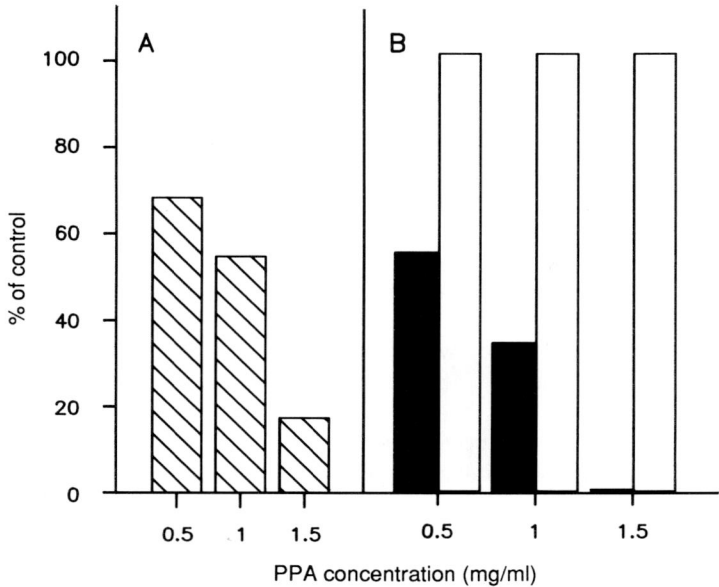

Fig. 7. The effect of partially purified autoinhibitor on both sexual cell fusion in *Dictyostelium discoideum* and calmodulin-dependent activation of $3':5'$-cAMP phosphodiesterase. Sexual cultures of *D. discoideum* were treated with various PPA concentrations (A) at 10 hr of development and binucleate levels were determined at 20 hr of development. (B) PPA was also added to *in vitro* assay mixtures containing either calmodulin-sensitive (■) or calmodulin-insensitive (□) PDE. The data in (A) are presented as a percentage of the binucleates present in equivalent untreated control cultures and in (B) as a percentage of the maximal PDE activation by calmodulin.

analogous to fertilization in higher eukaryotes, stimulation of a fusion-specific glycoprotein (gp166?) activates, via G proteins (Gα52?), phospholipase C liberating IP_3 and evoking release of Ca^{2+} from intracellular stores, and elevating the intracellular level of calcium. Extracellular calcium enters the cells via calcium channels and either directly or indirectly (via the replenishing of intracellular stores contributes to the increased level of intracellular calcium. This intracellular signal activates calmodulin, which in turn interacts with and activates those calmodulin-binding proteins that directly mediate cell fusion.

Cell fusion is terminated during zygote giant cell differentiation via two complementary pathways. The first is the down-regulation, via an unknown

FERTILIZATION   ZYGOTE DIFFERENTIATION

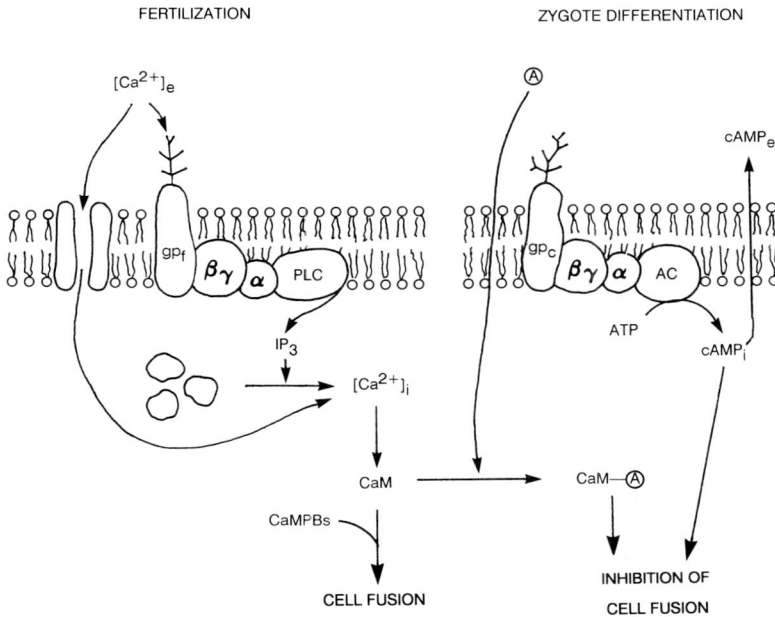

Fig. 8. A model for the initiation and termination of sexual cell fusion in *Dictyostelium discoideum*. $[Ca^{2+}]_e$, extracellular calcium; gp_f, glycoprotein receptor for cell fusion; $\beta\gamma$, G protein subunits β and γ; α, G protein subunit α; PLC, phospholipase C; IP_3, inositol 1,4,5-trisphosphate; $[Ca^{2+}]_i$, intracellular calcium; $[Ca^{2+}]_e$, extracellular calcium; CaM, calmodulin; CaMBPs, calmodulin-binding proteins; A, autoinhibitor of sexual development; gp_c, glycoprotein receptor for chemotaxis; AC, adenylate cyclase.

stimulus, of the cell surface machinery required for cell fusion. For example, gp166 and Gα52. In addition, zygote giant cells secrete the autoinhibitor, which interacts with and inhibits the function of calmodulin, thus inhibiting cell fusion. Finally, zygote giant cells chemoattract amebas via cAMP. cAMP, itself, inhibits cell fusion, but in this case, extracellular cAMP does not play a direct role, rather, the cAMP produced by the responding cells during chemotaxis inhibits cell fusion.

The progress to date in dissecting the calcium-dependent intracellular events during early sexual development in *D. discoideum* will allow the future examination, in detail, of intracellular mechanisms that mediate gamete formation and fusion in this simple eukaryote. Our investigations will focus on the purification and characterization of the autoinhibitor and those CaMBPs that are found associated specifically with gamete differentiation and fusion.

Purified autoinhibitor will be used to both prove that it inhibits calmodulin-dependent enzyme activation and to examine its calmodulin-binding characteristics. Purified CaMBPs will be used to both produce antibodies and to clone the genes for these proteins. The cloned genes will be used to produce deletion mutants to directly examine the role of these CaMBPs during gamete differentiation and fusion.

The ultimate goal of these studies is to apply the understanding gained with *D. discoideum* to investigations examining the role of calcium-dependent intracellular signaling during cell fusion in higher eukaryotes. Preliminary evidence has shown that PPA inhibits fertilization in sea urchins, thus it is quite possible that some of the mechanisms that regulate gamete fusion in *D. discoideum,* especially those that are calmodulin-dependent, also function in higher eukaryotes.

ACKNOWLEDGMENTS

This work was supported by research grants from the Bickell Foundation and the National Sciences, Engineering and Research Council of Canada.

REFERENCES

Abe, K., Orii, H., Saga, Y., and Yanagisawa, K. (1984). A novel cyclic AMP metabolism exhibited by giant cells and its possible role in the sexual development of *Dictyostelium discoideum. Dev. Biol.* **104,** 477–483.

Berridge, M. J. (1986). Inositol phosphates as second messengers. *In* "Phosphoinositides and Receptor Mechanisms" (J. W. Putkey, ed.), pp. 25–46. Alan R. Liss, New York.

Browning, D. D., and O'Day, D. H. (1991). Concanavalin A and wheat germ agglutinin binding glycoproteins associated with cell fusion and zygote differentiation in *Dictyostelium discoideum:* Effects of calcium ions and tunicamycin on glycoprotein profiles. *Biochem. Cell Biol.* **69,** 282–290.

Browning, D. D., Lewis, K. E., and O'Day, D. H. (1992). Zygote giant cell differentiation in *Dictyostelium discoideum:* Biochemical markers of specific stages of sexual development. *Biochem. Cell Biol.* **70.**

Browning, D. D., Poludnikiewicz, M. B., Proteau, G. A., and O'Day, D. H. (1993). Biochemical investigation of G-alpha subunits during cell fusion and chemotaxis in the sexual cycle of *Dictyostelium discoideum. Exp. Cell Res.* **250.**

Blaskovics, J. C., and Raper, K. B. (1957). Encystment stages of *Dictyostelium*. *Biol. Bull.* **113**, 58–88.

Bazari, W. L., and Clarke, M. (1981). Characterization of a novel calmodulin from *Dictyostelium discoideum*. *J. Biol. Chem.* **256**, 3598–3603.

Chagla, A. H., Lewis, K. E., and O'Day, D. H. (1980). Ca^{++} and cell fusion during sexual development in liquid cultures of *Dictyostelium discoideum*. *Exp. Cell Res.* **126**, 501–505.

Clarke, M., Bazari, W. L., and Kayman, S. C. (1980). Isolation and properties of calmodulin from *Dictyostelium discoideum*. *J. Bacteriol.* **141**, 397–400.

Cohen, P., and Klee, C. B. (1988). "Calmodulin." Elsevier, Amsterdam.

Cox, J. A. (1988). Interactive properties of calmodulin. *Biochem. J.* **249**, 501–505.

Europe-Finner, G. N., and Newell, P. C. (1984). Inhibition of cyclic GMP formation and aggregation in *Dictyostelium* by the intracellular Ca^{2+} antagonist TMB-8. *FEBS Lett.* **171**, 315–319.

Europe-Finner, G. N., and Newell, P. C. (1985). Calcium transport in the cellular slime mould *Dictyostelium discoideum*. *FEBS Lett.* **186**, 70–74.

Europe-Finner, G. N., and Newell, P. C. (1986a). Inositol 1,4,5-triphosphate induces calcium release from a nonmitochondrial pool in amebas of *Dictyostelium*. *Biochim. Biophys. Acta* **887**, 335–340.

Europe-Finner, G. N., and Newell, P. C. (1986b). Inositol 1,4,5-triphosphate and calcium stimulate actin polymerization in *Dictyostelium discoideum*. *J. Cell Sci.* **82**, 41–51.

Gilman, A. (1987). G proteins: Transducers of receptor generated signals. *Annu. Rev. Biochem.* **56**, 615–649.

Knudsen, K. A., McElwee, S. A., and Myers, L. (1990). A role for the neural cell adhesion molecule, NCAM, in myoblast interaction during myogenesis. *Dev. Biol.* **138**, 159–168.

Lewis, K. E., and O'Day, D. H. (1977). Sex hormone of *Dictyostelium discoideum* is volatile. *Nature (London)* **268**, 730–731.

Lewis, K. E., and O'Day, D. H. (1985). The regulation of sexual development in *Dictyostelium discoideum:* Cannibalistic behavior of the giant cell. *Can. J. Microbiol.* **31**, 423–435.

Lewis, K. E., and O'Day, D. H. (1986). Phagocytic specificity during sexual development in *Dictyostelium discoideum*. *Can. J. Microbiol.* **32**, 79–82.

Loomis, W. F. (ed.) (1982). "The Development of *Dictyostelium discoideum*." Academic Press, New York.

Lydan, M. A., and O'Day, D. H. (1988a). Developmental effects of the major ions found in a groundwater sample on sexual cultures of *Dictyostelium discoideum*. *Can. J. Microbiol.* **34**, 207–211.

Lydan, M. A., and O'Day, D. H. (1988b). The role of intracellular Ca^{2+} during early sexual development in *Dictyostelium discoideum:* Effects of LaCl$_3$, Ins-(1,4,5)-P$_3$, TMB-8, chlortetracycline and A23187 on cell fusion. *J. Cell Sci.* **90**, 465–473.

Lydan, M. A., and O'Day, D. H. (1988c). Different developmental functions for calmodulin in *Dictyostelium:* Trifluoperazine and R24571 both inhibit cell and pronuclear fusion but enhance gamete formation. *Exp. Cell Res.* **178**, 51–63.

Lydan, M. A., and O'Day, D. H. (1989). The autoinhibitor of cell fusion in *Dictyostelium* inhibits calmodulin. *Biochem. Biophys. Res. Commun.* **164**, 1176–1181.

Lydan, M. A., and O'Day, D. H. (1992). Production of ^{35}S-labeled proteins in *E. coli* and their use as molecular probes. *In* "Methods in Molecular Biology" (J. M. Walker and A. J. Harwood, eds.), Vol. 20, Humana Press, Clifton, New Jersey (in press).

Lydan, M. A., and O'Day, D. H. (1993). Calmodulin and calmodulin-binding proteins during cell fusion in *Dictyostelium discoideum:* Developmental regulation by calcium ions. *Exp. Cell Res.* **250**.

Machac, M. A., and Bonner, J. T. (1975). Evidence for a sex hormone in *Dictyostelium discoideum. J. Bacteriol.* **138**, 251–253.

Marshak, D. R., Clarke, M., Roberts, D. M., and Watterson, D. M. (1984). Structural and functional properties of calmodulin from the eukaryotic microorganism *Dictyostelium discoideum. Biochemistry* **23**, 2891–2899.

McConachie, D. R., and O'Day, D. H. (1986). The immediate induction of extensive cell fusion by Ca^{2+} addition in *Dictyostelium discoideum. Biochem. Cell Biol.* **64**, 1281–1287.

McConachie, D. R., and O'Day, D. H. (1987). Pronuclear migration, swelling, and fusion during sexual development in *Dictyostelium discoideum. Can J. Microbiol.* **33**, 1046–1049.

Morin, D. (1984). Fluorescent localization of calcium at the sites of spreading and attachment. *J. Exp. Zool.* **229**, 81–89.

O'Day, D. H. (1979). Aggregation during sexual development in *Dictyostelium discoideum. Can. J. Microbiol.* **25**, 1416–1426.

O'Day, D. H., and Lewis, K. E. (1975). Diffusible mating-type factors induce macrocyst development in *Dictyostelium discoideum. Nature (London)* **254**, 431–432.

O'Day, D. H., and Rivera, J. (1987). Lectin binding and inhibition studies reveal the importance of D-glucose, D-mannose, and N-acetylglucosamine during early sexual development of *Dictyostelium discoideum. Cell Differ.* **20**, 231–237.

O'Day, D. H., Szabo, S. P., and Chagla, A. H. (1981). An autoinhibitor of zygote giant cell formation in *Dictyostelium discoideum. Exp. Cell Res.* **131**, 456–458.

O'Day, D. H., McConachie, D. R., and Rivera, J. R. (1987). Appearance and developmental kinetics of a unique cell type in *Dictyostelium discoideum:* Is it the gamete phase of sexual development? *J. Exp. Zool.* **242**, 153–159.

Putney, J. W., Jr., Takemura, H., Hughes, A. R., Horstman, D. A., and Thastrup, O. (1989). How do inositol phosphates regulate calcium signaling? *FASEB J.* **3**, 1899–1905.

Saga, Y., and Yanagisawa, K. (1982). Macrocyst development in *Dictyostelium discoideum*. I. Induction of synchronous development by giant cells and biochemical analysis. *J. Cell Sic.* **55**, 341–352.

Saga, Y., and Yanagisawa, K. (1983). Macrocyst development in *Dictyostelium discoideum*. III. Cell fusion inducing factor secreted by the giant cells. *J. Cell Sci.* **62**, 237–248.

Stapleton, C. L., Mills, L. L., and Chandler, D. E. (1985). Cortical granule exocytosis in sea urchin eggs is inhibited by drugs that alter intracellular calcium stores. *J. Exp. Zool.* **234**, 289–299.

Steinhardt, R. A., and Epel, D. (1974). Activation of sea urchin eggs by calcium ionophore. *Proc. Natl. Acad. Sci. USA* **71**, 1915–1919.

Suzuki, K., and Yanagisawa, K. (1990). Purification and characterization of gp138, a cell surface glycoprotein involved in the sexual cell fusion of *Dictyostelium discoideum. Cell Differ. Dev.* **30**, 35–42.

Szabo, S. P., and O'Day, D. H. (1984). The low molecular weight autoinhibitor of sexual development in *Dictyostelium discoideum* inhibits cell fusion and zygote differentiation. *Can. J. Biochem. Cell Biol.* **62**, 722–731.

Szabo, S. P., O'Day, D. H., and Chagla, A. H. (1982). Cell fusion, nuclear fusion, and zygote differentiation during sexual development in *Dictyostelium discoideum. Dev. Biol.* **90**, 375–382.

Van Belle, H. (1981). R24571: A potent inhibitor of calmodulin-activated enzymes. *Cell Calcium* **2**, 483–494.

Wassarman, P. M. (1988). Zona pellucida glycoproteins. *Annu. Rev. Biochem.* **57,** 415–442.

Wakelam, M. J. O. (1985). The fusion of myoblasts. *Biochem. J.* **228,** 1–12.

Weiss, B., Sellinger-Barnette, M., Winkler, J. D., Schechter, L. E., and Progialeck, W. C. (1985). Calmodulin antagonists: Structure-activity relationships. *In* "Calmodulin Antagonists and Cellular Physiology" (H. Hidaka and D. J. Hartshorne, eds.), pp. 45–62. Academic Press, New York.

Index

ISBN 0-12-524155-0